DEPARTMENT OF THE ENVIRONMENT
DEPARTMENT OF ENERGY
SCOTTISH OFFICE
WELSH OFFICE

Commission on Energy and the Environment

Coal and the Environment

LONDON

HER MAJESTY'S STATIONERY OFFICE

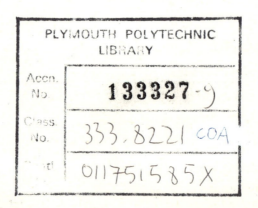

The Rt Hon Michael Heseltine MP
Secretary of State for the Environment

The Rt Hon David Howell MP
Secretary of State for Energy

The Rt Hon George Younger MP
Secretary of State for Scotland

The Rt Hon Nicholas Edwards MP
Secretary of State for Wales

Sirs

We were appointed in February 1978 to advise on the "Interaction between
Energy Policy and the Environment". Our first Report, "Review of Activities
1978–79", was submitted in July 1979. Our major effort has been concentrated on
a study of the longer term environmental implications of coal production, supply
and use in the United Kingdom looking to the period around and beyond the end of
the century, including likely new technologies and conversion to other fuels and
raw materials.

We now have pleasure in submitting the Report of that Study.

The Lord Flowers (Chairman)

R N Bottini

Sir Henry Chilver

Dr J G Collingwood

A G Derbyshire

Professor Sir Richard Doll

Professor Sir William Hawthorne

Professor F G T Holliday

Gerald McGuire

Mrs Naomi E S McIntosh

Mrs Veronica Milligan

Sir Austin Pearce

M V Posner

Sir Francis Tombs

D G T Williams

8 July 1981

Contents

PART II COAL PRODUCTION

Chapter 15　Coal conversion to substitute fuels

Chapter 16　Combustion technology

Chapter 17　Combustion residues of coal: sources, quantities, and control

PART IV   PLANNING AND CONCLUSIONS

List of Figures

List of Plates

Foreword

1.   The Commission on Energy and the Environment was appointed in March 1978 by the Secretaries of State for Energy, the Environment, Scotland and Wales. Our terms of reference are:

"To advise on the interaction between energy policy and the environment."

2.   The Commission which is a standing body was given a deliberately wide remit. When announcing our appointment to the House of Commons, the then Secretary of State for the Environment explained that we would have a wide diversity of interests. We would have to consider the environmental implications, nationally and globally, arising from the production and use in the United Kingdom of coal, oil, nuclear power, gas and electricity, and renewable sources. We would be concerned with pollution and planning (though not with specific planning cases) examining the interface between energy policies and land-use planning, and the implications of such policies for the natural world and the urban environment.

3.   As a Standing Commission we have the advantage that we are able to tackle a series of studies in a systematic way. We decided therefore that the major emphasis of our work should be placed upon analysis of the environmental implications sector by sector of the production and use of energy in the United Kingdom, looking to the experience of other countries where relevant.

4.   Nuclear power has been so much studied of late that we did not think it the right place to start. The environmental implications of its production have been examined in the Sixth Report of the Royal Commission on Environmental Pollution. To this should be added Mr Justice Parker's comprehensive Report on the Windscale Inquiry, and the expected inquiries into the proposed Pressurised Water Reactor and the Commercial Demonstration Fast Reactor.

5.   We concluded therefore that, in view of the central role coal seems likely to play in any future United Kingdom strategy, we should first examine in depth the environmental implications of increased coal production and use.

6.   The UK coal industry will need to adopt within the timeframe of out Study a wide range of new technology. This will have important implications for the training of the industry's manpower at management level and throughout the workforce. We have not dealt with these implications in our Report, although we have pointed out areas requiring further research. However, an important element of the follow-up to our work should be the early appraisal of the industry's future technical training and educational requirements. This appraisal should involve the Nationalised Industries, the unions, the universities and polytechnics, the relevant professional and technical institutions and Government Departments.

7.   Our terms of reference specifically excluded consideration of individual planning cases. Thus in examining the longer term environmental implications of increased coal production and use, we have punctiliously avoided judgements on the merits or otherwise of individual planning applications, and in particular of the Belvoir application on which a Ministerial decision is still awaited as we go to press.

8.   We invited evidence from various organisations with a special interest in the subject of our Study, and also from the general public. A list of those who submitted evidence is at Appendix 3. This evidence has been lodged in the libraries of the Department of Energy and the Department of the Environment. We visited areas of the United Kingdom to discuss particular problems with local authorities, industry and other local interests. We also visited research establishments to review current and prospective research and development activities. Details of our visits are given in Appendix 4.

9.   We leave until the end of this Report our general acknowledgement of the help we have received but we wish to record here our gratitude to the National Coal Board (N C B) for their unfailingly helpful and thorough response to our many inquiries during the preparatory phases of our work. We welcome the changes in its procedures for dealing with environmental matters introduced by the Board during the course of our Study. However, having received a great deal of evidence from a wide range of sources, we wish to stress that we reached our conclusions and recommendations independently of any of the interests involved.

10.   As we go to press, we have been informed that the Government intends to let the Commission fall into abeyance.

PART I

Introduction: Coal in the energy scene –
challenge and opportunity

Chapter 1 Introduction

THE COMMISSION'S REMIT

1.1 The Commission on Energy and the Environment was appointed in March 1978 by the Secretaries of State for Energy, the Environment, Scotland and Wales. Our terms of reference are:

"To advise on the interaction between energy policy and the environment".

This development was in accord with the mounting concern about the nature of the interface between energy and the environment in the wake of the energy crisis of 1973. In his report 'Energy and the Environment', published in 1974 on behalf of the Committee for Environmental Conservation, the Royal Society of Arts and the Institute of Fuel, Lord Nathan wrote:

"A policy for the development of energy resources must be created in the context of its environmental and social consequences . . . . Protection of the environment cannot be separated from the other objects of a policy for energy; the present skirmishing between environmental interests and the energy industries benefits neither side. . .".

1.2 We took as our starting point the Green Paper on Energy Policy, (Department of Energy 1978) since this represented the fullest available exposition of the then Government's thinking on energy policy. We considered the value of brief and immediate comments on the environmental consequences of the overall strategy outlined in the paper but decided that such comment would be of limited value.

1.3 We decided that the major emphasis of our work should be placed upon an analysis of the environmental implications sector by sector of the production and use of energy in the United Kingdom, looking to the experience of other countries where relevant. We concluded that in view of the central role coal seems likely to play in any future United Kingdom strategy, we should first examine in depth the environmental implications of increased coal production and use.

Terms of Reference of the Coal Study

1.4 We adopted the following terms of reference:

"To examine the longer term environmental implications of future coal production, supply and use in the United Kingdom, looking to the period around and beyond the end of the century, including likely technologies and conversion to other fuels and raw materials".

1.5 We have interpreted our remit as follows: to examine the environmental problems of a continued and expanded use of coal; to identify areas of concern; to suggest safeguards; and to present, as far as possible, an objective analysis of the interaction between the production and use of coal and environmental factors. This has two facets: firstly, to identify those circumstances where coal can be mined and used in an environmentally acceptable way; and secondly, to identify acceptable means to facilitate the further development of the use of coal.

1.6 Certainly it is the second of these which is the harder task – not least because many of the bodies presenting evidence to us addressed the past environmental disadvantages of coal, rather than the importance of coal in the energy future or the circumstances under which coal can be mined and used without environmental disadvantage. This was helpful to us in indicating what actions needed to be avoided; it was less helpful on the positive side in indicating what actions were necessary and desirable to promote environmentally acceptable coal burn when this is in the national interest.

Issues covered by the Study

1.7 It is important to emphasise at the outset that our report is a detailed study of just one of the major components of energy supply and use. It cannot, therefore, by itself answer such problems as what is the optimum fuel mix for the country to adopt. We are conscious that by its very nature this Study, limited to coal, does not enable us to comment on the relative advantages of coal compared to other possible options. We have therefore identified what seemed to us to be the critical environmental features of coalmining and use, and then analysed whether *on their own merits* they are likely to be acceptable, or could be made so.

1.8 The primary focus of our Study derives from our basic remit and the division of labour between ourselves and the Energy Commission, which was still in being when we embarked upon the Study. Thus it contrasts in scope and objectives with the wider study recently initiated in the Federal Republic of Germany under the sponsorship of the Federal Minister for Research and Development. The German study will analyse the entire complex of technical and economic requirements for increased coal use and for the related energy infrastructure, as well as the impact on man and the environment. The latter will include a comparative presentation of the environmental impacts of electricity generation from coal and nuclear power.

Timescale covered by the Study

1.9 The second important point to note from the remit is that the Study is directed to the longer term. We have attempted to look forward 20 years or more in order to obtain a broad overview of the likely role of coal in the country's future energy requirements, and then to decide the necessary environmental safeguards which may need to be instituted. We have thus been primarily concerned with the longer term prospects for coal, but have also taken account of present problems and past difficulties.

ENVIRONMENTAL IMAGE OF COAL

1.10 Our central concern has been to address the most sensitive and complex aspect of the relationship between energy policy and other objectives. The declaration issued at the end of the meeting of the seven Heads of Government at the Venice Summit in June 1980 stated:

> "Together we intend to double coal production and use by early 1990. We will encourage long term commitments by coal producers and consumers. It will be necessary to improve infrastructures in both exporting and importing countries, as far as is economically justified, to ensure the required supply and use of coal. We look forward to the recommendations of the International Coal Industry Advisory Board. They will be considered promptly. We are conscious of the environmental risks associated with increased coal production and combustion. We will do everything in our power to ensure that increased use of fossil fuels, especially coal, does not damage the environment."

1.11 This Report therefore deals with both environmental *and* energy issues. The link between the two is by no means clear, and is certainly not one way. It is much more complex than the conventional view of coal and the environment would suggest. That view is that coal is difficult and dangerous to mine, inevitably causing environmental stress where mining is undertaken; and that it is dirty and inconvenient to burn, contributing both air pollution and ash as residues to be absorbed into the environment.

1.12 Thus the conventional image of coal would have suggested that rigorous examination of the environmental impact of coal would inevitably throw up a picture highlighting the deleterious effects of coal extraction and use. Therefore, we conducted our examination with a view to ensuring that the development of the coal industry would not imperil the post-war progress on environmental protection as embodied in the Clean Air Acts, the Town and Country Planning Act and the Control of Pollution Act.

1.13 In the event, we have been reassured by the results of our investigations. We have indeed found features of coalmining and use, particularly that of opencast mining and spoil disposal, which pose serious problems. In general, however, excluding opencast mining, the worst features are associated with a pattern of coal production and use inherited from times past. We do not wish to minimise the scale and importance of this legacy and the urgency of measures needed to address it. However, what has particularly re-assured us is the progress reflected in the contrast between previous standards and those to which new projects are being planned and executed. Overall we regard the conclusions of the Study as hopeful.

1.14 At worst, the future expansion of the industry need not repeat inherited environmental problems; and at best, current procedures, if widely adopted, will clear the legacy of the past and make a future industry that is far more acceptable than the old. Certainly, over wide areas of coal mining and use, satisfactory techniques are currently available. It is in the N C B's interests to apply these techniques since consent for future activities will depend upon the use of the best methods to minimise environmental impact. We have been encouraged by the way in which the N C B is proceeding and, if it continues on its present path, there are reasonable grounds for taking the more optimistic view of future prospects.

1.15 We have been impressed by the extent to which the types of problem we have encountered were rarely unique to coal (although the actual practice of mining is, of course, individual – one of the few instances where the factory moves into the raw material rather than vice versa, and where the major operations are conducted underground). We have considered separately and in some detail subsidence and spoil disposal, two distinctive characteristics of deep mining. However, these apart, the environmental problems in developing a new mine are very similar to those involved in developing any new major industrial installation – for example, the location and siting of the plant, the design of the buildings, the transport of the raw materials and products, and the provision of a labour force. The requirements of planning and pollution control are likewise little different.

Character of Mining Communities

1.16 However, we have been struck from the outset by the unique character of mining communities. Mining has often been highly localised in areas which have become almost exclusively dependent on this single industry with very limited alternative opportunities for employment. With it have grown up extremely closely knit communities with an exceptionally developed sense of group identity reinforced by a strong tradition handed down from father to son. All this derives in large measure from the nature of mining, since miners depend very greatly on one another for protection and safety so that their lives rest on the cooperation, goodwill and even self-sacrifice of their fellows. The localised character of the industry and its strong sense of community tend to insulate miners to some extent from pressures of general public opinion. The effect of a legacy of dirt and danger often accompanied by harsh struggles cannot be underestimated.

1.17 It is against this background that two further problems can give rise to considerable apprehension: the socio-economic impact of new projects located in greenfield sites and the impact of pit closures on local communities. These are, of course, major problems in their own right with particularly sensitive regional dimensions. However, they are not unique to the coal industry, and their solution involves areas of concern much broader than coal and the environment.

1.18 It is a fact of life in an extractive industry that, while new pits continue to be opened up, other pits will be closed as they become exhausted or as reserves become so sparse and difficult to work that such pits become grossly uneconomic. This process will inevitably embrace a significant part of current capacity over the timescale of this Study. We fully recognise the acute sensitivity of the closure problem, particularly in areas such as South Wales confronted in the present recession by exceptionally high levels of unemployment. Within some localities in such areas there is almost total dependence on mining intensified by the contraction in the region of the steel industry. In this context, the legacy of the past cannot be underestimated. A vital element of the environment is the morale of people and their state of mind. This was graphically conveyed to us in evidence from the National Union of Mineworkers (N U M) which pointed out that, in the two decades before 1974, the British coal industry lost 70% of its collieries, 65% of its workforce and 45% of its output. The following extracts from the N U M evidence convey the strength of feeling bred of that experience:–

> "These figures do not, however, reveal the economic or social cost to individual mineworkers and their families, to traditional mining communities, and to the UK energy sector as a whole, of this period of decline. Members of the N U M became industrial gypsies as one colliery after another was prematurely closed. Whole families were forced to uproot and transfer to other areas where often there proved to be little difference in the long-term security of job prospects."

> "In the communities that were left behind there were the characteristic symbols of industrial decay in an ageing population, increasing land dereliction, deteriorating housing stock and the inability of elected local authorities, which by the closure of former collieries had been deprived of a major source of rate income, to undertake any effective measures of remedial action."

1.19 The industry itself must provide the machinery to determine which collieries are to close and the timing of closures, having regard to all the financial, marketing,

commercial and technical information about each colliery. The industry's Colliery Review Procedure provides for the essential full consultation with the unions who have rights of appeal to the Board at national level. However, the industry cannot solve the socio-economic impact of pit closures in isolation. We consider that these problems must be anticipated. They call for partnership between both sides of the industry, central Government and local authorities. The admirable work of the Welsh Development Agency (W D A), one of whose primary objectives is to attract alternative employment, demonstrates what can be achieved by tackling these problems on a regional basis. We fully understand that the unions might well interpret forward planning as an instrument for accelerating pit closures. However, the stakes are such, not least for the miners themselves, that we regard sensible planning based on advance dialogue between the interested parties as an essential component of effective and sensitive handling of this inevitably contentious issue. There would be major consequences for the industry and its workforce, and for the country, of failure to grapple with the realities of an extractive industry, whereby new capacity is needed as old capacity becomes either technically unworkable or grossly uneconomic, particularly if coal is to meet the challenge of its future prospects.

1.20 The socio-economic impact of new projects raises contrasting problems. Recent expansion has tended to be located adjacent to existing activity. Thus the developments currently in hand at Selby and planned for Belvoir lie close to existing coalfields and are to be serviced in large part by the transfer of men from existing collieries with some supplementary recruitment in greenfield areas. Even when the Board reaches the stage of developing greenfield capacity well away from any existing collieries it will still be necessary to employ a high proportion of skilled and experienced men to new recruits. The N C B 's current policy, as illustrated at Selby, is to absorb into the existing community the influx of men and their families required to service the mine. Such developments are bound over time to modify the traditional nature of mining communities. We discuss the wider socio-economic impacts of new projects located in greenfield sites in Chapter 6.

THE NATURE OF THE ENERGY PROBLEM

1.21 The communiqué after the Venice Summit meeting in June 1980 began with the words:

> "The economic issues that have dominated our thoughts are the price and supply of energy, and the implications for inflation and the level of economic activity in our countries and for the world as a whole. Unless we can deal with the problems of energy, we cannot cope with other problems".

1.22 The world energy problem is essentially a problem of the supply of oil in the short and medium term. The problem is not that the world is going to run out of oil by the turn of the century. The problem is rather that the world as a whole will have to adjust to a situation in which oil supplies will be declining, whereas formerly they were increasing at the rate of between 5% and 7% per annum. Moreover, the supply of oil will be unpredictable and vulnerable to political influence. Thus the world has to effect a transition from an economy in which oil was cheap, plentiful and certain to one in which energy in general will be expensive and oil will be increasingly scarce and uncertain. This transition has to be achieved in the next decade or so if a scramble for scarce energy resources towards the close of the century is to be avoided. It would be highly imprudent to weaken policies designed to reduce oil import dependence in the light of the current glut of oil on world markets.

1.23 The energy problem is not confined to the affluent industrialised nations. It is an even more serious problem for the developing countries. According to World Bank sources, 48 of the 74 developing countries which import oil are dependent on it for at least 90% of their commercial energy requirements; only 4 are less than 50% dependent. The energy situation must be seen in the wider perspective of prospects for world population, food production and raw materials. The Brandt Commission has reported that some 800 million inhabitants of the Third World are destitute. The Third World faces the need for massive increases in food imports. As oil becomes scarcer and therefore more expensive, the implications for agricultural production are critical.

1.24 Against this background, the central long term problem for Governments in the energy sector is to ensure that necessary adjustment is not unduly impeded; and to take such positive steps as are possible to facilitate the adjustment, in order to ensure that economic growth is not limited by shortage of energy or the economic strains imposed by its price instability. Failure to handle this transition successfully will have serious economic and social consequences and could endanger world peace.

1.25 The principal industrial countries have recognised that the reduction of present dependence on oil is central to the restructuring of their energy economies. Meeting in Venice in June 1980, the seven Heads of Government agreed that it was essential to break the link between economic growth and oil consumption. To that end they agreed that, by 1990, oil should be reduced to no more than 40% of total primary requirements. Beyond that they agreed to strengthen energy conservation programmes, accelerate the development of alternative energy resources and in particular increase the use of coal and nuclear power in the medium term. In the long term they agreed that it was necessary to develop new energy technologies. This commitment reconfirmed earlier agreement at the Tokyo Summit to increase coal use, production and trade as part of the overall strategy.

1.26 It is against this background that the development of U K energy policy has to be set. The objectives of U K energy policy have been traditionally defined as ensuring that there should be adequate and secure energy supplies; that these supplies should be efficiently used; and that these objectives should be achieved at the lowest practicable cost to the nation. These objectives need to be considered in the light of our international commitments and obligations; the need to ensure that energy policy does not inhibit economic growth; and, in particular, the prospects of scarcer and dearer oil.

1.27 The United Kingdom is fortunate in being richly endowed with indigenous energy. We have significant reserves of oil and gas, very large reserves of coal, and considerable experience of nuclear power which already produces some 13% of our electricity. But our supplies of oil and gas from the North Sea are of limited life. We could become net importers again before or during the 1990s. Thus we have the same long term interests as those of our partners in the industrialised world who are deficient in energy. Since we are a trading nation, the health of our economy is inextricably linked to the economy of the industrialised world as a whole. Given prospective conditions in the world energy market, there is likely to be advantage in meeting as much of our energy requirements as we can economically from secure indigenous sources and in diversifying our sources of supply.

1.28 In order to diversify our sources of supply there is a strong case for developing both coal and nuclear power. To present our energy policy options as "coal versus nuclear" is misguided; nor does it render a service to the coal industry. We cannot comment in this Report on the content of the nuclear programme. However, we cannot overlook the implications of its magnitude for the demand for coal. A modernised U K coal industry is compatible with the enhanced nuclear component of our energy supplies as envisaged by successive U K Governments. Moreover, it may well be that even the most recent lower estimates (22 Gigawatts (G W)) of the likely nuclear contribution by the turn of the century may prove incapable of achievement.

1.29 Our appreciation of the market prospects for coal in the longer term reinforces this judgement. We discuss in Chapters 13 and 15 the prospects for a major re-entry of coal into the industrial market and, as we move into the next century, the new prospects which will open up based on new technologies in both production and use. In addition, coal exports should be looked at anew. A modernised coal industry could bolster the U K contribution to the collaborative international effort to reduce oil import dependence. U K coal constitutes the greatest single fuel resource in Western Europe. The U K coal industry should aim at greater commercial penetration of the huge Community energy market.

1.30 The challenge to both Government and the N C B is to convince the workforce and the customer of the reality of the industry's future prospects, notwithstanding the short term difficulties arising from the current recession. Our Study has underlined the extent to which the future prospects of the industry will be determined by its progressive modernisation aimed at bringing onstream new, low cost, high productivity capacity. This will involve the expansion of long life capacity in existing areas and the introduction of greenfield production. The modernisation of the industry is the more important since, over the timescale of our Study, the costs and efficiency of energy supply will become of increasing importance in terms of the international competitiveness of U K industry. This factor has been heavily underlined in the recent debate about energy pricing to U K industry. The cost of U K coal will be critical, for example, in determining future U K electricity prices.

THE INTERACTION BETWEEN COAL AND THE ENVIRONMENT

1.31 We have been struck from the outset by the inherent uncertainties of the energy scene. It is the massive uncertainty of the future which makes forecasting such a hazardous business. The unexpected cannot be predicted, but it will surely happen. We have not attempted to generate independent forecasts. Instead, we have proceeded, taking the Green Paper (Department of Energy 1978) as our starting point, by examining the environmental implications of a range of possible levels of future coal production and coal burn.

1.32 We have undertaken a detailed analysis of a number of major topics: the environmental effects of deep and opencast mining operations; storage, handling and transport; the environmental implications of the markets for coal, including coal conversion to other forms of energy; the environmental consequences arising from the combustion of coal; and the implications for planning policy and procedures.

1.33 Three inter-related issues have arisen repeatedly during the Study, concerning the adequacy of *current best practice*; the relationship of this concept to the *best practicable means* formula; and the implications of *multiple standards* in an industry characterised by striking contrasts between past and present practices.

1.34 In most areas we are satisfied that the best available modern techniques are environmentally adequate. There is encouraging evidence that N C B best practice incorporates these techniques in new developments. Thus the problem which we have faced is not so much that of identifying the scope for improved standards of *current best practice*, but of securing the wider application of such standards. Economic pressures on developers might cause them to retreat from current best practice or to modify it.

However, once a particular feature of current best practice has been implemented through good design or competent operation, there will be considerable public pressure to ensure that that feature is repeated in future designs or operations. In considering applications for new developments, local planning authorities will usually expect current best practice to be used; and in their total evaluation of a project they can be expected to subject to critical scrutiny any case for not using it.

1.35 However, as we consider at length in Chapter 9, concern about spoil disposal has featured predominantly in the evidence submitted to us. Even here we have been reassured that the current techniques of low profile tipping and progressive restoration are an immense improvement upon the traditional, uncompacted, unrestored, conical spoil heaps. Nevertheless, the extent to which even these improved methods are acceptable depends substantially on local circumstances. In particular, however good the operation of the techniques, in some areas there could still be problems over the sheer scale of the future spoil disposal requirement.

1.36 A further complication with the concept of current best practice derives from the reality of *multiple standards* which arise from the varied age structure of the N C B 's installations. This reflects the contrast between the modern techniques embodied in the latest projects, and those available in the past. We would expect a continuing improvement in the standards applied to new projects as new techniques of design and operation evolve. However, in general, it is neither feasible nor economic to redesign existing projects so that there is complete uniformity of standards. The physical constraints of existing layout of plant and machinery are one major reason why this is so. A similar complication arises with new extensions to old projects – again the constraint offered by the design of the old project may in practice limit the extent to which the very latest sophistication in design, plant or operational techniques can be deployed in them compared with the greenfield situation. Every effort should be made to apply improved techniques to existing projects, but greater constraints will apply. We have drawn specific attention to the scope for such efforts – for example, in the physical appearance and tidiness of some older pits.

1.37 The reality of multiple standards is acknowledged in the related but distinct concept of *best practicable means* (Bpm) enforced by the Inspectorates which we discuss more fully in Chapter 4. Fundamental to the operation of Bpm is flexibility in adapting general standards to local circumstances. Bpm does not necessarily mean "all technically possible means". In some cases such a requirement would involve unsupportable costs. New standards are not usually applied immediately to existing plant. Industry cannot be expected to re-equip their plant with the latest control equipment too frequently, particularly when this forms an integral part of process design. Existing equipment must be allowed a reasonable economic life unless it is grossly ineffective.

1.38 In contrast, we have used the term current best practice to relate only to the highest standards which have been used in current operations. It is not embodied in any legislation, nor is it enforced by any agency comparable to the Alkali Inspectorate.

1.39 Whether standards are judged on the basis of current best practice or on the basis of Bpm, the apportionment of costs is critical. The following quotations from the Organisation for Economic Co-operation and Development (O E C D) publication, 'The Polluter Pays Principle' (O E C D 1975), encapsulate some important elements of why such a principle is required and why its operation is an efficient means for allocating costs.

1.40 "Environmental resources are in general limited and their use in production and consumption activities may lead to their deterioration. When the cost of this deterioration is not adequately taken into account in the price system, the market fails to reflect the scarcity of such resources at both the national and international levels. Public measures are thus necessary to reduce pollution and to reach a better allocation of resources by ensuring that prices of goods depending on the quality and/or quantity of environmental resources reflect more closely their relative scarcity and that economic agents concerned react accordingly."

1.41 "The notion of an 'acceptable state' decided by public authorities, implies that through a collective choice and with respect to the limited information available, the advantage of a further reduction in the residual social damage involved is considered as being smaller than the social cost of further prevention and control. In fact the Polluter Pays Principle is no more than an efficiency principle for allocating costs and does not involve bringing pollution down to an optimal level of any type, although it does not exclude the possibility of doing so."

1.42 The polluter pays principle has been an integral part of pollution control policy in the U K for a long time. It states that the polluter should bear the expense of pollution control measures. Governments consider it conducive to economic efficiency because the prices of goods reflect the environmental costs incurred in their production; and they also justify it on grounds of equity, as those responsible for pollution then bear the costs of its control. The principle has been long accepted by international bodies, including the E E C and the O E C D. They apply the additional precept that measures designed to secure pollution control "should not be accompanied by subsidies that would create significant distortions in international trade and investments".

1.43 In principle it is a statement of where in public life the costs of any necessary preventive or remedial actions should fall. In practice, however, where the polluter bears the cost of pollution control, he will usually transfer it to the product, and so ultimately it will be passed to the customer. In this sense, the polluter pays can be interpreted as the consumer pays. The alternative to the polluter pays would be for the taxpayer to pay. In such circumstance, pollution control would be subsidised. This applies in the current system of central government grants for derelict land clearance to remove pollution inherited from the past.

1.44 In the case of the coal industry, the polluter pays principle means that the costs of environmental control should in general be borne by the N C B, and be incorporated in the price of coal. Likewise, the costs of environmental control in coal-burning power plants should be borne by the Central Electricity Generating Board (C E G B), and be incorporated in the price of electricity. There are, however, two limits to this. Firstly, the scale of the costs which are imposed for environmental reasons needs to be considered against the decreased competitiveness of coal in the market place. Thus, there is a trade-off between environmental values on the one hand and energy policy dictates on the other. It is probable that public perception of this trade-off will fluctuate in the time span covered by the Study. It will not therefore be capable of permanent solution. Nevertheless, for the purposes of our Study, we have not consciously limited our recommendations to those which can be accommodated under existing cost structures. We have taken as our touchstone what is necessary to achieve an environmentally satisfactory use of coal. The fact that in general our recommendations will not substantially increase the cost of coal is not a reflection of any arbitrary cost limit we have set, but more of the high standards that can be achieved with modern best practice.

1.45 Secondly, this principle, that the costs of necessary environmental protection should be reflected in the price of coal, should in equity only relate to current activity. It would not be appropriate to expect N C B, given its present remit, to clear up the dereliction caused by the activities of past coal entrepreneurs over which it had no control. There is no argument for the remedial cost for that past pollution to be placed upon the current price of coal. Under these circumstances, where it is not possible to make the polluter pay, it is appropriate for central Government to undertake that task, which is placed on all taxpayers who, in a sense, pay for the past shortfall in environmental standards.

1.46 The more difficult question is whether an existing agency should be required to remedy pollution which it has itself caused in the past, but at a time when there were no statutory requirements for it to be controlled. If these requirements change we consider that any enhanced obligations should be made effective from the date of such change and not be made retrospective. However, there is an obvious need to secure that such pollution is cleared. One particular example of this relates to spoil heaps, which in the past were permitted under the General Development Order (G D O), but which now may require improvement. This problem is considered at length in Chapter 9.

1.47 It will be clear from this brief introduction that the main body of the Report will encompass wide ranging issues of public policy. However, the evidence we have received has heavily underlined the extent to which problems causing public antipathy towards coal are a function of detailed issues which may not be of major importance in themselves but which cumulatively give rise to considerable concern. We have therefore sought deliberately not to confine our attention to a wide sweeping analysis of the major policy issues, but also to consider in the appropriate detail the specific matters which together comprise a significant element in public perception of the interface between coal and the environment. While risks to health must command priority attention, we have also addressed the complex of impacts which, although they have a less direct impact on physical well-being, nevertheless in total impinge significantly on the quality of life of those most closely exposed to the production and use of coal.

Chapter 2   The coal industry – decline and renaissance

2.1 In this Chapter we review the fluctuating fortunes of the coal industry during the post-war period and the changes in coal policy brought about largely by external events bearing upon the industry's prospects. We then review the dramatic changes in the energy scene resulting from the events of 1973 and since. We see an understanding of the protracted post-war period of contraction and decline as essential to an appreciation of the nature of the adjustment needed by the industry to meet the challenge and opportunity arising from its changed prospects.

2.2 The massive oil price increases imposed by the Organisation of Petroleum Exporting Countries (O P E C) in 1973 brought to an abrupt end the era of cheap and plentiful energy. This sudden price increase and the subsequent further escalation of oil prices confronted oil importing countries, industrialised and developing alike, with a far reaching and complex challenge to reduce oil dependence.

2.3 As a result of these developments, the U K coal industry now faces a fundamental change in its long term prospects, notwithstanding the short-term difficulties posed by the depth of the current recession. The value of the U K 's substantial coal reserves has been transformed. To seize the opportunities opened up by these potential prospects constitutes a massive challenge to the industry. We discuss more fully the nature of that challenge in the next Chapter. However, in order to understand the magnitude of the adjustment involved, full account must be taken of the fluctuating fortunes of the industry since it reached its peak in 1913, when the U K was the world's major coal producer and output was 292 million tonnes, compared with 125 million tonnes in 1980/81. Even on the basis of more pessimistic forecasts of future coal demand, the scale of this adjustment remains formidable since lower growth of coal demand will place an even higher premium on the modernisation of the industry if it is to lay the foundations for later growth. The fluctuating fortunes of the industry are encapsulated in the title of the book by I Berkovitch 'Coal on the Switchback' (1977).

2.4 The decline of the industry set in during the First World War and continued after an immediate post-war boom throughout the inter-war period. The situation did not improve during the Second World War. In response to the fuel crisis of 1942, the Government used its emergency powers to take control of the industry and created the Ministry of Fuel and Power to coordinate fuel and power supplies as well as to promote efficiency in their distribution and consumption. An examination of British coal production by the Reid Committee in 1945 (Coal Mining 1945) identified the industry's major problems as undermechanisation, lack of technical development, the fragmentation of the industry in small and unplanned mines and the level of productivity which had shown little increase since 1913. This was the situation inherited by the National Coal Board when it assumed responsibility for coal production on 1 January 1947.

THE SCALE OF DECLINE

2.5 The scale and character of the industry's decline since 1947 is demonstrated in Table 2.1.

Figure 2.1   U K Energy consumption

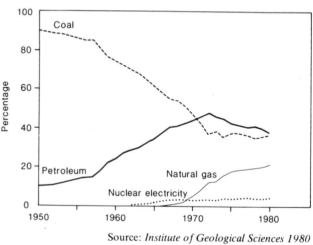

Source: *Institute of Geological Sciences 1980*

Markets

2.6 The declining role of coal in total energy consumption in the U K during the first decades of nationalisation is indicated in Figure 2.1. The decline in individual markets, in spite of a considerable upsurge in total energy demand, is indicated in Figure 2.2. With the exception of power stations, all markets and particularly general industry, railways and gasworks, reflected a very substantial reduction in coal consumption. A complex of factors contributed to this continued decline. The most significant was the abundance of cheap oil, mainly from the Middle East, which in addition to its price advantage offered greater convenience in use and handling. Moreover, the Clean Air Acts of 1957 and 1968 encouraged both industrial and domestic consumers to switch to other fuels. Technical developments also contributed to coal's decline. Railways had developed diesel and electric traction. In the gas industry, oil gasification was developed and natural gas discovered. Considerable advances were made in fuel burning efficiency throughout industry. Improvements between 1948 and 1970 in the thermal efficiency of power stations alone meant that the electricity supply industry was burning about 17 million tonnes less than would have been the case without these improvements. Lastly, nuclear power entered the power station market.

2.7 The fall in coal markets would have been even more drastic but for measures taken in the late 1960s aimed at improving the market situation. These included a tax on fuel oil, a ban on coal imports together with a virtual ban on the conversion of coal-fired power stations to oil.

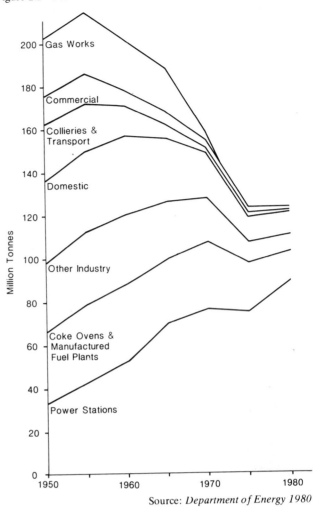

Figure 2.2   Markets for coal 1950–80

Source: *Department of Energy 1980*

Table 2.1   *The Coal Industry in the U K*

	1949 (Calendar Year)	1959 (Calendar Year)	1968/69	1978/79	1980/81
(a) Total Output (million tonnes)	218·2	209·4	163·2	119·9	124·9
(b) Number of pits (at end of period)	912	737	317	223	211
(c) Colliery Manpower (thousands)	712·5	658·2	336·3	234·9	229·8
(d) Colliery output per manshift (tonnes)	1·19	1·37	2·16	2·24	2·32 (calculated on a different basis)
(e) Cost per tonne £ (deep-mined collieries)	2·21	4·06	4·75	24·87	34·9
(f) Inland Consumption (million tonnes)	198·9	189·6	167·3[1]	119·9[2]	120·9[3]
(g) Coal as a percentage of U K inland energy consumption	91·3%	75·9%	53·3%[1]	35·3%[2]	36·9%[3]

Source: *N C B*

[1] Calendar Year 1968
[2] Calendar Year 1978
[3] Calendar Year 1980

Manpower

Figure 2.3   The coal industry 1900–1980

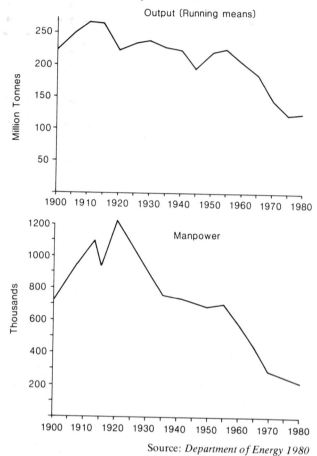

Source: *Department of Energy 1980*

2.8   The decline of the industry in terms of output and manpower is indicated in Figure 2.3. Since 1947, the total number of collieries has declined by 747, and total N C B colliery manpower has fallen by about 472,500, a scale of rundown unparalleled in any other sector of British industry. Manpower problems were exacerbated by the need, while contracting the labour force, to recruit young workers for craftsman training in an industry which was becoming increasingly mechanised.

2.9   In the years immediately following nationalisation, there was indeed a chronic manpower shortage. At the same time, rapidly increasing productivity in the pits being mechanised continued to be offset by low and falling productivity in the older, poorer pits, with a limited life. Overall, productivity rose only by 15·6% between 1947 and 1957.

2.10   By 1959, the long-term nature of the rundown was reflected in the N C B's Revised Plan for Coal. This represented a significant turning point in the development of the industry. It attempted to grapple with the problems of decline in contrast with the expansionist policies of the post-nationalisation period. The Revised Plan provided for a decrease in manpower to 626,000 by 1975 compared to the actual level of 658,200 in 1959. In the event manpower fell to 465,600 in 1965.

2.11   This further decline was accompanied, however, by a considerable rise in productivity. Between 1957 and 1967/68 output per manshift increased by just over 57% in sharp contrast to the 15·6% improvement in the previous decade. This resulted from a combination of factors including increased mechanisation, economies of scale arising from greater concentration of output and rationalisation through closure of the most unprofitable pits. The total number of pits fell from 737 in 1959 to 504 in 1965.

2.12   This increase in productivity was achieved in spite of a number of negative factors. There was a considerable increase in the average age of the workforce, which in turn increased the rate of involuntary absenteeism. A recruitment problem persisted in some areas in the face of policies the overall aim of which was to reduce the labour force. Low morale in the industry and lack of confidence in its future intensified the recruitment problem.

CHANGES IN POLICY

2.13   Coal policy since nationalisation has reflected the fluctuating fortunes of the industry which have been in large measure a function of external events bearing upon the industry's prospects. The relative optimism of the immediate period following nationalisation was embodied in the 1950 Plan For Coal. This proposed a £635 million programme of capital investment over a 15 year period. Output was to rise from 219 million tonnes to 224 million tonnes by 1965. The N C B's financial performance was considerably inhibited by an informal pricing policy agreement with the Government. The N C B was encouraged to meet its capital requirements by Exchequer borrowing rather than by price increases and self-financing.

2.14   In 1956 the N C B published a revised plan 'Investing in Coal'. The 1965 output target was confirmed at 224 million tonnes, but capital requirements were drastically revised upwards – £715 million more than anticipated in 1950. In contrast, the N C B's Revised Plan For Coal in 1959 attempted to grapple with the long-term nature of the downturn and thus provided for reduced output levels.

2.15   However, the White Paper in 1965 (Ministry of Power 1965) and the 1967 Fuel Policy paper (Ministry of Power 1967) sought to moderate a continued rundown. The White Paper noted that while coal was currently dearer than oil the situation might change; coal capacity, once lost, would be difficult to recreate; and there were practical limits to the rate at which the industry could contract economically. Thus, the Coal Industry Acts of 1965 and 1967 contained a number of measures designed to help the industry cope with the worst financial effects of decline.

2.16   Such then was the prelude to the latest and most dramatic change in the industry's prospects resulting

10

from the fundamental changes in the energy scene since 1973. By the end of the 1960s the industry, like its counterparts in Western Europe, had struggled through decades of very rapid decline. Output had fallen in 1969 to 156 million tonnes. The labour force had declined from 602,000 to 311,000. The number of collieries in operation had fallen to 304. A change of policy was required to deal with the many investment, production and manpower problems. Investment in new capacity had almost stopped during the 1960s: now a massive investment in new capacity was essential virtually starting from scratch. In 1973 over half the total of deep mined output came from collieries which were over 70 years old, the newest colliery being Kellingley where production had begun in 1965. Currently the average age of operating collieries is 86 years. Figure 2.4 illustrates the operations of the coal industry since the War with reference to year-on-year capital investment by the N C B.

Figure 2.4   N C B and subsidiaries: additions to fixed assets 1948–1980 (1980 prices)

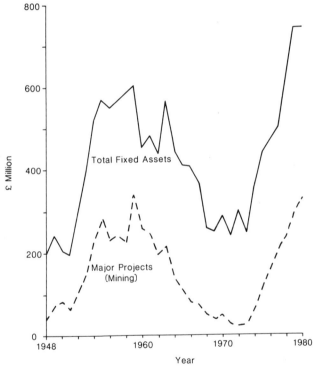

Source: *N C B*

DEVELOPMENT SINCE 1973

2.17 In 1974 the N C B put forward 'Plan For Coal' (N C B 1974) which was aimed at halting the decline with a view to eventually reversing it. The essential feature of this plan was that the rundown of outdated, uneconomic capacity would continue, but would be more than balanced by large investment both in new pits and in major improvements at existing pits. New capacity was to be gained in three ways: by major reconstruction projects at long-life collieries with sufficient reserves to sustain larger capacity; extension at short-life collieries where access could be proved to

new areas of reserve; and by the sinking of new mines which were planned to provide half of this new capacity. A new approach to manpower policy was also required. Although pit closures would continue as an inescapable feature of a long established extractive industry, now the emphasis was on recruitment, accompanied by the redeployment of skilled miners. By 1985 the coal industry would be producing at about the same rate as in the early 1970s but with much better productivity and working conditions.

2.18 Plan For Coal was adopted by the Labour administration in 1974 and endorsed by the then Conservative Opposition. In accordance with the plan, the level of investment in the industry has built up very rapidly; in real terms, it has more than trebled since 1973 and this year (1980/81) it will be about £810 million including deferred interest and leasing. The Public Expenditure White Paper published in March 1980 provides for investment to be kept at this level in real terms over the years to 1983/84.

2.19 There has been serious slippage in carrying out parts of the Plan for Coal. In particular, it has proved necessary to allow more time for drawing up development plans for new collieries, and for full consultation with the local interests concerned and for the operation of the planning process. It is expected that the original Plan for Coal objectives will be achieved by the late 1980s – later than originally expected. The benefits of the investment made since 1973 are however now beginning to come through. Total output, including opencast and licensed mines, was 3·3 million tonnes greater and output per manshift 2·4% higher in 1980/81 than in 1979/80.

The Coal Industry Act 1980

2.20 We understand from evidence provided by the Department of Energy that, in the view of the present Government, coal's potential can be realised only if it can be economically produced at competitive prices. Ensuring that the industry is fully competitive is therefore the necessary counterpart of the policy of promoting coal as a major energy source. The Government consider that, with the benefits of investment now accruing, the N C B, like the other nationalised industries, can be expected to achieve financial viability, rather than need Government support for its operations indefinitely.

2.21 The concept that nationalised industries should pay their way taking one year with another was, of course, inherent in the original Nationalisation Acts, in coal's case the Coal Industry Nationalisation Act of 1946. The White Papers on Nationalised Industries, issued by different Governments, have also consistently emphasised that there can be misallocation of resources if the industries fail to find an economic return on the resources entrusted to them. The rapid growth of financial assistance to the N C B is a relatively recent development. The N C B last broke even before operating grants in 1976/77; the big increase in operating support has taken place only since then.

2.22 We understand that the Government has been discussing with both sides of the coal industry the way in which the industry should develop over the next few years. The Board will continue to receive social grants, which include the Government's contributions to the Mineworkers' Pension Scheme for pre-1975 Pensioners, and payments towards the cost of pit closures, as well as grants towards their operation and revenues.

2.23 We also understand that the Government will continue to provide finance for investment, and in the Coal Industry Act 1980 the N C B's borrowing limit was raised from £2,200 million to £3,400 million, with provision for a further rise to £4,200 million. Output can be maintained or expanded only by continuous investment in new faces or pits. Because of the natural process of exhaustion of pits, and the long lead time of much coal investment, N C B's capital expenditure is very substantial. Since 1974, the N C B have undertaken major investment in new and modernised deep mined capacity, starting some 200 major projects at a total cost of £2,500 million. The twin drifts at the new mine at Kinsley, which will provide access to an estimated 12 million tonnes of reserves, began production in August 1979. The Selby coalfield, which will cost some £900 million at today's prices, and should create a complex of workings which will bring 10 million tonnes per annum of coal to the surface at a single rail terminal, is on course to begin production in 1983. By 2000, virtually all deep mined output will come from new and modernised productive capacity.

The right starting point for planning capital investment must be an assessment of potential demand in the market and the industry's ability to supply economically. The size of investment programme which is appropriate in any year or period will depend, therefore, on several factors, in particular on the likely future supply/demand balance for coal, and the need for new and replacement capacity. We discuss this more fully in the next Chapter.

2.24 We have reviewed the protracted period in which the dominant theme has been contraction and decline in order to highlight the nature of the adjustment needed by the industry, both its management and its workforce, if it is to meet the challenge and opportunity afforded by its longer term prospects. We discuss in later chapters the socio-economic implications of the legacy of past decline and of the creation of a modernised industry. The very landscape of the coalfields reflects the co-existence of old and new. It ranges from the ravages wrought by past practice in an area such as Barnsley to the achievement of the N C B at Selby, the design of which could well become a model for coal developers the world over. Similarly, the mining industry reflects both its history and its beckoning future in the contrasting nature of its mining communities. The tight knit communities of the Welsh valleys grew up when the industry was highly labour intensive and when entire towns depended almost exclusively on this single industry. These contrast strikingly with the new developments serviced by commuting miners from surrounding villages made up of mixed communities. In the process of change, there remains considerable contrasts in existing practice between the old and the new which are reflected in a diversity of both landscape and communities.

Chapter 3 Prospects for coal

3.1 In this Chapter we examine the place of coal in the U K energy scene and its future prospects. We take as our starting point the Department of Energy's Energy Projections 1979. We take account of criticisms that the Department's approach is too supply oriented, and underline that one of the most controversial elements in the energy debate concerns the scope for conservation. In our analysis of the energy policy implications of the Department's forecasts, we give particular attention to the controversy about the role of coal production and coal consumption over the timescale of the Study, which we consider prudent to use in the context of our primary concern with environmental issues.

ENERGY PROJECTIONS

3.2 The Department of Energy states that its projections are necessarily based on certain broad long term assumptions about economic growth, technical developments and movements in energy prices, all of which are very uncertain. The projections do not purport to be predictions of what will happen nor prescriptions for what should happen. They are intended rather to provide a broad quantitative framework for the consideration of possible energy futures and policy choices. Nor, as we understand, do they imply a Government commitment to particular levels of energy production; these will depend on the way the market develops and on decisions that will be taken progressively as the Government's appreciation of possible future needs and supply prospects develops.

3.3 The two main scenarios which formed the basis of the Department's projections published in 'Energy Projections 1979' examined the possible demand for energy associated with annual economic growth rates averaging (i) 2% and (ii) 2·7%. These indicated that total U K energy demand might rise from 377 million tonnes of coal equivalent (mtce) in 1979 to 410–435 mtce in 1990 and 445–510 mtce in 2000. On the supply side, it was estimated that by the end of the century potential indigenous production could be in the range of 390–410 mtce, including 137–155 million tonnes of coal. Taken in conjunction with the estimates of future demand, the Department estimated that there could thus be a net import requirement, especially of oil and gas, of some 35–120 mtce by the year 2000. This requirement would be growing as we move into the next century. A summary of the Department's

published projections of energy demand and indigenous supply for the year 2000 (based on Energy Projections 1979) is shown in the Table below, along with comparable figures for 1979.

Table 3.1   Energy projections for 2000

		mtce
	1979	2000
Total U K Demand (including non-energy and bunkers)	377	445–510
of which Coal Demand	130	128–165
Indigenous Supply		
Coal	122	137–155
Gas	58	62–65
Oil	130	100
Nuclear and Hydro	16	88–95
Total U K indigenous supply	329	390–410
Net fuel imports	47	35–120

Source: *Department of Energy*

3.4 On these assumptions, and with North Sea supplies peaking round about the mid-1980s, coal would have a vital role to play in meeting the U K 's future energy requirements by 2000. Beyond 2000, the Department's projections indicated that the need for coal would become greater still; they showed that whilst demand for primary fuels might rise by some 6–11 mtce a year after that date, supplies of oil and gas from the North Sea could be declining at about 6 mtce annually. On this basis the potential gap between U K energy demand and indigenous supplies would therefore grow wider, unless coal, conservation and nuclear energy could together make it good.

3.5 We understand that the best advice available to the Government is that the renewable sources of energy are unlikely to make a significant and economic contribution to our energy supplies before the next century. They could, however, begin to make a more important contribution thereafter. A widely based research and development programme is under way. Expenditure last year on renewables amounted to £9 million and in the current year will be about £12·6 million. Until the extent of the possible economic contribution from renewables is clear, the official view is that it is necessary to work on the basis that the U K 's assured long-term energy supplies consist of coal and nuclear power. Together with conservation, all three are at the centre of the Government's energy policy, and we accept that all three are essential elements in ensuring a robust U K energy strategy for the longer term.

3.6 Since Energy Projections 1979 was published, the Department of Energy has considered, as a sensitivity on the 2% and 2·7% economic growth rate scenarios, the possible effects on energy demand of gross domestic product (G D P) growth rates averaging about 1% per annum up to 2000. In this low growth variant, rates of growth somewhat below 1% are assumed in the period up to 1985 with rates higher than 1% during the remainder of the period. We understand that this is not to imply that the Department regards the possibility of very low economic growth as a secure foundation for planning designed to ensure that the price and availability of energy are not constraints on the economy. The Department has suggested that, even with this low growth, demand for coal might be around 115 million tonnes in 2000, within a total primary energy demand of about 400 mtce.

DEMAND

3.7 Possible U K demand for coal by sector is illustrated in the following Table, on the basis of average economic growth assumptions of 2% and 2·7% taken from Energy Projections 1979 and of the low growth case mentioned in paragraph 3.6.

Table 3.2   Coal demand by market in 2000

			million tonnes
Economic growth	2%	2·7%	1% (low growth case)
Power stations	66	78	65
Coke ovens	16	19	12
Other industry (including collieries)	39	45	32
Domestic (including manufactured fuels)	3	3	3
Other	3	5	3
Substitute Natural Gas (S N G)	1	15	0
Total	128	165	115

Source: *Department of Energy*

3.8 Our view of these projections is that, with the deterioration in economic prospects that has taken place since the Department of Energy proposed them, the likely outcome may lie somewhere between the 2% growth case and the 1% growth variant. However, such projections are surrounded by massive uncertainty. Moreover, extrapolation of trends in the depth of a recession can become self-fulfilling prophecy. An energy policy firmly based on low growth might well prove insufficiently robust; some over-provision of energy supply is preferable to the risks attendant upon

energy shortage. However, severe opportunity costs would clearly be involved if the insurance margin were to be excessive.

3.9 The Department's projections allowed for the possibility of up to 40 GW of installed nuclear capacity by the year 2000. We have severe doubts about the achievability of such a nuclear capability over this period on grounds of practicability. We therefore asked the Department if it could provide estimates of the possible implications for the coal demand projections of a nuclear capability limited to 22 GW by the year 2000 for the 2% growth case and the 1% growth variant. The Department estimated that for the former, power station coal burn might increase by 33 million tonnes, and for the latter it could increase by about 10 million tonnes. The Table below summarises these adjustments and their possible effect on projected coal demand in 2000.

Table 3.3   Coal demand in 2000

		million tonnes
	2% growth case	1% growth variant
Department of Energy Projections	128	115
Estimated coal demand for Nuclear capability reduced to 22 GW	33	10
Revised projections	161	125

Source: *Department of Energy*

3.10 The allowance made for energy conservation in the Department's projections (14–20%) will call for a sustained and substantial conservation effort. We accept the Department's assessment that the renewable energy sources will not be making a significant contribution to total energy supply by the turn of the century, though they might thereafter.

3.11 It should be noted that our revisions to the coal demand projections in Table 3.3 above include no allowance for the effects of higher world oil price in 2000 than that assumed. Similarly, we make no allowance for the possibility of a smaller contribution by Combined Heat and Power (C H P) to total energy demand than that assumed in the Department's projections. Bearing in mind these qualifications, our revisions are very close to the range published by the Department for the 2% and 2·7% economic growth cases (128–165 million tonnes). Furthermore they underline the continuing importance that coal is expected to have in meeting the U K's future energy requirement.

PRODUCTION

3.12 The N C B estimate that operating reserves (reserves fully proved in respect of thickness, quality and mining conditions and which are either accessible to existing mines or have been proved sufficiently to identify new mines) are 7 billion tonnes – equivalent to some 47 years production at 150 million tonnes of coal a year. Total resources could be as much as about 190 billion tonnes (coal-in-place in seams over 0·61 metres (2 ft) thick and less than 1,220 metres (4,000 ft) deep, after allowing for the coal which has already been worked). The N C B estimate that about 45 billion tonnes of this coal may ultimately become recoverable national reserves (the proportion of known coal in place which is considered to be recoverable under current economic conditions and using current mining technology (Moses K 1980)).

3.13 These N C B estimates have been the subject of considerable public debate. In response to evidence submitted by the Institute of Geological Sciences (I G S), we invited the I G S and the N C B, in association with the Royal Society, to review the extent and definition of U K coal reserves and to report to us their findings.

3.14 After lengthy deliberations, a highly technical report was made to us, which summarised separately the views of the N C B and of the I G S. We had hoped for a clearer and more precise outcome; and we are disappointed that a greater measure of overt agreement could not be reached between two such responsible bodies. Nevertheless, some benefit appears to have resulted from our initiation of this exchange of views: discussions between I G S and N C B are to continue with a view to a possible revision of the N C B 's Production Instruction on operating reserves, which would ensure a greater uniformity of application across the country. We have noted from the report that there is a great deal of coal-in-place in the UK. The problem lies in assessing how much of this is likely to be retrievable under present or likely future circumstances, taking account of possible changes in both technical expertise and economic constraints. Even taking these into account, it seems clear to us that there are very substantial reserves of coal for future exploitation.

3.15 In our view, the scale of U K coal reserves satisfies any reasonable current concern about the future physical availability of coal. We have not been concerned to define with precision whether U K coal reserves will last, for example, for 200 or 300 years, or any intermediate duration. However, we attach considerable importance to the N C B 's policy of maintaining, on a continuing basis, operating reserves sufficient to meet 50 years of current production. The N C B/I G S joint analysis does not, in our view, throw doubt upon the ability of the industry to sustain operating reserves at this level for as far into the future as has any practical meaning.

3.16 In addition to the coal reserves on land, it is known that there are substantial coal deposits under the North Sea. Preliminary surveys made in conjunction with oil and gas drilling indicate deposits of coal at distances up to 100 kilometres from the coast, in areas where the sea depth varies from less than 25 metres to 200 metres. Coal seams up to 15 metres thick have been found, located at depths between 600 metres and 3000 metres beneath the sea.

3.17 These deposits represent a long term resource, for which new extraction techniques would obviously be needed. There are major problems both in gaining access to the seams, and in extracting coal from them. Access could be possible from extensions to shore based workings, from sea-level operations, or from sea-bed based operations. Options for extraction could include automated manned mining systems, remote mechanical extraction, hydraulic mining, or in-situ decomposition for example by gasification, microbiological decomposition or chemical methods. All this seems too far off, technically and economically, to justify any investigation by us, and we therefore make no recommendations. However, we hope that research on these ideas will continue in establishments and universities, because new methods of extraction could be important both under the sea and also on land. We suggest that North Sea coal might be regarded as the equivalent for the coal industry of the fusion programme for the nuclear industry – a vast possibility at present not useable for which the required technology can be foreseen but is unproven.

Deep Mined

3.18 U K deep mined production in 1980/81 was 109·6 million tonnes (the figure includes a very small proportion of tip coal), produced from 211 N C B collieries employing 224,800 miners. A significant feature of N C B deep mined production is its concentration in the Yorkshire-Derbyshire-Nottinghamshire coalfield which had 99 collieries (47% of U K total), in March 1981, 109, 100 miners (48%) in year ending 1980, and produced in 1980/81 60·4 million tonnes (55%). The breakdown of operating collieries by N C B Area for 1980/81 in these counties was:

Table 3.4 *Operating collieries in the Yorkshire–Nottinghamshire–Derbyshire coalfield (1980/81)*

South Yorkshire	18
North Yorkshire	16
Barnsley	18
North Derbyshire	11
North Nottinghamshire	14
South Nottinghamshire	12
Doncaster	10
Total	99

Source: N C B

At 1 April 1974, this coalfield had 4,924 hectares of active colliery spoil tipping and 2,099 hectares of derelict spoil heaps (not all of which were attributable to coal mining). The scale of spoil in this region has created areas of considerable industrial dereliction, the implications of which we address in Chapter 9. Over the next 20 years, the concentration of deep coal mining in Yorkshire–Derbyshire–Nottinghamshire is likely to increase, whilst there will be some continuing decline in the older traditional mining areas of South Wales and North East England.

3.19 A further significant feature of deep mined production is the wide variation in production costs. Figures taken from the N C B Report and Accounts for 1980/81 (N C B 1981a) indicate that average production costs in 1980/81 range from £27·9 per tonne in the North Nottinghamshire area to £51·4 per tonne in South Wales, an average of £34·9 per tonne for all deep mining areas. Although production costs per tonne at individual pits are not available, it is clear that they would indicate a much wider range of costs. To the extent that high production costs and poor environmental characteristics are associated with older, less efficient pits, the transition of coal mining to a modern industry, concentrating its activities in pits employing new technologies, would be beneficial in terms of both environmental impact and production costs. This clearly has major implications for the workforce which we have discussed in Chapter 1 and which we return to in Chapter 10.

3.20 The Board are now spending some £19 million a year on their exploration programme. The map we reproduce later at Figure 6.1 illustrates some concentration of effort to the North and East of the Yorkshire/Nottinghamshire coalfield which moves further South to new coalfields, as at North East Leicestershire, the South West Midlands and further South into Oxfordshire.

3.21 Coal is an extractive industry and output can be maintained only by continuous investment in new faces and pits to replace reserves which are worked out. In 1978 the N C B estimated that without further major investment their capacity would fall to some 80 million tonnes per annum in 2000. Output would be slightly less than this. Projects already in hand would raise capacity to some 95 million tonnes per annum; but even assuming a contribution from opencast in line with the present target of 15 million tonnes per annum, it is clear that further substantial investment is needed. It would in fact be needed even to meet the 115 million tonnes demand projected in the low growth case described above, to say nothing of the demand in the two main cases of Energy Projections 1979.

3.22 Although the N C B are investing heavily in the replacement of obsolete capacity and the improvement of productivity, they do not expect to reach their Plan

For Coal (N C B 1974) target production level of 120 million tonnes deep mined coal until the late 1980s. Allowing for some further delay in realising their planned investment, especially in new mines, total production including opencast might therefore be 127–138 million tonnes in 1990. Output in 2000 might, we consider, be 137–155 million tonnes.

3.23 On the Board's latest forecasts, existing mines which have been developed and modernised will produce 86 million tonnes of coal in 1985 and new mines some 4 million tonnes. The N C B's principal major projects are firstly the development of the Selby coalfield, which will cost £900 million at today's prices and should produce 10 million tonnes a year; and secondly, the development of new mines in the Vale of Belvoir, which is the subject of a public inquiry. In addition, there are many projects for the modernisation and reconstruction of existing pits.

3.24 The N C B argue that although theoretically the introduction of replacement and new capacity could occur rapidly at the end of the century, there is danger in relying on this method of meeting demand because of the long lead times, and potential shortages of trained managers and miners to carry out the required expansion quickly enough, and also because such a rapid increase of capacity might prove to be at the expense of the environment. A steady build up is therefore essential both to ensure that demand can be met and to provide adequate opportunity to take environmental factors into account. We endorse this general approach, particularly since a crash programme would be highly prejudicial to environmental considerations. We accept that the low growth case will place an even heavier premium on bringing low cost high productivity capacity onstream. We are not, however, proposing a guaranteed market for coal. Incremental decisions will need to take full account of the continued competitiveness of coal and of the optimum level of reliance on indigenous energy sources.

3.25 Costs of N C B deep mined coal have risen at about 5% per annum in real terms in the eight years since the 1972/3 rise in oil prices. This compares with an average annual rate of about 1% per annum in the previous 8 years. The main reasons for this acceleration were the rate of increase in miners' wages coupled with constant levels of productivity taking the period as a whole. The rate of increase of costs over the last 5 years has dropped significantly to about 2·5% per annum, as a result in part of recent increases in productivity. Prices have also been rising rapidly since 1973, although the rate has varied between markets. In the power station market, the average rate since 1973 has been about 8% per annum in real terms, but in the domestic market only 3·5% per annum. In the last 5 years, the rate of price increase has slowed to 3·5% per annum in the power station market. Coal has improved its competitive position against oil in spite of these price rises. The coal/oil price ratio in 1973 was about 1 but has fluctuated between 0·6 and 0·8 in the years since then. In 1980 it stood at 0·65. Figure 3.1 compares the rate of increase in the real price of fuels used by industry since 1970. Whereas fuel oil prices have increased at almost twice the rate of coal price increases, gas is now more competitive against coal than it was in 1970.

Figure 3.1   Industrial sector: 'real' (1) fuel price indices (1970=100)

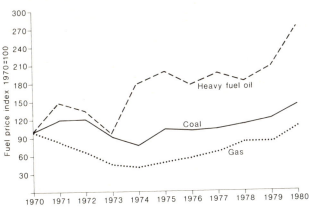

(1) Fuel price indices deflated by the Wholesale Price Index of materials (excluding crude oil and carbonising coal) purchased by all manufacturing industry

Source: *Department of Energy*

Opencast

3.26 Opencast production in 1980/81 by the N C B 's Opencast Executive was 15·3 million tonnes. A further 1·06 million tonnes was produced by licensed private operators. The map we reproduce later at Figure 11.1 illustrates the distribution of opencast operations.

3.27 Opencast coal is an important part of the N C B 's overall production, and its production costs are much lower than those of deep mining. In 1980/81, the N C B made an operating loss of £134·9 million on 109·6 million tonnes of deep mined production (before grants and interest), and a profit of £156·5 million on 15·3 million tonnes of opencast production. The annual production target of 15 million tonnes was first agreed in 1974, and has been subsequently endorsed on a number of occasions, most recently in answer to a Parliamentary Question on 3 July 1980.

3.28 Opencast coal can be produced economically, is normally of good quality, supplies some essential markets with which deep mine supplies cannot cope (this represents at least 10% of total opencast output), and in many instances is used to upgrade such supplies to an acceptable quality (30%). However, while substantial benefit to the landscape after restoration can result from some opencast operations, residual and continuing environmental damage is the inevitable result of many opencast operations. This damage must be weighed in the formulation of policy. We consider this issue in detail in Chapter 11 in the light of the extent and character of the evidence submitted to us.

IMPORTS

3.29 Imports supply only a small part of the U K 's total requirements for coal, having varied between 2% and 6% of coal consumption over the past six years:

Table 3.5   U K coal imports and consumption 1975–80

million tonnes

	1975	1976	1977	1978	1979	1980
Total coal consumption	122·2	123·6	124·0	120·5	129·4	123·5
Total coal imports	5·1	2·8	2·5	2·3	4·4	7·4
Of which:						
Steam coal	4·2	1·4	1·0	0·7	1·8	4·5
Coking coal	0·8	1·3	1·2	1·4	2·4	2·5
Anthracite	0·1	0·1	0·3	0·2	0·2	0·4

Source: *Department of Energy*

3.30 It must be stressed, contrary to popular belief, that the great majority of current N C B supplies are at present competitive with imports when compared on a delivered coal cost basis. This is the case even allowing for operating grants to the industry which are lower than those received by the industries of other important E E C coal producing countries. The N C B 's investment programme, aimed at reducing costs in existing capacity and opening new low-cost mines, should enable the Board to maintain its market position. For instance, on current expectations, coal from the new Selby mine should cost only 50–60% of the present average cost per tonne of N C B coal. However, changes in the exchange rate will exercise a major influence on the competitiveness of U K coal. The high value of sterling in the early part of 1981 worked to the disadvantage of the N C B 's position; but any fall in the value of the pound of course strengthens the Board's competitiveness.

3.31 There are in any case serious doubts about the availability and cost of coal imports in the long-term. World demand for coal imports is expected to grow rapidly over the next 20 years as a result of the increasing cost of oil. The World Coal Study (Coal Bridge to the Future 1980) has predicted that world requirements for imported steam coal will increase 10–15 fold between 1979 and 2000. Meanwhile, it must be added that coal demand is not growing as expected. The increase in supply needed to meet growth in demand will depend critically upon the ability and commitment of major coal exporters – Australia and the U S A principally and to a lesser extent Poland and South Africa – to expand their exports at the pace required.

3.32 The U S A and Australia certainly have massive reserves of coal and extensive opportunities for exploitation at relatively low cost. There are, however,

17

problems for mine developers in both countries and in particular the rate at which new mines are developed could be severely curbed by tightening environmental requirements, heavy costs for additional infrastructure and escalating labour costs and needs. Development costs are heavy. Mining and transport companies are tending to ask increasingly that overseas customers should commit themselves to long-term contracts for coal from new mines before development begins. Difficulties may also arise about the political acceptability of adopting the role of major energy exporter, especially when this involves confronting additional environmental problems.

3.33 The South African Government is planning to expand coal exports substantially in the long-term but this programme of expansion, while not being free from a number of problems which the U S A and Australia face, is inevitably also subject to a particularly high degree of political uncertainty. Exports from Poland are not now expected to rise significantly above their 1979 volume, or indeed to return to that level in the light of recent events. The sudden change in at least the short-term prospects of supplies from Poland illustrates vividly the vulnerability of imports to internal developments in the four major exporting countries.

3.34 Uncertainty over the future course of world coal prices is perhaps even greater than uncertainty on availabilities. By the end of 1980, prices at Rotterdam had risen to about $65 per tonne for steam coal and $75 per tonne for coking coal. Growing demand pressures, coupled with supply difficulties due to congested port and inland transport facilities in the U S A, are widely expected in the trade to result in further price increases during 1981, perhaps of over $10 per tonne.

3.35 Longer-term price movements cannot be predicted with any degree of certainty. If demand for coal outstrips supply, then coal prices can be expected to rise to the equivalent level of prices for fuel oil, the main competitor in the steam coal market. If supply exceeds demand, coal prices will be determined by supply costs and would be less likely to rise to the equivalent oil price. But a market condition close to this second scenario has existed in the U S A home market for many years and nevertheless prices have risen because production costs have risen; during the period 1973–78, for example, mining costs rose on average by 6·5% per annum in real terms, not far off the British record.

3.36 In view of the great uncertainties which characterise the world coal market, there is clearly an argument for ensuring that the U K's coal supplies, which are vital to our overall energy strategy, should be supplied from our own coal industry. But on the other hand, protection does traditionally encourage costs to rise. We assume that successive Governments will

resolve this dilemma by allowing some continuing flow of imports, and by allowing that flow to increase when home cost pressures seem particularly great. And we assume also that, as a result of such policies, the share of imports in total will, around the turn of the century, be not more than 10% of U K coal consumption. This leads us, as a working assumption for our Study of environmental effects, to the range of U K coal production discussed in Paragraph 3.45 below.

ENERGY POLICY IMPLICATION

3.37 The Department of Energy has stressed in its evidence that U K prospects, as they emerge from the Projections, need to be considered against the uncertainties inherent in the energy scene. Given prospective conditions in the world energy market, there is likely to be advantage in meeting as much of our energy requirements as we can economically from secure indigenous sources and in diversifying our sources of supply. Therefore, in the view of the Department, what emerges from the Projections is the need for a flexible strategy to cope with the uncertainties of the future; the need to take out insurance on a broad front and not to foreclose prematurely any supply option; the need to ensure a smooth transition from the short to medium term period of possible energy surpluses to the longer term when the supply position will be tight and getting progressively tighter; and the need for our energy policy in the longer term to be developed on three main components: coal, nuclear power, and energy conservation.

3.38 We accept that all forecasts are inevitably surrounded by great uncertainty and that therefore energy strategy cannot be translated into a blueprint. We see the role of the Department of Energy's projections as providing one among several possible technical and objective views of the future, and a quantitative backcloth against which Ministers can consider individual policy decisions. The Department of Energy claims no monopoly of wisdom in this difficult area. Continuing dialogue with experts outside Government and the fuel industries can make an important contribution, particularly in the difficult area of demand forecasting.

3.39 We attach particular importance to the Government's role in stimulating public debate and wider understanding of the broader framework within which particular energy policies contribute to national objectives. The inter-relation of fuels and the repercussions of decisions in one field on others are such that sectoral strategies can only be determined to a limited extent separately from each other. The public acceptability of the requirement for major projects as a function of these sectoral strategies would be considerably enhanced by a more systematic explanation by Government of the national policy framework. We return to this major theme in Chapter 21.

3.40 The Department's critics would argue that its approach is too supply orientated. In particular, they believe that the demand figures are substantially over-estimated and that we could achieve far greater energy savings than those projected, even though the Department has made a substantial and explicit allowance for energy conservation in its projections. This line of attack is perhaps best exemplified by Gerald Leach in his book 'A Low Energy Strategy for the U K', (Leach G, et al 1979) in which he argues that similar levels of G D P growth to those used by the Department could be achieved with little or no growth in energy demand. However, there is little evidence to show that such levels of energy savings could be achieved without a much more dirigiste and detailed energy policy than successive U K governments have been prepared to contemplate. Moreover, in a world increasingly dependent on coal and nuclear power, which require large fixed investments with long lead times, the presumption must be that a shortfall in supply could not easily or quickly be made good. The consequences of a relatively minor shortage in the supply of one fuel were demonstrated in the Summer of 1979 with the reduction in Iranian oil output. The need is for the development of robust policies which would not continually require alteration in response to external pressures, because the lead times for investment must be counted in decades, not the mere months or years of some other industrial activities.

3.41 Nevertheless, one of the most controversial elements in the current energy debate does concern the scope for conservation. It raises the fundamental question whether the current balance between investment in energy supply and in energy efficiency is right. We would see considerable merit in a major study evaluating the risks inherent in a supply orientated policy compared with a policy emphasising efficiency in energy use. Such a study would need to encompass, inter alia, the implications for the fuel mix of energy pricing policy, oil and gas depletion policy and high or low electrification. It would also examine the environmental impacts of energy conservation, not all of which are necessarily advantageous. We would have proposed to carry out such a study ourselves had we not been allowed "to fall into abeyance".

3.42 Secondly, there is controversy about the role of coal imports. The study by Professor Robinson and Miss Marshall entitled 'What future for British coal' (1981), which appeared very late in the course of our own work, is notable chiefly for bringing into sharp relief the crucial importance for the industry of its own cost structure. The fact that most commentators see an extremely bright future for the world coal producing industries does not necessarily imply that coal mining in any one country will prosper. World trade in coal is bound to increase very substantially by the end of the century, and it is unrealistic to suppose that, under any government, U K customers will persist, for a whole run of years, in paying substantially more for coal than they would have to pay if they bought on the world market. The Robinson and Marshall study draws attention to the high costs that would result from open-ended commitments to subsidise high cost collieries in the U K, and also, more controversially, to the high costs which might be incurred even in modern collieries if protection from imports were to lead to excessively high wages and sluggish management. Such high costs could come home to the taxpayer, in the shape of a charge for subsidies that would enable the industry to charge prices competitive with low cost imports; or come home to the customer directly in the form of high prices. But our view on costs, investment, and guaranteed markets must be stated clearly here. We have been told that rising productivity, the introduction of new capacity, and the full use of existing long-life capacity could yield a steady improvement in the N C B's cost structure, and we accept this. Moreover, we believe that most western governments will continue to find it sensible to favour indigenous supplies, and to be unwilling to jeopardise sensible long-run developments by allowing customers to play the market by allowing short bursts of low cost imports during temporary recessions. And, perhaps most important of all, we recognise the social unacceptability of repeating in the last decades of the century the sequence of creation and then destruction of mining communities in the way described in Chapter 1.

3.43 However, this does not lead us to the view that imports could or should be banned, nor constrained within guaranteed quantitative limits. It is not for this Commission explicitly to discuss at length the balance between the advantages of output guarantees to the N C B and the disadvantages to the consumer or taxpayer of being locked in to the output of an industry to some extent insulated from competition. But we are aware that this policy issue is one of balance, and we note particularly the challenge which clearly confronts the industry, both its management and its workforce. The interests of the taxpayer and the energy consumer on the one hand, and the long-term prospects for employment and real wages in the mining industry on the other, can only be reconciled by investment in industrial modernisation and realistic wage settlements which allow coal costs to retain and increase their competitive edge over other forms of fuel. These two sets of interests cannot be reconciled by policies which lead the N C B to produce the last possible tonne from obsolescent, high-cost capacity – the burden which such a policy would place on the consumer or on the taxpayer could not in practice be endured – and to persist in such a course would certainly mean that investment in modernisation would be the casualty.

3.44 But the issue of imports remains. Our view is that it is indeed reasonable for governments to do their best to provide markets for coal which is produced at productivity levels, and at real cost levels, which bear some clear and measurable relationship to the expectations of both parties (the producers and the government) when the original plans were made. It is most unlikely, and probably against the public interest from most points of view, that such agreed plans could be made to constitute an unbreakable guarantee – that would not be practical politics – and would very likely be bad economics. On the other hand, a regime in which the future markets of the Coal Board were completely uncertain would rationally reduce output and investment, managerial effort and productivity performance over the long run, and would probably over the short run (which will last for a decade or more) very materially increase costs to coal consumers who in that short run would have no alternative to N C B supplies.

3.45 We conclude therefore that neither the prospect of very much greater conservation (or of a particularly poor G D P performance), nor of vast supplies of cheap imported coal, is sufficiently strong to remove the presumption that we make about coal consumption and indigenous coal production in the U K by the turn of the century, if the environmental issues can be so managed as to constitute no substantial barrier. The range of coal consumption that we feel it prudent to examine is between 110–170 million tonnes per year; and the range of production is perhaps best characterised as lying between 100–150 million tonnes per year. We would stress that these figures do not constitute forecasts but represent a broad enough range of possibilities for the purposes of our Study of environmental issues. However, even if the lower levels of production and consumption are considered, it will be essential (perhaps particularly essential) for the Coal Board to invest in new capacity, on a fairly large scale, which makes Chapters 5–12 of this Study of particular importance. And throughout our Study we have kept in the front of our minds the consideration that only continued progress with improved and higher productivity processes can allow the Coal Board, its workforce, and its customers to prosper.

Chapter 4   Coal: The framework of environmental control

4.1 In the preceding Chapters we have reviewed the historical background and the future prospects of the coal industry. In subsequent Chapters we examine the environmental impact of coal production and use. We review in this Chapter the framework of planning and environmental legislation within which the N C B has to carry out its operations, and the evolution of that framework. This has two distinct elements. The first is the town and country planning system which has as a basic objective the reconciliation of competing demands on the use of land resources, but which also has an important role in the protection of the environment particularly in the control of development. The other is the array of powers which central and local government have to control the release of pollutants from industrial and domestic sources to the atmosphere, land, sea and water. The 1960s and 1970s saw a particularly marked increase in the concern of the government and the wider public about the environment. This growing concern was part of a wider international development. Much environmental legislation has been adopted in the wake of the United Nations Conference on the Human Environment held in Stockholm in 1972. The European Community adopted an environment programme in July 1973 and since joining the Community the United Kingdom has started to apply Community legislation embodying environmental standards.

TOWN AND COUNTRY PLANNING

4.2 At about the same time as the NCB was being established, the foundations of the present planning system in England and Wales were being laid down in the Town and Country Planning Act 1947 (since superseded in part by the 1968 Act and consolidated in the 1971 Act). Planning legislation in England and Wales has been paralleled in Scotland by the Town and Country Planning (Scotland) Acts. The equivalent in Scotland of the 1971 Act covering England and Wales is the Town and Country Planning (Scotland) Act 1972. There are some differences between the two systems, but they are not differences of principle. The 1947 Acts made a new start in the field of planning: development plans were to be prepared for every area in the country showing in outline the way in which each area was to be developed over the next 20 years; almost all development was made subject to planning permission; the development value of land was nationalised leaving

owners with virtually no rights except to continue using their property for its existing purpose; the number of planning authorities was reduced by transferring planning responsibilities from district to county and regional councils.

Development Plans

4.3 The basis of U K town and country planning is the development plan which, broadly speaking, contains a local planning authority's main objectives for the use of land resources in its area over a period of years. In 1968 a new type of development plan system consisting of "structure" and "local" plans was introduced to separate strategic from local issues; to reduce administrative delays; to emphasise positive planning for the creation of a pleasant environment; and to enable the public to play a greater part in the planning process. Structure plans are now prepared by county planning authorities in England and Wales and by regional planning authorities in Scotland for approval by the appropriate Secretary of State. They consist of a written statement, illustrated diagramatically, which sets out and justifies the main planning policies for the area; it provides a framework against which more detailed decisions can be made. Planning authorities may subsequently prepare local plans although there are now provisions whereby they can be prepared in advance of structure plans in some cases. Local plans consist of a written statement and a map setting out the authority's proposals for the development and other use of land for the area, defining precisely the area of land affected.

Development Control

4.4 With certain exceptions all development (which in planning law includes most forms of construction, engineering, and mining and any material change in the use of land or existing buildings) requires the prior consent of the local authority. They have considerable discretion in deciding whether or not to grant planning permission. The planning authority must take account of the development plan but they may take other material considerations into account and can approve an application which does not accord with the plan. Where the planning authority consider a proposal would be a substantial departure from the plan they must give the public an opportunity to make representations, and inform the Secretary of State who can then choose to call in the application for his own decision. The planning decisions of the authority can be one of three kinds: unconditional permission, permission subject to such conditions as they think fit, or refusal. The exceptions to this are those types of development for which planning permission is already deemed to be granted by general provisions in the General Development Order. These powers are of particular importance to the N C B and we discuss them more fully in paragraph 4.10.

Central and Local Government

4.5 Town and country planning is largely a function of local government. Nevertheless the Secretary of State does have an important role in overseeing the working of the town and country planning system, particularly where issues arise that go beyond the area of an individual planning authority and are of national or regional significance. In a wide range of matters central Government approval is necessary for proposals made by a local authority: structure plans have to be approved by the Secretary of State; and, if a local planning authority informs him of a major departure from a development plan he may direct that permission be refused or call in the application for his own decision. Anyone who receives an adverse decision from a local planning authority on a planning application may appeal to him. He also has the power to call in any case which he considers sufficiently important for his own decision. Once the Secretary of State has issued his decision he cannot review it and its merits cannot be challenged except in the High Court on points of law.

Town and Country Planning: Environmental Considerations

4.6 A major objective of the town and country planning system is the production of a satisfying environment. A local planning authority describes its policies for achieving this in its development plan. It implements them through its own activities and by controlling the activities of others by means of development control. Local planning authorities have very wide powers to decide whether to grant a planning application and if so what conditions they attach to it; in the terms of the legislation they may grant permission subject to "such conditions as they think fit". "Amenity" is a key concept in this aspect of town and country planning. It has no precise legal definition and is an all-embracing term which takes account of all the factors which establish the general quality of the environment and life in a particular area. Its meaning and importance may best be conveyed by a statement once included in a publication by the Ministry of Housing and Local Government: "Anything ugly, dirty, noisy, crowded, destructive, intrusive, or uncomfortable may 'injure the interests of amenity' and, therefore, be of concern to the planning authority".

4.7 Local authorities therefore have very wide scope for establishing conditions, often in consultation with the developer, which control the siting and design of a development and the manner in which activities associated with it are to be carried out. There are no specific standards of amenity or design against which local authorities have to assess planning applications. It is their responsibility to decide whether the proposed development is acceptable in the context of the locality where it is to take place and the general environmental standards prevailing at the time. This is an important point. Over the years higher standards have come to be applied to improve and protect the environment and this has been reflected in the operation of the development control system. However, it is an important general principle of that system that conditions cannot be imposed retrospectively unless a local authority pays full compensation to the developer for the cost of doing so. Thus, although improved standards of design and environmental protection may develop over time, this does not mean that, generally speaking, planning permissions can be continually updated to take account of them. At any one time a range of standards of design and operation may be found connected with any particular activity.

Town and Country Planning: the N C B

4.8 The reconciliation between economic and amenity interests in mineral working is an obvious matter of concern for planning authorities. Their responsibility generally stated is to ensure that mineral working is carried out with proper regard for the appearance and other amenities of the area and, if the working inflicts too great an injury to the comfort and living conditions of the people in the area or to amenities generally, it can be limited or even prevented. Powers to control working stem from the definition of development which includes "the carrying out of mining . . . operations in, on, over or under land". Further, the tipping of waste constitutes development. Generally speaking therefore mineral workings, ancillary buildings, depositing of waste and the construction of means of access to sites require planning permission.

4.9 However, a number of special provisions apply to the operations of the N C B. In the first place, opencast coal working has taken place under legislation separate from the Town and Country Planning Acts. It began during the war under emergency legislation and continued under this until 1958. The Opencast Coal Act of 1958 laid down a special method of control operated by the then Minister of Power, now the Secretary of State for Energy, which was subsequently amended by the Coal Industry Act 1975. We give full details in Chapter 11.

4.10 The other major exception to the general principle that the N C B's activities require planning permission from the local planning authority is the power it has under the General Development Order

1977 (S I 1977 No 289) in England and Wales and in Scotland under the Town and Country Planning (General Development) (Scotland) Order 1975 (S I 1975 No 679). As we explained, this gives deemed planning permission for a number of activities and there are both historical and practical reasons why this is so. The Town and Country Planning Act 1947 established the basic principle that a change in the use of land required planning permission but the continuation of an existing use did not. As a result the continued working of mines begun before 1 July 1948 is "permitted development" and, therefore, did not and still does not require specific planning approval. Moreover, any extension of these activities does not require planning permission. The same also applies to the tipping of waste on a site used for that purpose on 1 July 1948 whether or not the superficial area or height of the tip is extended. Local authorities may not at present impose any requirements for waste management or after-care on the N C B for spoil tipped before 1974, but they can require the N C B to submit for approval a tipping scheme of waste deposited after 1 April 1974. These rights also enable the N C B to recommence tipping at a disused tip without requiring further planning permission. The N C B may also carry out ancillary developments in connection with coal industry activities subject to the county planning authorities approving the detailed plans and specifications for the erection, alteration and extension of buildings and other structures. The main purpose of this provision, which is similar to that available to other industrial developers, is to allow a certain degree of flexibility in carrying out work on sites in connection with the use for which planning permission has been given. Otherwise numerous planning applications would be required for development which would not significantly alter the impact of the site on the surrounding area.

4.11 The Government has recently introduced the Town and Country Planning (Minerals) Bill which, as we complete our Report, has not yet received Royal Assent. It will introduce some significant changes into the planning law which affects mineral operators including the N C B. The Department of the Environment has also issued a consultation document

Figure 4.1

HOW PLANNING PERMISSION IS GRANTED CURRENTLY

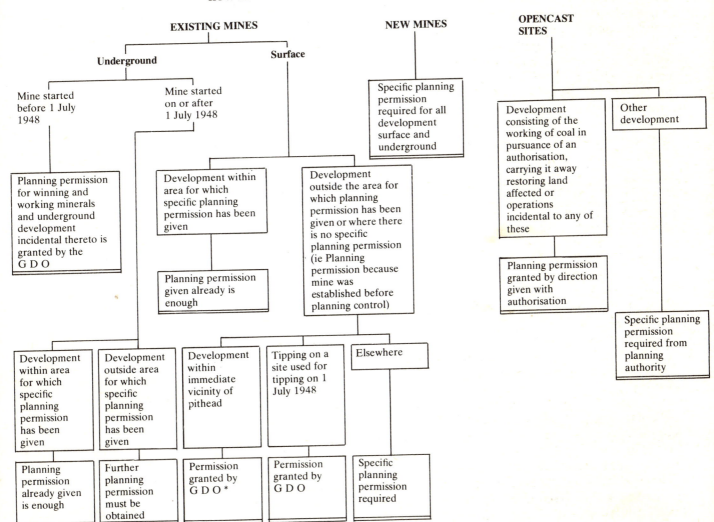

*The approval of the planning authority is required for certain details of buildings

Source: N C B

(1979a) on its proposals to amend the G D O which will also have major implications for the N C B. We discuss the details of these proposals in the relevant Chapters of the Report. Figure 4.1 shows in summary form how the existing planning system affects the N C B's activities. We examine the operation of the planning system in a wider context in Chapter 21.

POLLUTION CONTROL

4.12 To complete the account of the framework of legislative control we now describe the way in which the release of specific pollutants and their effect on the environment is controlled in the United Kingdom. The two systems are not entirely separate: careful planning can be important in ensuring the effectiveness and in minimising the cost of satisfactory pollution control because it ensures that siting decisions can take account of the possible pollution problems arising from too close an association of industrial and residential development. The pollution control machinery has, however, particular relevance to the combustion of coal.

The Definition of Pollution

4.13 Pollution is generally defined in terms of effects. For example, the definition of air pollution adopted by the United Nations Economic Commission for Europe (E C E) is: "The introduction by man, directly or indirectly, of substances or energy into the air resulting in deleterious effects of such a nature as to endanger human health, harm living resources and ecosystems and material property and impair or interfere with amenities and other legitimate uses of the environment". This definition allows for the fact that in many cases the environment has a certain capacity to assimilate "pollutants" by converting them to less harmful substances and that deleterious effects may not occur until the concentration of the substance exceeds some threshold level. For example, sulphur dioxide, a widespread pollutant, is emitted in some measure by the combustion of all fuels, and by the reduction of many non-ferrous ores, but over the world, as a whole, natural emissions – from volcanoes, decaying vegetation and forest fires – are about equal to man-made emissions. Indeed sulphur is an essential element for all living organisms and so there are natural global cycles involving its release to the atmosphere or to water, absorption by rain, oceans or surface materials and re-assimilation into growing material. At low concentrations sulphur dioxide cannot therefore be regarded as a pollutant.

The Objectives of Pollution Control

4.14 The broad objectives of pollution control are usually accepted as being the maintenance of human life and health both now and in the future, and of the life support systems necessary to sustain mankind. The control of pollution to the external environment

presents a number of particular problems to the policy maker. A pollutant is normally transferred by a combination of air, freshwater, land and sea and concentrations of individual pollutants in the environment at large are much lower than those that occur in the working environment. However, they may also be present in combination with consequent risks of interaction. Given these considerations, the likely number of people to be affected ultimately, by what means and by what combination of pollutants, is sometimes very difficult to assess. They will also vary according to the general geographical and industrial character of the locality.

4.15 The aim of pollution control policy in the United Kingdom therefore has been to try to reduce emissions progressively in a way that is consistent with economic and technical feasibility and with what at any one time is thought to be an acceptable ultimate objective for the locality in question. In attempting to achieve that objective the United Kingdom has followed the general principle that the discharge of polluting substances should be looked at in the light of local circumstances, because the danger of any pollutant depends on its toxicity, the circumstances of its release, the likelihood that it will reach a target and the impact it may have on the target. Executive responsibility has wherever possible been left to local agencies operating within the framework of broadly drawn legislation. Another important principle is the "polluter pays principle" which we discussed in Chapter 1.

Emissions to air

4.16 Control of air pollution in the United Kingdom is provided for under the Alkali Acts, the Clean Air Acts, the Control of Pollution Act 1974 and the Public Health Acts which together cover emissions from industrial and domestic sources and until recently no explicit standards for ambient concentrations had been set. Emissions have been controlled at source. However the E E C Council of Ministers recently adopted a directive on concentrations of smoke and sulphur dioxide in the atmosphere which prescribes mandatory maximum values for the ground level concentration of these pollutants to be met throughout the Community by 1 April 1983 (European Communities 1980). It also provides a non-mandatory lower set of values, which are intended to serve as reference points for the longer term improvement of air quality and for the setting of targets in special zones if necessary. The following paragraphs give a brief factual description of the legislation. A comprehensive review of the arrangements for controlling domestic and industrial air pollution was undertaken by the Royal Commission on Environmental Pollution (R C E P) and the results published in its Fifth Report, 'Air Pollution: An Integrated Approach' (R C E P 1976a). The Government's conclusions on this Report are still awaited. We discuss the control of air pollution from coal use in Part III of this Report.

Alkali Etc Works Regulation Act 1906

4.17 The emissions from certain chemical and industrial processes have long been subject to special control. The first was the production of alkali which gave rise to huge emissions of hydrochloric acid gas. The Alkali Act of 1863 was subsequently extended to cover other processes and was consolidated in the Alkali Etc Works Regulations Act of 1906 which remained the governing statute until it was partly subsumed by the Health and Safety at Work Etc Act 1974. There are now almost 60 processes, involving some 2,300 works and 3,300 operations registrable under the Act in England, Wales and Scotland, including the major industrial coal burning processes such as electricity generation, gas and coke works, cement and lime producers and the ceramic industry.

4.18 H M Alkali and Clean Air Inspectorate of the Health and Safety Executive (H S E) are responsible for the administration of the control of air pollution from registered works in England and Wales, and in Scotland similar responsibility lies with H M Industrial Pollution Inspectorate, who operate in this respect as agents of the Health and Safety Commission (H S C). The Act specifies emission limits for only four chemical processes, and for all other processes the requirement is that Bpm shall be used to prevent the emission of noxious or offensive gases and to render what is discharged harmless and inoffensive. "Bpm" is not defined in the Act, but other related legislation defines "practicable". The Clean Air Act 1956 gives a definition that essentially describes the interpretation of Bpm by the Inspectorates. This is:

> " 'Practicable' means reasonably practicable having regard, among other things, to local conditions and circumstances, to the financial implications and to the current state of technical knowledge, and 'practicable means' includes the provision and maintenance of plant and the proper use thereof."

4.19 For some processes for which a statutory emission limit is not provided, emission limits are laid down by the Inspectorates after discussions with the industry concerned, and adherence to these limits is regarded by the Inspectorates as evidence that Bpm are being used at the relevant points. These limits are altered from time to time as new techniques for control are developed or because of changing circumstances. The obligation to comply with Bpm is a continuing one and regular inspections and tests are carried out by the Inspectorates to ensure that they are met. In general, the Inspectorates are consistent in their approach although each case is considered individually.

Clean Air Acts 1956 and 1968

4.20 The other major sources of man-made emissions to the atmosphere are domestic sources and industrial sources not covered by the Alkali Acts. Control of these emissions is the responsiblity of local authorities.

Before the 1950s they were empowered by certain Public Health Acts to prevent nuisance from smoke and similar pollution. However, in the 1950s it became clear that urban pollution containing a high proportion of smoke and sulphur dioxide had been the direct cause of premature deaths among chronically sick and old people: it is estimated that the London smog of 1952 resulted in 4,000 such deaths. It was recognised that more systematic and widespread control measures were needed. As a result the Clean Air Act of 1956 was passed; this provided for the introduction of "smoke control areas"; replaced and extended the provisions relating to smoke nuisances in the Public Health Act; brought under control certain emissions from industrial combustion processes not within the scope of the Alkali Act; and dramatically improved air quality in urban areas. The provisions were revised by a further act in 1968. It is most helpful to look at the controls relating to industrial and domestic sources separately.

4.21 The provisions of the Acts relating to industrial sources of emissions established some prohibitions but generally gave local authorities a number of discretionary powers. Dark smoke may not be emitted from any trade or industrial premises or from the chimney of any building. The Secretary of State can relax these provisions by order and this has been done to accommodate essential lighting up and soot blowing and to exempt certain activities subject to conditions where dark smoke is unavoidable.

4.22 The discretionary powers available to local authorities concern controls over new furnaces. The local authority must be informed of any proposal to install a new furnace. New furnaces other than domestic boilers with a certain maximum heating capacity, have to be capable of being operated continuously without emitting smoke when burning fuel of the type for which they were designed. The local authorities also have discretionary powers to ensure that certain types of new furnaces are equipped to prevent the emission of grit and dust and also have powers to approve the height of any new chimney built to serve furnaces of this type. Operators of furnaces may appeal to the Secretary of State against the decision of local authorities in these cases. The Secretary of State may prescribe limits on the rates of emission of grit and dust from certain furnaces and also exempt classes of furnaces from the need to fit equipment to prevent the emission of grit and dust. In effect these powers reflect the general principles of local enforcement and the examination of the emission of substances to the environment in the light of local circumstances. However, there is a high degree of uniformity in the application of some of these controls, partly because guidance is given in Circulars and Regulations issued by central Government and partly because local authorities seek advice from the Inspectorates.

4.23 The other major power given to local authorities by the Clean Air Acts is the power to make smoke control orders prohibiting the emission of smoke from buildings, including dwellings, in all or part of their district. Specific buildings may be exempted by the order. Householders may burn fuels which have been specified in regulations by the Secretary of State as being smokeless or may install coal burning appliances which similarly have been specified as being smokeless. Grants are available to cover part of the cost of conversion. The initiative for making smoke control orders lies with local authorities although central control is maintained over the certification of fuels and appliances as being smokeless. Section 8 of the 1968 Act also enables the Secretary of State to require the introduction of smoke control areas.

Control of Pollution Act 1974

4.24 Part IV of this Act empowers local authorities in England and Wales to carry out investigations into air pollution by enabling them to obtain information about the emissions to the atmosphere from any premises other than private dwellings. Any information obtained by the local authority has, with certain exceptions, to be kept in a register open to the public. Section 76 of the Act gives the Secretary of State for the Environment and the Secretaries of State for Scotland and Wales new powers to limit or reduce air pollution by making regulations to control the sulphur content of fuel oil for furnaces or engines.

Water pollution and noise: The Control of Pollution Act 1974

4.25 We have already referred to the provisions of the Control of Pollution Act 1974 in respect of air pollution. The Act also contains important provisions on water pollution and noise pollution. The control of the discharge of water from coal mines has been an integral part of the development of coal mining. The Control of Pollution Act 1974, when Part II comes into force, taken together with earlier legislation, and particularly the Rivers (Prevention of Pollution) Acts 1951 and 1961 and the corresponding Scottish Acts of 1951 and 1965, should ensure that all discharges of trade and sewerage effluent made to rivers, the sea, specified underground waters and land are subject to the control of water authorities in England and Wales and, in Scotland, the river purification authorities. The control of pollution in rivers takes account of the different uses to which water is being put as quality

requirements will vary considerably. Standards are set for particular discharges to achieve the desired quality set for a river in the light of particular circumstances.

4.26 Noise has only recently come to be treated as a matter of environmental concern. The Control of Pollution Act strengthened local authority powers which had first been introduced by the Noise Abatement Act 1960 to require the abatement of a noise nuisance, including the execution of any necessary works. This legislation is directed at the control of noise in the environment generally and it has always been left to specific legislation to deal with noise from road traffic, air traffic and within industrial premises. There are no statutory standards relating to neighbourhood noise. A new power in the Control of Pollution Act enables local authorities in England and Wales to establish noise abatement zones within which the authority registers typical noise emissions from classified premises and uses these as reference levels to prevent noise increasing and against which noise reduction action can be taken, having regard to the best practicable means. This has not yet been implemented in Scotland. The Act gives no indication of what local authorities might regard as acceptable. As we have already explained planning permissions may include conditions relating to noise.

Conclusions

4.27 The United Kingdom has adopted to date a pragmatic approach to environmental control. Statutes have not generally specified standards either for particular emissions or for general environmental quality. Authorities have been given discretion to set their own local standards although in practice they often work to fairly uniform standards or within widely accepted limits. The system has encouraged the progressive improvement of the environment over time. Control over particular pollutants can be tightened as new technology becomes available or if new evidence becomes available of their effect on people and the environment which justifies stricter controls. A policy of environmental management cannot set out to abolish polluting substances entirely because of the cost in terms of other sources of human welfare. And it has always been recognised that pollution only arises when substances are present in sufficient concentrations to be liable to cause damage. U K practice has concentrated on areas where there is a need for control for health reasons and, for the rest, where effective control can be achieved at an acceptable cost.

PART II

Coal production

Chapter 5: Deep mining

INTRODUCTION

5.1 In Part II we examine in detail the environmental effects of both deep and opencast coal extraction. In this Chapter we look briefly at methods of deep mining, since some understanding of mining operations is necessary to appreciate the nature and range of environmental impacts. In Chapter 6 we look briefly at the N C B's approach to the opening up of new deep mines, and at the extent of their exploration programme. The environmental effects of deep mines themselves are next examined, dealing first with the more minor effects, of surplus water disposal, noise and dust (Chapter 7); and then concentrating in the following two Chapters on the major problems of subsidence and spoil disposal (Chapters 8 and 9). We then take into account the effects of past mining activities and the resulting dereliction, and consider how to ensure that in future dereliction will not result when particular mining activity ceases (Chapter 10). Chapter 11 discusses all aspects of opencast mining. Chapter 12 draws together our overall conclusions on coal production.

UNDERGROUND MINING OPERATIONS

5.2 The type of mining techniques employed can make a substantial difference to the impact of a mine on the surrounding community. Throughout this century the industry has experienced a change from hand methods of extraction to mechanisation. This has had a substantial benefit in terms of increased output per man-shift, and in improvements in working conditions for those below ground.

Access to deep coal

5.3 Historically there have been two ways whereby access has been obtained to deep mined coal. The simplest way of all has been by "drift" or "adit"; this is a sloping or horizontal tunnel driven in early days into the side of a hill where coal outcrops. The alternative way was by a vertical shaft. In primitive mining in the shallow coalfields this took the form of a short shaft negotiated by rope or ladder leading down to a chamber. The miner hollowed out the sides of the chamber for as far as was safe. Once the limits of safety were reached, a fresh shaft was sunk and the coal was extracted from a further chamber. Viewed in section such a chamber was usually shaped like a bell, hence its

name of bell pit. In later and deeper mines, the shafts were lined with supports and mechanised winding gear installed.

5.4 These two basic methods of access are still in use today. The vertical shaft is probably the most familiar form of coal mine, and is particularly needed for the deeper mines. However, drift mines are still important, and can have considerable advantages over the vertical shaft; the cost of transporting men and materials is considerably lower, largely because conveyor belts rather than vertical winding can be used. The removal of coal from the new Selby mine will be through a drift. However, drift access is not always feasible or economic, depending upon the depth of the mine, on the characteristics of the overlying rock, and the configuration of the coal seams.

Methods of working

5.5 There are two major methods of working coal inside a coal mine, although many variants have been used and in some cases continue. The older method is called stall and pillar (or room and pillar, or bord and pillar). In this system coal is won by driving a network of narrow roadways ("stalls") through a coal seam. Substantial pillars of coal are left between these stalls to support the roof. In old workings there was frequently sufficient space for only one man, or at most a small team, to hew coal in each of the narrow stalls. The slow rate of advance by this narrow work often resulted in excessive congestion in the workings, a low percentage of extraction, and difficulties in ventilation.

5.6 Almost all output in the U K is now produced by longwall methods, which originated in Shropshire in the 17th century. In the past these involved a large number of men working together along the line of a face, and removing the whole of the coal. The roof was supported by a row of props which were moved forward as the coal face advanced. Pack-walls built with stones from the roof supported the waste area behind the coal face; the coal won from the face was transported along the line of the face to the main roadway and thence to the pit bottom. The coal was obtained from the face by undercutting the seam with a pick. The coal then either collapsed under its own weight, or was broken down by use of wedges or later by explosives (see Plate 5.1). Early attempts at mechanisation involved undercutting the coal by machine rather than by hand. Loading of the coal on to the transport system was, however, still undertaken by hand during these early stages of mechanisation.

5.7 The modern mechanised longwall face (Figure 5.1) differs in some important respects from these early techniques. In the first place, the coal is generally cut by a power loader machine which shears the coal from the face instead of under-cutting it. The shearer consists of a drum rotating on its horizontal axis which has picks around its barrel. It rotates at high speed, cutting into the coal as it passes along the face. The coal which is cut is pushed away from the shearing head by a plough on to an armoured chain conveyor. This conveyor carries the coal to the ends of the face, back along the access roadways to the main roadway, and thence to the bottom of the pit. At each pass along the face the machine cuts coal off the face to a depth normally of up to one metre. This method of cutting coal results in a high proportion of small sized coal coming from the face, in contrast to the undercut methods. However, much of the coal is now sold at power stations, where pulverised coal is a necessary feed into modern boilers. Thus the small size of the coal coming from this type of machine is no real disadvantage.

drawn forward, and then returned to the support position for the roof. This is done successively along the face as the shearing machine passes by. The roof area behind the face is left unsupported, and usually collapses almost immediately after the withdrawal of supports. It is this process of the collapse of the roof strata, or "caving" that results in subsidence at the surface (see Plate 5.2).

5.9 There are two ways in which the longwall system can be operated. In advance mining, the face moves forward into the coal away from the main roadway. In the retreat system, the access roadways at either end of the face are first driven at right angles from the main roadway up to the boundary of the area to be worked. The coal is then produced from that boundary working backwards towards the main roadway. Retreat mining now accounts for about 24% of major longwall face output. By working a face between roadways that have already been driven, retreat mining removes the need to synchronize the advance of roadway headings with the face, making production simpler and more efficient than on conventional advancing faces. It also gives advance warning of faults or other difficult mining conditions which the face may encounter. Average output per day from the 131 retreat faces working at

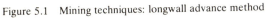
Figure 5.1   Mining techniques: longwall advance method

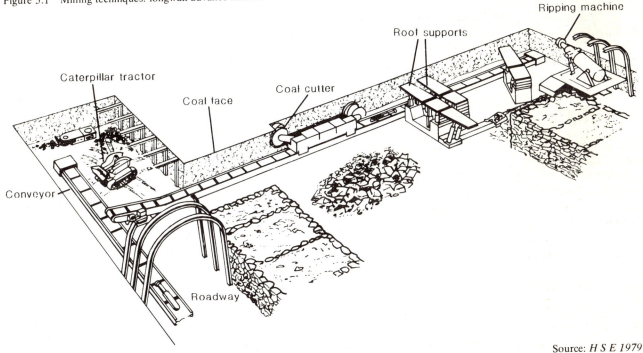

Source: *H S E 1979*

5.8 The coal face can advance at up to 25 metres per week or even more. This speed means that manual advance of the roof supports is too slow for safety. Hydraulically powered supports have therefore been developed which hold up the roof by hydraulic pressure, and are moved forward by the same pressure. The average face is about 200 metres long, and there are approximately 170 sets of support for this length. As the face moves forward these supports are lowered,

the end of March 1981 was 24% higher than from advancing faces. However, the retreat system cannot be used universally. It can involve financial penalties, as there is considerable initial expense involved in driving the access roadways before productive output of coal is derived from the face. There are places where the convergence of the strata on the preformed access roadways would be excessive, entailing constant repair work. There may be need to leave wide pillars of coal unworked between faces in order to minimise this convergence. This sterilises a substantial proportion of possible output. The use of handwon methods, together with mechanised systems, to extract coal from the pillars protecting gate and trunk roads, could result in both greater employment generation, and increased coal recovery. However, detailed assessment of this possibility would be needed, not least from the safety aspect.

5.10 The coal faces themselves may be 50–250 metres long. The seams worked vary in thickness from 0·5–3 metres, and are typically about 1·5 metres thick. The maximum depth below the surface at which they are worked is normally 1,000 metres, and the minimum depth 100 metres. To be economic, a coal face normally requires a life which enables an advance through a minimum of about 1,000 metres of coal, because of the high cost of installing the coal face machinery.

5.11 There are three major differences to be seen on the surface resulting from the modern techniques of longwall mining, compared with the traditional stall and pillar methods. The first results from the higher productivity of the longwall faces. Clearly with high productivity faces, a more extensive range of surface buildings will be required to operate the mine and to deal with the greater output. Secondly, the machines used to shear the coal from the face and to drive the roadways forward cannot be operated to discriminate as sensitively as the old hand methods between coal and spoil. Thus a much higher proportion of spoil to coal is now produced than formerly. Thirdly, the longwall mining method with early caving offers the possibility of a predictable and short term effect of subsidence at the surface. The earlier methods of stall and pillar working give a continued support to the roof after the coal is extracted; but over time the pressure of the overlying strata can weaken the pillars, which later collapse, thus causing less predictable subsidence for many years.

Conditions of mining

5.12 It goes without saying that conditions of mining have improved immeasurably over the years. No longer are young children required to drag, on their knees, and in passageways sometimes as little as 16 inches (40 centimetres) high, sledges of coal weighing 2–5 hundredweights (100–250 kilograms). The past legacy of dirt and danger often accompanied by harsh struggles cannot be over-estimated.

5.13 The introduction of the power loading machines in the 1960s has done much to ease one of the most tedious and burdensome tasks. Even as recently as the 1950s it had been possible to observe that "we must realise how difficult it is to make a man enthuse about throwing 12 or more tons of coal a distance of 7 feet or more, in about 6 hours of working time, often in dust

30

laden, warm, humid atmospheres, and in cramped positions – and keep this up daily for 40 years or more" (Sales W H, quoted in Berkovitch I, 1977).

5.14 In spite of modern improvements, mining is still an uncomfortable and unpleasant job. We have observed this at first hand. We were struck by the use of heavy water jets needed to keep down dust, the all-pervasive sour smell, the seeping oil, and the sweat-streaked blackened bodies of the miners, "To a miner comfort is a luxury. Dust, firedamp, high temperatures and high relative humidities, intense noise, low lighting luminous levels, cramped and extremely dirty working spaces, besides the risks of sudden firedamp emission, water insurges, explosions mean that a coalmine must be designed to have the best health, safety and welfare conditions possible". (Croome D J 1978).

5.15 Although conditions vary from mine to mine, the difficulties even now have been described recently as follows: "Miners are ordinary people coping with extraordinary working conditions. They descend in cages to depths of as much as 3,000 ft; they travel as far as five miles from the pit bottom to the coal face by man-riding facilities and by foot; they walk along roadways which are made for transporting coal not people, and which are narrow, uneven, have very low roofs and are dangerous because they are dominated by conveyors, machines and the chains and cables which serve them. At the face, miners work in environments often no bigger than the knee hole in a typist's desk, on their sides, on their knees, sometimes in water, often in high temperatures, always in dust and in the dark and, since the introduction of machines, with great noise". (Allen V L 1981).

Figure 5.2   Coal mines: accidents from all causes and total rate per 100,000 manshifts 1947–79

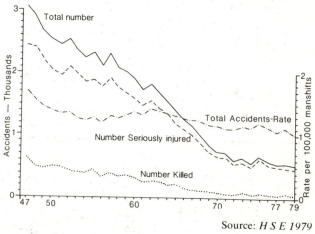

Source: *H S E 1979*

Health and safety

5.16 Together with the general improvements in working conditions underground, there has also been a considerable reduction in accidents at the face. This reflects a general reduction in the number of accidents

Plate 5.1 (*above*) Getting coal by hand, Frog Lane Colliery, near Bristol, circa 1905. Before nationalisation 97% of coal was won by hand-mining and this picture shows typical hand-getting methods using picks and shovels.
(*Photograph by courtesy of the National Coal Board*).

Plate 5.2 (*left*) Mechanised mining, Riddings Drift Mine, Barnsley, Yorkshire. The power-loader (right) slices coal from along the 300 metre face taking a depth of coal of a metre at each pass. After each cut the shearer moves forward on top of its armoured flexible conveyor (running from foreground to background) and the hydraulic powered roof supports (left) move forward to support the roof.
(*Photograph by courtesy of the National Coal Board*).

Plate 5.3 (*left*) Park Mill Colliery, Clayton West, Yorkshire. This aerial picture shows the proximity of many older collieries to residential areas.
(*Photograph by courtesy of West Yorkshire County Council*).

Plate 5.4 (*above*) Six Bells Colliery, Abertillery, South Wales. Sunk in 1890 at the height of the Victorian coal boom in South Wales. Such pits were sunk in sparsely populated areas, and villages to house the miners were built as close as possible to the pithead.
(*Photograph by courtesy of the National Coal Board*)

Plate 5.5 (*below*) Betws Drift Mine, South Wales. This, the most modern Welsh colliery, was completed in 1979. The surface buildings have been designed to be as unobtrusive as possible and are landscaped in a rural setting. In 1980 the design received a premier award from the Business and Industry Panel for the Environment.
(*Photograph by courtesy of the National Coal Board*)

Plate 5.6   Colliery buildings and yard, Barnsley. This picture shows the adverse
environmental effects of untidiness.
(*Photograph by courtesy of Barnsley Metropolitan Borough Council*)

Plate 5.7   Thurcroft Colliery buildings and yard, Yorkshire. In contrast to the picture above
this shows the benefit of a neat and well maintained appearance.
(*Photograph by courtesy of the National Coal Board*)

throughout the industry as illustrated in the graph at Figure 5.2. The graph at Figure 5.3 shows the changes in the principal types of accidents. The reduction over the years in accidents caused by falls of ground is clearly shown. However, in part the overall reduction in numbers of accidents can be attributed to the reduction in quantity of output over the years, in part to the increases in productivity whereby fewer men can produce greater output, and in part to a decreasing workforce.

Figure 5.3 Coal mines: number of accidents by major cause 1947–79

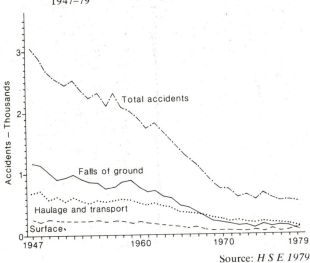

Source: *H S E 1979*

5.17 Changes in the incidence rates of accidents indicate the underlying safety trends more clearly. Figure 5.2 shows the incidence rates of all reportable accidents per 100,000 man-shifts between 1947 and 1979. It indicates a generally falling trend up to 1956, and from 1962–79, but with an increase in the late 50s and early 60s. The incidence of accidents is now the lowest over the entire period – over 30% lower than in 1962. Accidents could also be expressed per million tonnes of coal output; but this would ignore changes in productivity over the period, and so would give a less clear picture of long term changes in mine safety as they affect the individual miner.

5.18 The principal characteristics of accidents over the years have changed. There has been tendency for accidents that are caused by uniquely mining features, for example, falls of rock, underground explosion, suffocation, to reduce, but for there to be some increase in those associated with machinery. Thus there has been a very significant reduction in the accident rate caused by falls of ground between 1947–79. However, taking coal haulage and transport as a particular feature of the progress of mechanisation, there has been little significant change in the rate of accidents over the last 20 years.

5.19 Figure 5.4 also shows an erratic incidence rate of accidents at the surface over the same period, with no apparent tendency to diminish. Indeed, in 1979 the

total number of reportable accidents was the highest recorded since 1967, and represented 20% of all reportable accidents at coalmines. This is despite the extensive modernisation of colliery surfaces which has resulted in a reduction in the labour involved and has made many tasks less onerous. Of all surface accidents, 22% occurred in coal preparation plants, and the majority of accidents occurred while machinery was being maintained.

Figure 5.4 Coal mines: rate of accidents by major cause 1947–79

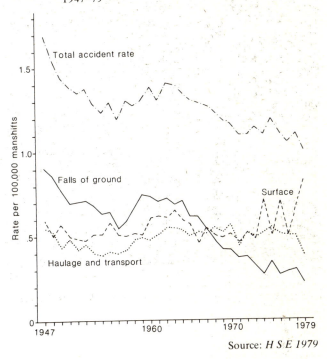

Source: *H S E 1979*

5.20 Nevertheless, despite the reduction in underground accident rates, it is important to remember that mining takes place in a hostile environment. Table 5.1 sets out the type and location of all incidents in 1979. Of all fatal and serious accidents, 89% and 79% respectively occurred underground, although only 21% of fatal accidents and 25% of serious accidents actually occurred at the working face. However, the total number of accidents caused by falls of ground is the lowest ever recorded, and the H S E view this as a "justification for investment in new equipment and the application of sophisticated mining techniques at the coal face" (H S E 1979).

5.21 In addition to these accidents which are unique to coalmining, the industry also carries with it the risk of conventional industrial accidents involving tools, machinery, transport, conveyors and electricity. In 1979 44% of all fatal accidents and 32% of all serious accidents underground were caused by transport and machines. The H S E in their Annual Report for 1979 have expressed concern over accidents in underground transport which constitute 28% of all reportable accidents. Although the number of such accidents (14 deaths and 130 seriously injured) is the lowest ever recorded, the H S E have pointed out that if the general trend of improvement is to be sustained, reliance can no longer be placed on outdated and over-burdened transport systems to meet the needs of the industry. They are of the view that there is clearly a case for more emphasis on safety in underground transport operations, since 55% of underground transport accidents result from lack of discipline, bad operator practice and illegal man-riding.

Table 5.1   Coal mines: Percentage of fatal and serious (reportable) accidents by type and location 1979

Location	Percentage of fatal accidents	Percentage of serious accidents
Underground		
At the face:	21	25
Falls of ground	6	16
Transport	4	5
Machines	11	4
Elsewhere underground:	68	54
Falls of ground	4	2
Transport	27	22
Machines	2	1
Miscellaneous	29	28
Shaft	6	1
Total underground	89	79
Surface (at collieries)		
Transport	3	6
Machines	4	2
Miscellaneous	4	13
Total surface	11	21
Total, all accidents %	100%	100%
Total, all accidents, number	46	473

Source: Derived from H S E 1979, Appendix 3

Figure 5.5   Trend of pneumoconiosis cases (new) 1958–79

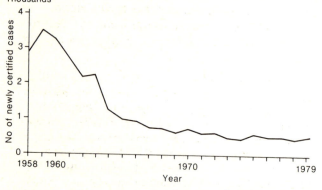

5.22 Mining causes its own health hazards, of which pneumoconiosis is the best known. However, as the graph at Figure 5.5 shows there has been a general downward trend in new cases over the past 20 years. The number of newly certified cases diagnosed by the Pneumoconiosis Medical Panel of the Department of

Health and Social Security in 1979 was 476. Although this represents an increase of 11% on the previous year, most of the first certificates now apply to men in the older age groups, of whom almost 60% have since left the industry.

5.23 Pneumoconiosis is caused by exposure to dust over a long period so that these figures do not reflect health conditions in modern mining. Although the problem of respirable dust control has been increased by the mechanisation of mining, especially in the 1960s, dust control techniques and other protective measures have been developed and have resulted in an overall reduction in the levels of exposure to airborne dust. According to the H S E "comparisons of the prevalence and progression of pneumoconiosis at certain collieries from surveys made since 1959 indicate a further fall in prevalence, and provided there is no deterioration in current dust control methods, continued decline in the incidence of pneumoconiosis can be expected. Nevertheless, there is no room for complacency and efforts not only to improve the methods of dust suppression but also the monitoring of the environmental conditions, must continue". (H S E 1979).

5.24 Coal mining is still a hazardous activity compared with many others. Table 5.2 demonstrates that the accident rate in coal mining is higher than construction activity, which is normally regarded as a high risk occupation.

Table 5.2   Accidents to employees in selected industries

Industry	Deaths per 100 000 employees at risk		Accident per 100 employees at risk	
	Average 1970–74	Average 1975–79	Average 1970–74	Average 1975–79
Agriculture	14·0[1]	11·4	2·1[1]	1·6
Coal Mines[2]	25·0	20·9	24·1	19·2
Manufacturing	4·3	3·4	3·6	3·5
Construction	18·9	14·4	3·6	3·4

Source: H S C 1980
[1] Based on partly estimated figures of employees at risk.
[2] Excluding opencast.

Table 5.3   Comparative risks of deep mine and opencast coal production

Year	Incidence rate of accidents per 1000 at risk			
	Fatal		Serious	
	Deepmine	Opencast	Deepmine	Opencast
1972	0·22	0·17	1·8	1·90
1973	0·3	0·38	2·1	3·05
1974	0·19	0·69	1·9	2·26
1975	0·25	0·18	2·3	0·9
1976	0·20	0·73	2·1	2·2
1977	0·16	0·63	2·0	2·4
1978	0·25	1·06	2·0	1·42

Source: H S E

5.25 It is frequently supposed that deep mining is more hazardous per employee than opencast. In fact in general this is not so. Table 5.3 demonstrates that the incidence rate of both fatal and serious accidents has usually been higher for opencast mining than for deep mining over the period 1972–78. However, it should be noted that the deep mining industry employs about 250,000 people, compared with somewhat less than 10,000 in the opencast sector. This difference of scale is reflected in the number of accidents – for example 63 fatal accidents at coal mines occurred in 1978, compared with only 9 at opencast coal sites. This qualifies any conclusions based on comparisons between these two sectors.

SURFACE OPERATIONS

5.26 Certain surface facilities are essential at all collieries. Thus surface structures will always be needed at the access to a mine, although it will take the form of winding gear for a shaft, and conveyor gear for a drift. Pithead administration buildings will be common features, including operation rooms, baths, and canteens. Stockyards for materials and machinery to be used in the mine will also be a normal requirement (see Plate 5.3).

5.27 Many collieries will also incorporate a substantial coal preparation plant for cleaning, sizing, and separating dirt from the coal, although in some cases one preparation plant can process the output from several nearby collieries, and in others, the coal produced is sufficiently clean not to require sophisticated preparation plants. Where substantial coal preparation plants are needed, lagoons may also be necessary for the settlement of tailings.

5.28 Similarly, spoil heaps close to collieries are not an inevitable feature at all pitheads. Where collieries work thick clean seams, relatively little spoil is produced (for example, Daw Mill in Warwickshire), or none at all (as expected at Selby). Other solutions for spoil disposal are also possible, which can minimise the need for spoil heaps close to the pit; we consider these in detail in Chapter 9.

5.29 There are three further factors which combine to produce a great variation in the appearance of surface operations and plant at different pits. First, there is an increasing sophistication and specialisation in large scale deep mining; and thus not all pits are now individual operational units independent of each other. Each pithead may no longer have to provide a fully comprehensive range of all the surface facilities which are necessary for mining operations. Some pits in Yorkshire and elsewhere have been inter-connected underground in a way which enables the processing of coal and disposal of waste at a common location. This eliminates the need for each pithead to make its own separate and local provision for spoil disposal. At Selby, all the output from the mine will be taken up one drift to the surface, although there will be five other shaft sites in the mine area for access of men and materials to the workings. Thus the surface facilities for coal preparation, stockpiling and transport are not needed at the five shaft sites.

5.30 Secondly, there is great variation in the age of mines, which range from those over 100 years old to modern installations. The appearance of surface facilities reflect very strongly the standards of the time when they were first provided (see Plates 5.4 and 5.5). Thirdly, even within a group of mines of the same period, their present appearance can vary widely according to the standards set by local management. During our visits we became very aware that surface facilities at some mines presented a very neat and well maintained appearance, whilst at others there was seemingly little concern with the external appearance of the installations (see Plates 5.6 and 5.7).

5.31 On the whole we have concluded that the appearance of surface installations does demonstrate a marked improvement from times past, partly as a result of the efforts of the NCB, and partly as a result of the activities of local authorities in securing environmental improvement in the vicinity of pits. We have been struck by the appearance of several modern collieries compared with the dismal squalor of pits exhibited in old photographs, and described by writers over the last 100 years, and indeed within our own memories.

5.32 The archetypal view is of the "pit, piled up in the bottom of a hollow, with its squat brick buildings, raising its chimney like a threatening horn, [seemingly] to have the evil air of a gluttonous beast crouching there to devour the earth" (Zola, E 1885). Orwell describes "The monstrous scenery of slag heaps, chimneys, piled up scrap iron, foul canals, paths of cindery mud criss-crossed by the prints of clogs" (Orwell G 1937). He describes the spoil heaps as examples of "an ugliness so frightful and so arresting that you are obliged as it were, to come to terms with it":

"A slag heap is at best a hideous thing, because it is so planless and functionless. It is something just dumped on the earth, like the emptying of a giant's dustbin. On the outskirts of the mining towns there are frightful landscapes where your horizon is ringed completely round by jagged grey mountains, and underfoot is mud and ashes and overhead the steel cables where tubs of dirt travel slowly across miles of country. Often the slag heaps are on fire, and at night you can see the red rivulets of fire winding this way and that, and also the slow moving blue flames of sulphur, which always seem on the point of expiring, and always spring out again. Even when a slag heap sinks, as it does ultimately, only an evil brown grass grows on it, and it retains its hummocky surface. One in the slums of Wigan, used as a playground, looks like a choppy sea suddenly frozen; 'the flock mattress', it is called locally."

5.33 More prosaically the Reid Committee (Coal Mining 1945) reported in 1945 that:

"It is an unfortunate fact that the appearance of many colliery yards and offices often makes an impression of dirt and disorder which must have contributed materially to the critical attitude of the general public towards the coal industry . . . much remains to be done to improve and brighten colliery premises." (Paragraph 579).

They recommend that:

"no plans for a new mine, nor, where the surface arrangements are affected, for the remodelling of old mines, should be started, before taking the advice of an architect, as well as of an engineer, on the design of the surface layout as a whole; and every effort should be made by planting trees and shrubs, and laying down grass in open spaces to make the colliery surface attractive."

Our investigations have made it clear that this recommendation has in large measure now been implemented, although progress has by no means been uniform region to region.

5.34 Techniques of design of surface installations have improved immeasurably since the early years of this century. We have noted a clear general trend of significant improvement, although in some localities this has been less pronounced than it should have been. However, at the same time, public expectations have also risen. The general improvement which we have noted cannot therefore be interpreted as necessarily signifying approval or acceptance of the appearance and design of modern surface plant in all circumstances. Whether or not any individual project is acceptable will depend very much on local circumstances and the merits of the case.

5.35 We note that relatively few areas now exhibit the stark characteristics described by Orwell. However, we are much concerned that such areas do still exist. During our visits to Yorkshire and Wales we saw coalmines where the surface plant, although no doubt improved since the early years of the century, still does not present the appearance which we would have expected for the 1980s. Particularly in areas such as Barnsley which have a high concentration of pits, the lack of attention to the appearance of colliery plant can create a cumulative impression of great desolation. We make recommendations in subsequent Chapters designed to ensure that such features are speedily improved.

Chapter 6 New deep mine developments

INTRODUCTION

6.1 Coal mining is an extractive industry so the N C B have a continuing need to invest in new capacity simply to maintain current levels of output. Additional investment is also required to meet any planned increases in production such as those agreed by the N C B, the unions and the Government in 1974 and set out in 'Plan for Coal' (N C B 1974). It must be remembered that during the previous two decades the U K coal industry lost 70% of its collieries, 65% of its work-force and 45% of its output. Following the oil crisis of 1973, the industry had for the first time for a number of years to prepare not only to maintain production but to plan to increase it up to and beyond the end of the century. In this Chapter we examine how the Board plans the introduction of new capacity.

6.2 New capacity is acquired not only by opening up new greenfield sites but also by investment in existing long life mines to make greater extraction of reserves from these mines both economic and technically possible. The Board estimated that to meet the target production figures set out in 'Coal for the Future' (Department of Energy 1979), that is, 150 million tonnes per annum of deep mined coal by the year 2000, 4 million tonnes of new capacity must be opened every year from 1984, 3 million tonnes from new deep mines and 1 million tonnes from existing mines. The 1% G D P growth case provided by the Department of Energy (discussed in Chapter 3) would suggest that these target figures are optimistic. However, if U K coal is to remain competitive, a lower growth in coal demand increases the premium on the need to introduce low cost, high productivity capacity. In any event, there would be considerable danger in relying on a crash programme towards the end of the century rather than a steady build-up of this capacity. The opening of new capacity involves long lead times of about 10 years. If its introduction were delayed, there could thus be a shortage of U K coal, forcing reliance on imports at a time when world coal prices were likely to be much higher than now; there could be a shortage of trained engineers, surveyors, and miners to carry out the required rapid expansion; most importantly, there is a danger that any crash programme could only be achieved at the expense of the environment. A steady build-up of capacity would be consistent with the need for robust policies and would also provide adequate opportunity to take environmental factors into proper account. We therefore endorse the need for a steady build-up of new capacity but we recognise that each particular proposal must be judged on its merits in the light of competing claims for investment, and likely impact on the environment and employment.

6.3 In the following sections we examine how the Board open up new capacity. We deal firstly with expansion of existing mines and secondly with new greenfield sites. We then look in a little more detail at the Board's development at Selby and the proposals for possible development at Belvoir (North East Leicestershire Prospect) as these are the two major developments that the Board have embarked on in the last decade.

DEVELOPMENT AT EXISTING MINES

6.4 Under existing legislation if the N C B wish to extend a colliery which was brought into operation after 1 July 1948, they need to obtain planning permission in the normal way from the local planning authority, or on appeal from the Secretary of State. If the area where they propose the extension is already covered by the original planning permission granted for the colliery no further permission will usually be necessary.

6.5 As we explain in Chapter 4, if a mine was in operation on 1 July 1948, planning permission is deemed to have been granted under Article XX of the General Development Order 1977 (S I No 289) in England and Wales and Article XVII of the General Development Scotland Order 1975 (S I No 679). This, briefly, gives a general grant of planning permission for:

"(i) The winning and working underground, in a mine commenced before 1 July 1948, of coal . . . and any underground development incidental thereto:
(ii) any development required in connection with coal industry activities . . . and carried out in the immediate vicinity of a pithead."

However, where the erection, alteration, or extension of a building is proposed, the prior approval of the local planning authority is required, although this approval can be withheld only on limited grounds.

General development order

6.6 In the evidence to us concern was expressed lest the N C B should continue to retain these existing use rights, particularly as, under present powers, it appears that the N C B can use the underground working rights and the ancillary development rights together, to proceed with surface developments which open up new access underground. Such developments can result in new areas being affected by subsidence; surface buildings can be erected and further waste can be deposited without express planning permission. The local authorities and the Royal Town Planning Institute (R T P I) consider that such matters are sufficiently significant to a locality for such developments to be brought within the normal planning controls.

6.7 In practice, the N C B have only rarely used their G D O rights to open up new mines, but the Kinsley Drift mine in Yorkshire, which was opened in 1979 is one such example. The N C B consider that if a seam can be worked more efficiently from one point on the surface than another because of underground mining considerations, then such an operation should be allowed to take place without formal requirements for permission. Moreover, they point out that under Article 4 of the G D O there is a safeguard available in that the Secretary of State, or the local planning authority, may require that specific planning permission should be sought for any development for which permission would otherwise be granted by the G D O. On the other hand, the use of Article 4 powers often carries with it the obligation to pay compensation, which would not be the case with a normal planning permission.

6.8 We have noted that the main purpose of these provisions is to allow the N C B a degree of flexibility in carrying out work on sites where they have planning permission for general use, without requiring further planning applications for developments which will not alter significantly the impact of the site on the surrounding area. We would not wish to see the N C B placed in a less favourable position than other mineral operators, or than industry generally, by their removal. In principle these are common to all industrial, commercial and mining operations. It is sensible and economic for any concern to be able to make necessary adjustments to plan and site layout to improve the operation of its legitimate activities without undue hindrance.

6.9 However, the environmental impact of such operations must be taken into account in the exercise of these rights. The effects of a new mining development on the surrounding area will last for 20 years or more, which is a long time for uncontrolled activities. Moreover, the majority of sites which are operated under the G D O provisions without express planning permission are concentrated in the Yorkshire coalfield; and this area also suffers a substantial concentration of dereliction from past mining, with consequential adverse effects upon the environment and the possibility of attracting other industry. If the local planning authorities had to grant planning permission, this would allow for a greater degree of planned and co-ordinated development of the mining industry in their areas. It would serve to ensure that the optimal environmental standards were applied to these mines.

6.10 We are aware that the Government are currently proposing some changes, both by amending the G D O, and by the introduction of the current Town and Country Planning (Minerals) Bill. These changes will tighten up the existing provisions in some important respects, although it is not proposed to withdraw the N C B's existing rights for the working of coal underground, together with underground development incidental to it, in a mine commenced before 1 July 1948.

6.11 The Government's approach is to accept that the N C B require a degree of flexibility to cope with unforeseen difficulties in their underground working of a colliery, and that it would be unreasonable to ask the N C B to apply for a separate planning permission to change direction underground in such cases. The N C B's rights relating to developments above ground after the proposed changes will be identical, to all intents and purposes, to the rights of other mineral operators, and to those applied to industry generally under Class VIII of the Order.

6.12 We acknowledge that the Government's proposed amendments meet most of the concerns expressed on this issue. However, we consider that it should be made clear that any surface development which is proposed in order to open up a new access underground should require an express grant of planning permission. There seems to be some doubt as to whether the Government's own amendments would achieve this end. However, it could be done quite simply. If the deemed rights to erect or extend buildings were specifically to exclude any such buildings or structures designed to open up new access to underground workings, this should ensure that planning permission would be required for the opening up of all new shafts or drifts. We recommend, therefore, that the Government incorporates an amendment to the General Development Order 1977 to this end.

DEVELOPMENT IN GREENFIELD SITES

6.13 Before opening up new mines in greenfield sites the N C B require planning permission. However, before planning permission is sought for a particular proposal the Board are involved in a long process of internal planning from the basic geological exploration up to the detailed proposals for the siting of pits and the restoration of spoil heaps. The N C B have been most helpful in explaining these processes to us.

Exploration and Development

6.14 From 1974 the Board have built up their exploration programme from almost nothing to the extent that they are now spending about £19 million per year and the rate of drilling has reached about 100 deep boreholes per year. These boreholes currently require permission from the local planning authority and the results of the drilling are published. Of the 100 boreholes roughly 20 are for existing collieries, for hazard evaluation or for colliery extension; 70 are for prospects in the later stages of development; and 10 are for prospects in the preliminary stage, of which about 3 are wildcat boreholes (that is, holes bored on a speculative basis), and the remaining 7 are drilled on the basis of geological information available from different sources.

The Ladder of Exploration

6.15 During the course of our discussions, Mr Moses, Director of Planning and Major Projects at the N C B, has developed and expanded the concept of the "Ladder of Exploration". This clarifies what stages of planning any prospect must go through before mining can begin. The following paragraphs both define the stages and describe the Board's exploration programme to March 1981 in relation to each stage (the geographical location is illustrated on the map at Figure 6.1).

I *THE POTENTIAL PROSPECTS STAGE:* Areas at this stage are simply those in which coal is believed to exist, usually by knowledge gained from some source other than coalmining exploration. There are very many places in this category.

II *THE PRELIMINARY EXPLORATION STAGE:* Areas at this stage are those where coal is known to exist, but further proving is needed to establish the boundaries of the coal. The most promising prospects at this stage of exploration are:

Canobie (located north of Carlisle). Four boreholes were completed in 1954–56, two in 1979 and three in 1980. 44 kilometres of seismic survey were carried out in 1980. Further drilling and a detailed seismic survey would be needed to establish the boundaries of the coal: this work would take several years.

Firth of Forth (located beneath the Firth between Musselburgh and Buckhaven). Twelve boreholes were completed prior to the preparation of Plan 2000 and a marine seismic survey of approximately 285 kilometres was carried out in 1979. A large and relatively costly programme of offshore boreholes and seismic surveys would be needed next.

Yorkshire Block 5 (located North West of York). Eight boreholes and 73 kilometres of seismic survey are completed, with a further four boreholes and 40 kilometres of seismic survey planned.

Yorkshire Blocks 9, 10, 11, 12 and Carrlands (located east of the Hatfield, Markham and Rossington "takes"). Twenty boreholes have been completed together with some seismic work on the edge of Hatfield.

Oxfordshire (to the north and east of Witney). Four boreholes plus one I G S hole have been completed. Further holes and seismic work are required to establish the structure of the field.

III *THE INTENSIVE (OR SIGNIFICANT) EXPLORATION STAGE:* This is the stage when the boundaries of the coal are known and results from earlier stages justify intensive exploration to determine the geological and chemical characteristics of the coal.

Prospects at this stage are:

Amble (located at the northern end of the Northumberland coalfield offshore from Amble, and immediately north of the Ellington/Lynemouth "take"). Twenty-seven offshore boreholes and 454 kilometres of marine seismic survey have been undertaken. A Feasibility Study could follow the results of further exploration.

The East Staffordshire Prospect (centred on Lichfield and located to the south and east of the Lea Hall "take"). Twenty boreholes and 230 kilometres of seismic survey have been completed. Further holes are proposed.

Till (located mostly to the east of the River Trent and to the north and south of Gainsborough). Thirteen boreholes have been completed and three more are proposed. The prospect would require considerable further exploration to define a worthwhile project area.

Oxford (Banbury) (located to the south of the town). Twelve boreholes and one I G S hole have been completed, together with a limited amount of seismic work. Further exploration is needed.

East Nottinghamshire (to the east of the Harworth Colliery "take"). The presence of a significant prospect is suspected from the knowledge of the reserves at the nearby colliery. Only limited work has been carried out to date and further boreholes are needed.

IV *THE FEASIBILITY STUDY STAGE:* If the results of the earlier exploration are encouraging, a prospect moves to this stage during which an engineering team will be set up to determine possible areas for mining access, potential output levels, and to make an assessment of the economics of extraction. The broad environmental impacts of the possible mining activity will be taken into consideration. Feasibility Studies are under way or under consideration as follows:

South Warwickshire (located south and west of Coventry and north of Leamington). Fifty-two boreholes and approximately 430 kilometres of seismic

Figure 6.1 Exploration programme potential prospects

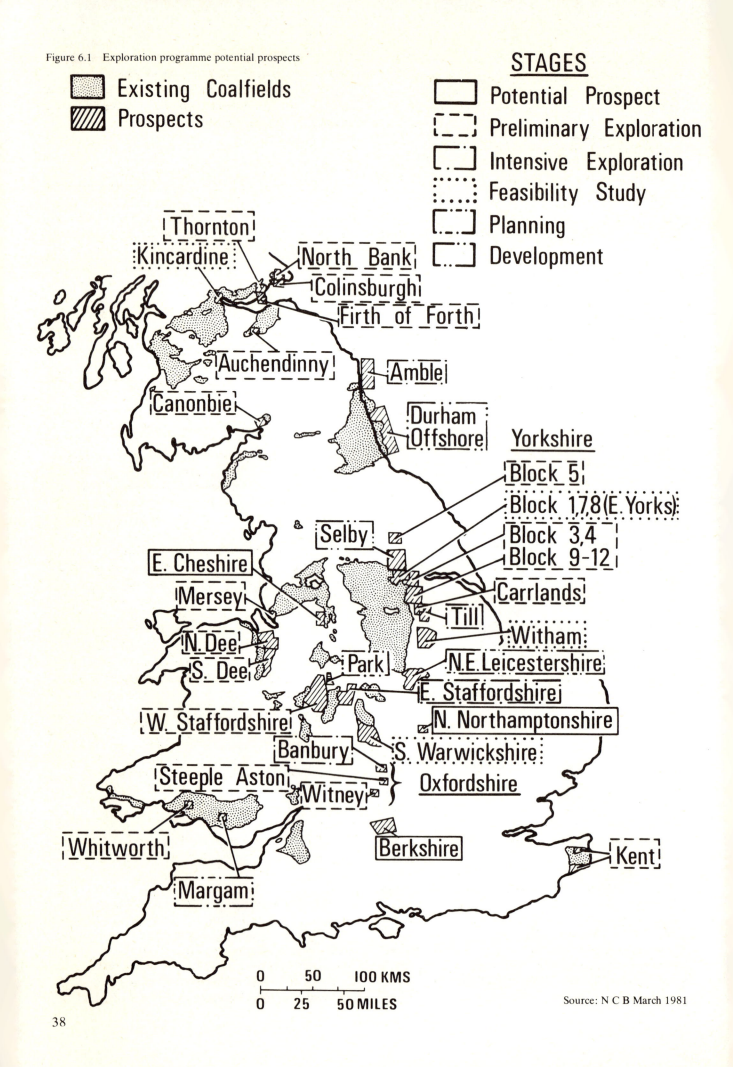

STAGES

Existing Coalfields
Prospects

Potential Prospect
Preliminary Exploration
Intensive Exploration
Feasibility Study
Planning
Development

Thornton
Kincardine
North Bank
Colinsburgh
Firth of Forth
Auchendinny
Amble
Canonbie
Durham Offshore
Yorkshire
Block 5
Block 1,7,8 (E.Yorks)
Block 3,4
Block 9-12
Selby
E. Cheshire
Carrlands
Mersey
Till
Witham
N. Dee
S. Dee
Park
N.E. Leicestershire
E. Staffordshire
W. Staffordshire
N. Northamptonshire
Banbury
S. Warwickshire
Steeple Aston
Oxfordshire
Witney
Whitworth
Berkshire
Kent
Margam

0   50   100 KMS

0   25   50 MILES

Source: N C B March 1981

38

work have been completed. A Feasibility Report is nearing completion although further boreholes will be required to define reasonable areas of consistent workable section within the variable Thick Coal complex.

East Yorkshire Blocks (located around Snaith and Howden). Thirty-four boreholes and some 165 kilometres of seismic survey have been completed. A Feasibility Study may well be started shortly.

Witham (which is located between Newark and Lincoln). Forty-one boreholes and a limited amount of seismic surveying have been completed. Further work is necessary before a Feasibility Team can be set up for a full study.

Kincardine (Hirst basin located to the west of Kincardine and Longannet). Some twenty-one boreholes have been completed, and more are planned. Additional drilling may be necessary to prove faulting.

V *THE PLANNING STAGE:* After consideration of the Feasibility Report the Board may decide to support a specific proposal which, with the support of the Department of Energy, is moved to the planning stage. During this stage the details of mining the coal are worked up into a planning application for outline planning permission to work the coal from access points discussed with the local planning authority. This stage includes the time after submission of the planning application until the application is accepted or rejected. Prospects at this stage are:

North East Leicestershire Prospect. A decision by the Secretary of State is awaited.

Park (Staffordshire). The Board have decided not to proceed at this stage with this prospect because of the high chlorine content of the coal and so the planning application has been withdrawn.

Margam (West Glamorgan). The Board have decided not to proceed at present with this prospect following the change in coking coal demand. A planning application has not been made.

VI *THE DEVELOPMENT STAGE:* A prospect reaches this stage if planning permission is granted. This stage takes the prospect to production. Planning permission for detailed aspects of the prospect (such as building designs) will still be necessary. The only greenfield prospect currently at this stage is Selby.

6.16 Projects may move up or down the ladder from any stage in the light of progressively refined appraisal. The withdrawal of the planning application for the proposed new development at Park (see paragraph 6.15 V above) is an example of this. On the other hand, the more developed assessment of the East Yorkshire Prospect has suggested that this might be a much more viable proposition than it was thought to be in September 1979.

6.17 The concept of a ladder of exploration explains not only what the N C B are doing at each stage but also what information they have available. It has therefore significant implications for consultation and planning procedures which we discuss fully in Chapter 21.

6.18 The Board have only two developments at the most advanced stage of planning and construction (since the withdrawal of the Park application), the North East Leicestershire Prospect, better known as Belvoir, and the development at Selby. We have examined in more detail the Board's approach to the planning and development of these proposals which illustrate two contrasting solutions to the problem of getting coal out from below ground. The difference arises from the interaction of geological, economic, engineering and environmental factors which combine to produce very different visual impacts on the existing landscape.

Selby

6.19 In their evidence to the Commission the N C B described fully how the impact of the developments at Selby and Belvoir on the surrounding area was taken into account in the detailed planning of the proposals. Some members of the Commission paid a visit to the work in hand at Selby and were particularly interested in the visual impact of the development. When the Board sought permission to develop Selby in 1974 the application covered the largest single area ever to be dealt with under the planning legislation – namely an area of about 110 square miles. Full production is expected by the end of the 1980s, when about 4,000 miners will be employed, producing about 10 million tonnes of coal every year. The mine will comprise five independent collieries dispersed over the field. These will consist of two shafts to provide input and extract ventilation. One shaft will be for man-riding and the other for equipment and materials. The coal won by each mine will be transported underground to conveyors working in a central drift (that is, a tunnel sloping at a gradient of 1 in 4 from the surface). This will come to the surface, by that particular combination of good luck and good judgement which makes a brilliant design, in a disused railway siding miles away from the collieries with plenty of room for coal treatment and storage, and access to a rail link. The formation of the seam is such that the coal will be free of dirt and there is therefore no requirement for waste tipping.

6.20 The one winding tower (the other shaft will be a headframe only) can be kept relatively low (29·5 metres only) because it will not be required to handle coal. The other surface buildings for ventilation plant, welfare, canteen, administration and workshop and storage functions will be laid out differently at each site according to the local terrain and access opportunities.

Debris from the shaft sinking will be disposed about each mine to make planted mounds which will screen the works, car parking and equipment storage areas. Although each building element is a standard design, the variations in grouping and careful choice of materials for external walls, roofs and planting, will reflect the context of each site, and should ensure that each colliery will be comfortably assimilated into its locality with no more impact than a new farm with a couple of tall silos. What might have become a devastating intrusion in a virtually flat agricultural terrain can in fact be looked forward to as a series of well dispersed new points of interest. It is no accident that excellent relations exist between the Area Deputy Director of Mining who leads the project, and his team of consultant architects, landscape architects and engineers.

Belvoir

6.21 At Belvoir the N C B are seeking planning permission to extract coal from under some 100 square miles of land. The coalfield lies roughly within a triangle formed by Nottingham, Grantham and Melton Mowbray. The area is sparsely populated with an essentially agricultural economy with the exception of industrial estates at Melton Mowbray and Langar and the British Steel complex at Asfordby. The coalfield is divided by the Harby Hills, a marlstone escarpment which crosses the area from north-east to south-west. The escarpment is wooded and topped by a ridge of trees leading up to Belvoir Castle. To the north and west lies the plain known as the Vale of Belvoir and to the south and east is a limestone plateau of high landscape quality. Although by national standards the area is not one of outstanding beauty or scientific interest, it is one of the more attractive locations of the East Midlands and the castle receives around 250,000 visitors per year.

6.22 The N C B proposal would mean the extraction on completion of the project of some 7·2 million tonnes of coal per year from three new mines employing 3,800 mineworkers. The Board estimate that the first two mines (producing 5·2 million tonnes of coal per year) could be in full production 8 years after planning permission had been granted, with the third following some years later. Unlike Selby, however, the Belvoir field would produce spoil (at a ratio of 2·56 tonnes of coal to 1 tonne of spoil) and tips are proposed by the Board for each of the three mines. The tipping proposals are discussed in more detail in Chapter 9.

6.23 The Board told us that from the first they had accepted that the three sites preferred for mining might prove unacceptable for environmental reasons. Consultants had therefore carried out what amounted to a major environmental impact analysis, adopting sieve techniques in the first instance, to find the most suitable location for the three mine sites. Three main groups of factors were taken into account as follows:

(1) Operational factors – road and rail access and availability, conflict with other major services such as power lines, water and sewerage installations etc.

(2) Terrestrial factors – ground slopes and shapes (slopes in excess of 1 in 25 or land of irregular shape can cause construction problems) wildlife, vegetation, high agricultural land values, areas of land unsuitable for development.

(3) Activity/Land Use – proximity of mine buildings and dirt tips to settlements (aim to keep buildings 1 kilometre and dirt tips 0·5 kilometres from existing villages) landscape value (aim to keep away from Sites of Special Scientific Interest, Areas of Outstanding Natural Beauty and other statutory designations such as conservation areas).

A series of maps analysing these factors on a grid of half kilometre squares was produced to enable their combined effect to be assessed. A second stage analysis was then carried out to take into account localised environmental impacts at the possible mine sites which emerged. This involved an assessment of the effects that some of the major features of deep mining would have on a basically rural area such as disposal of mine waste, dust, water, and subsidence. In addition the visual impact of the proposed mines, the resultant noise, dust, loss of landscape value, loss of agricultural value, and traffic intrusion were taken into account. Three sites were finally chosen for mines at Hose, Asfordby and Saltby. These sites differed from the original sites preferred by the N C B on mining grounds.

6.24 As we have made clear when discussing our remit in Chapter 1, judgements on individual planning cases fall outside our terms of reference. However, from the information available to us we have been struck by the extent to which geology can result in projects of contrasting visual impact on the landscape. Thus the Belvoir proposals would result in a visual impact very different from that at Selby. Our understanding is that this would arise for the following reasons:

(a) mining activity at Belvoir would be concentrated at three sites on the surface rather than six;

(b) the combination of mineral handling, access for men, equipment and materials, ventilation and coal treatment at each mine would inevitably produce at Belvoir a bulkier concentration of surface structures occupying a larger area of ground;

(c) owing to their multiple function and the higher duty that would be required of the shafts, the winding towers would have to be higher (45·8 metres for service towers and 58·3 metres for coal towers compared to 29·5 metres for the service tower at Selby);

(d) a large quantity of waste would have to be disposed of whereas at Selby, once the shafts are sunk and the drift is driven, there will be none.

Of course, visual impact is only one aspect amongst others of any new proposal, but the fact that the same method of assessment of the same set of factors can produce such different results for purely geological reasons, in the first instance, underlines the need for the earliest possible consultation, continuing as a project develops, so that all concerned can understand and accept its evolution.

6.25 Throughout this and subsequent phases of planning the N C B kept closely in touch with the local authorities concerned and those living in the area and consulted a large number of statutory undertakers, Government and associated organisations, other national organisations, such as the Royal Institute of British Architects, Royal Fine Arts Commission and National Farmers Union and local organisations, for example, the Vale of Belvoir Protection Group, and the Grantham Canal Restoration Society. In all over 45 different bodies were consulted. During the course of the consultation prior to the submission of the planning application, the N C B set up with the local authorities an Environmental Working Party. This was one of a number of working groups set up with the authorities and other interested bodies.

6.26 The N C B therefore undertook by any standards a very thorough environmental impact assessment in order to try to minimise the environmental effects of the proposed mines. Such doubts as we do have about the way in which the Board take into account environmental factors in building up their programme of new developments relates to the earlier stages of the "ladder of exploration". As already stated, we discuss these aspects fully in Chapter 21.

Effects of new developments on existing communities

6.27 In the preceding paragraphs we have looked largely at the way in which the N C B takes into account physical impacts when planning new developments. However, as further proposals for new mines in non-traditional areas are likely to come forward before the end of the century, it is important to make some assessment, however rough and ready, of the magnitude of the disruption likely to be caused thereby to the existing communities of the new coalfields. This is particularly important, as we continue to emphasise, because of the out-dated image that still persists of the mining industry. There is little practical evidence on which to draw, since developments of new coalfields on the scale of Selby and the Belvoir proposals have not taken place for many years.

6.28 We have already reported that it is the N C B's policy to absorb the influx of men and services required for new mines into the existing community. We now look at examples of what this means in practice for employment and other necessary services such as housing and education.

Employment

6.29 Most of the N C B's recent expansion has tended to be located adjacent to existing activity. Thus the development at hand in Selby and the proposed mines at Belvoir lie close to existing coalfields and are to be serviced in part by the transfer of men from existing collieries with supplementary recruitment from the locality of the new developments. For practical reasons it will also always be necessary for the Board to employ a high proportion of skilled and experienced men in relation to raw recruits. The Board estimate that the three mines at Belvoir would employ on completion about 3,800 mineworkers and 300 office staff. Due to manpower wastage in the industry this would involve employment of something like 8,000 men over a period of years, as follows:

4,600: transfers from collieries in the Nottingham-Leicestershire area and from further afield.

1,100: re-entrants into the mining industry.

1,200: juveniles (under 18) for training; most of these would be local youths and the maximum intake would be 170 a year.

1,300: young men with no previous mining experience; the bulk would probably be local men attracted to the industry and the maximum intake would be 160 a year.

In addition over the 13 year construction period, a further 1,200 workers would be required. The number would peak after about 6 years. About half those employed in surface construction would probably be local men. Finally, it is estimated that the development would create a further 1,900 jobs in induced employment, mainly in the service sector. The development would therefore create some new employment for the locality over time, although in the longer run it may also increase the numbers seeking employment from the families of the mineworkers.

Housing and other local authority services

6.30 The N C B no longer build houses for their own employees, but as with the provision of other local authority services such as schools, health and personal social services, work closely with the relevant local and health authorities. At Selby proposals have been agreed with the local authority for in-filling and small scale development of new houses within the surrounding villages so that the numbers in any one village will be no more than the local community might be expected to absorb. As Selby is a small district authority the provision of rented council accommodation is to be supplemented by the provision of some housing by local housing associations. The Board are also expecting a number of people to purchase their own property in the area, and others to commute from the traditional mining areas located further South. The provision of other supporting services, in particular of schools, is being monitored by joint standing committees of the authority and the Board. The local education authority are currently satisfied that they will be able to meet the additional requirements for school places in the area. The University of York have also been involved with the Board and other relevant authorities in the monitoring of developments at Selby.

6.31 In addition to working with the statutory authorities on the provision of relevant services to cater for the incoming population, the N C B have been making attempts to help integrate the new population with the existing community. Through the Coal Industry Social Welfare Organisation they plan to provide social clubs and other amenities for the benefit of the whole community and not just their own employees.

Local Authority Finance

6.32 A particular point has been raised by the Board and local authorities about the effect of new developments on the provision of central government finance to local authorities through the Rate Support Grant in England. The local authorities concerned are able to obtain rates from the new development as from any other business within their area. However, under the Rate Equalisation Scheme the amount of new rate so raised is deducted from the local authorities Rate Support Grant allocation. The additional services the local authority has to provide as a result of the new development will probably be taken account of eventually in an adjustment to the "needs" factor which central government takes into account when distributing the Rate Support Grant amongst local authorities. We are told there are time lags in the system because of the data used. It is possible that the authority concerned has to provide additional services without additional funds unless the rates are increased for that purpose. We have not examined this question in any detail as the operation of the Rate Support Grant system is outside our remit. We cannot therefore make any recommendations, but we suggest that the Department of the Environment (and, if necessary, the Scottish and Welsh Offices) should examine this aspect of the distribution of the Grant.

CONCLUSIONS

6.33 There has not yet been sufficient experience with the introduction of major new mining projects to be able to draw firm conclusions from the above material. However, in our view the introduction of a modern mine into a new coalfield, both in terms of the scale of increase in the population and the associated house building rate need have no greater impact than that associated with any medium size industrial development. It could be argued that the degree of change is likely to be in keeping with that experienced by many areas of the country where new development has taken place within the last 20 years or so. However, the acceptability in such areas of new mining projects will be determined by those impacts which are unique to deep mines and are not therefore associated with other forms of industrial development. We examine these in the following Chapters.

Chapter 7   Environmental impacts of deep mining operations

INTRODUCTION

7.1 We now turn to the environmental effects of deep mining operations. Chapter 5 – Deep Mining – illustrates some of the most significant impacts inherent in modern methods of extracting coal underground. In our judgement, well-supported by the evidence, the two most important impacts are subsidence and spoil disposal. We examine these subjects in more detail in the next two Chapters. Here we concentrate on other effects as follows:

- disposal of surplus water
- noise
- dust
- impact of tipping and colliery operations

DISPOSAL OF SURPLUS WATER

7.2 The disposal of surplus water is an integral part of mining operations. Such water has a number of sources including water pumped from underground, surface drainage from stocks and buildings, drainage from tips and effluent from coal preparation plant. In addition problems can be caused by overflows of polluted water from abandoned mines into rivers and other water courses.

Surplus water resulting from mining underground

7.3 Most mine drainage waters come from the water bearing rocks lying between or over the coal seams. The extraction of the coal releases water from the strata into the mine workings and this water has to be pumped to the surface if mining is to continue. In the newer coalfields the coal measures are overlain by heavily water-bearing rocks which are frequently used for public and industrial supply.

7.4 The largest volumes of minewater have been produced in South Wales, Scotland and North East England, and the least in the Midlands (see Table 7.1). The amount of water pumped varies greatly from mine to mine. For example – for shallow workings the ratio of water pumped to saleable coal produced can be as high as 30 to 1; for deep workings it can be as low as 0·001 to 1.

7.5 Waters which enter the workings from the strata can become contaminated with coal and clay particles picked up from the mine roadways and with the soluble oxidation products of iron pyrites present in the coal

Table 7.1   Volume of mine drainage waters pumped by the N C B – 1977

N C B area	Volume (megalitres per day)	Tonnes water per tonne saleable coal
Scotland	140	4·6
North East	210	5·2
North Yorkshire	17	0·7
Doncaster	7	0·3
Barnsley	60	2·8
South Yorkshire	44	2·1
North Derbyshire	15	0·7
North Nottinghamshire	3	0·1
South Nottinghamshire	7	0·3
South Midlands	37	1·4
Western	36	1·1
South Wales	213	8·1
Total	789	2·3 (average)

Source: N C B

and exposed during mining. The oxidation products include ferrous, ferric and aluminium sulphates and sulphuric acid. The quality of water may vary with depth. At the greatest depths saline waters may occur which have ten times as much salt as sea water, although the deep waters from the coalfields of Wales and Scotland contain almost no salt.

7.6 Control underground is achieved both by leaving an adequate thickness of strata between the workings and any substantial body of water or aquifer, and by allowing any drainage to accumulate in underground sumps and pools. Storage may be for a few hours, or months or even years, and has the advantage of allowing suspended matter such as coal and shale particles to be deposited before water is pumped to the surface.

7.7 Although most drainage from underground can be discharged directly into rivers, streams, estuaries or the sea, some kinds of drainage require treatment to improve quality. Suspended particles such as coal or shale are removed by sedimentation, if necessary assisted by flocculants, in earth-walled lagoons. However lagoons have a major impact on the environment which we discuss in Chapter 9. Iron compounds may be removed chemically and saline water (about 50 megalitres per day) has to be treated in specially constructed plants before discharge. About

88% of mine water is discharged directly into rivers and streams, about 7% is used for industrial purposes and about 5% used for public supply.

Surface drainage from colliery premises and coal stocks

7.8 Most surface drainage is caused by rain. Drainages from the colliery plant are basically similar to drainages from other buildings and can be discharged directly into drains or sewers without problems. Drainages from roads, yards or sidings may be contaminated with coal or shale particles. Where necessary these are collected and treated by sedimentation to standards required by trade effluent consents. Drainage from coal stockpiles is similarly treated by sedimentation in lagoons to meet trade effluent standards. Some coals which produce acidic and ferruginous discharges after prolonged storage, in so far as possible, are not stockpiled.

Drainage from spoil tips

7.9 Drainage from spoil tips can cause problems by either seepage from the "toe" of the tip or from surface run-off. Older tips which were not as heavily compacted as more modern tips tend to have no problems with surface run-off. In periods of wet weather or storms the run-off from modern tips can be so rapid that large quantities of suspended matter can be carried into rivers and streams. These may not cause difficulties if the receiving water courses are in spate at the time but can cause problems otherwise. The N C B argue that provision to cope with such irregular occurrences is out of proportion to the benefits gained and are in discussion with the water authorities to try and find a satisfactory technical compromise.

Coal preparation effluents

7.10 Very large proportions of the "run-of-mine" coal have to be treated in preparation processes which involves immersion in water before they are of a quality acceptable by the C E G B and other users. The volume treated has increased by about 60% since the beginning of the war. The preparation process involves the use of large volumes of water, about 2·5 tonnes of water per tonne of coal, much of which is circulated in the preparation plant, to remove the fine particles of unwanted shale and sandstones from the coal. These particles are concentrated into a suspension of tailings. Previously large quantities of tailings were discharged directly to rivers and streams. Now most preparation plants have closed water circuits and the tailings are either filtered out in pressings or allowed to settle in specially designed lagoons. The separated water is then used either for further process work or discharged as a trade effluent. Surge lagoons are provided to contain any discharges containing suspended solids if there are any problems with the preparation plant. There have been problems of discharges from preparation plant but the N C B expect that as modern preparation plants replace older units, all processes will be closed water circuits.

44

Legislative controls

7.11 The evidence to the Commission, including that from the National Water Council, suggests that although localised operational problems can arise, the disposal of surplus water from working mines is such an integral part of deep mining that the water authorities in England and Wales and the river purification authorities in Scotland can in general cope and will probably have adequate powers when Part II of the Control of Pollution Act 1974 is brought into force.

7.12 Under existing legislation in England and Wales (the Rivers (Prevention of Pollution) Acts 1951 and 1961), the N C B do not need approval to discharge water raised or drained from any underground part of a mine to a stream in the same condition in which it is raised or drained from underground unless an order is made by the Secretary of State. However, process waters such as those from the coal preparation plant are trade and sewage effluent and require consent from the water authority before this can be channelled to form a discharge. There is some doubt as to whether these powers actually apply to surface run-off or leachate. The Department of the Environment is currently considering suggestions from the National Water Council aimed at clarifying the law on contaminated surface water and leachate to enable them to ensure that they may be properly controlled.

7.13 Under Part II of the Control of Pollution Act 1974 consent will be necessary for any discharge of trade or sewage effluent from working mines and there will be no further distinction made in respect of water discharged in the same condition as it was drained from the mine. Part II of the Act has yet to be fully implemented. Improvement Programmes will be drawn up according to the degree of pollution caused and to quality objectives relating to the use of the receiving water concerned. Over the years the N C B have established close co-operation with the water authorities in England and Wales and the River Purification Authorities in Scotland in seeking practical solutions to water pollution problems. The N C B is also a member of the Water Research Centre and maintains technical collaboration with the Natural Environment Research Council (N E R C), and its Institute of Geological Sciences (I G S). On proposals for new developments such as Selby and Belvoir the water authorities work with the Board on the planning stage of the proposals to ensure that effects on water services, supplies and drainages are taken into account.

7.14 We do not foresee therefore any major new problems arising in respect of the disposal of water from working mines. The major difficulties arise from leachate from spoil tips and from coal stocks, particularly where the latter are stored for longer periods due to shifts in market demand as we saw in South Wales. We are satisfied that the N C B and the water authorities can and do work together to sort out the most appropriate technical solutions. However, it would be helpful to the water authorities to have the additional powers provided under Part II of the Control of Pollution Act 1974 available to them and we recommend that the Government brings these into force as soon as possible.

Discharges from abandoned mines

7.15 On the other hand, the evidence suggests that pollution from discharges of water from abandoned mines can be a locally severe problem particularly as in both England and Wales there is no clear legal liability to pay for the costs of removing or preventing the pollution caused, especially where land has changed hands. In Scotland, where the court has identified a guilty party, the offender must pay for removing the pollution. However, once the provisions of the Control of Pollution Act are brought into force, discharges from abandoned mines will no longer be an offence and it will not be possible to require anyone to pay for removing pollution caused by it.

7.16 As explained above, while a mine is in operation the inflows to the workings are immediately cleared by pumps and there is little time for contact between the water, loose materials, and exposed faces of the workings. Normally the discharge is not severely polluting. However, when a mine is abandoned the workings progressively become flooded. At some places, after a time, the water rises to the surface and breaks out as an uncontrolled flow. Unfortunately, due to the much longer retention time (possibly for a number of years) and greater contact area with the coal workings, the water can absorb a considerable amount of metals in a soluble form including iron and aluminium. It also develops a highly acidic character. When released to the atmosphere, where it can absorb oxygen, and with dilution of the acidic strength by other more alkaline waters, the metals are thrown from solution and deposited as oxide sludge commonly known as ferruginous or ochre sludges. Where older workings have also been "heated" or are on fire a much greater uptake of metal may occur. The resulting pollution can be very serious, can persist for years and can be costly to deal with.

7.17 A particularly instructive example is provided in Scotland by the pollution of the River Girvan in Ayrshire from 1979 onwards. Iron-contaminated waters from the N C B's colliery at Dalquharran which was closed in 1977 began overflowing into the river at the rate of about 0·5 million gallons a day. The pollution has had an adverse effect on local industry which had been relying on the river for a supply of process water of good quality. The river was a prime salmon fishery and is now completely inhospitable to any fish life between the point of pollution and the sea. Consulting engineers employed by the Scottish Development Agency recommended the installation of a treatment plant at the colliery at a cost of about

£400,000. The Agency were unable to find the funds. The Clyde River Purification Board however took the N C B to court under the Rivers (Prevention of Pollution) (Scotland) Act 1951 for causing polluting matter to enter the river. Recently three Appeal Court judges found in favour of the River Purification Board and directed a sheriff to overturn his previous decision and convict the N C B of polluting the Water of Girvan. The court ruled that the over-flowing mine had been caused by the Board's establishment of the mine and their decision to cease pumping water when the mine closed. The responsibility for clearing up the river now rests on the Board. The decision has significant implications for the Clyde area where there are some 34 kilometres of polluted waterway and for the Fife area, which is even more badly affected.

7.18 The position in England and Wales is in need of clarification as here it is not at present an offence to cause water raised and drained from a mine to enter a stream. From the evidence there appear to be local problems particularly in South Wales, Yorkshire and Northumberland. However, there are no firm data available on the extent of pollution caused by discharges from abandoned mines, on the significance of the problems or on the cost of remedial measures. The situation will vary in each case depending on the nature of the discharge and the quality objectives for the receiving waters. For this reason Section 50 of the Control of Pollution Act 1974 (which is in force) provided water authorities with the powers to investigate the extent of the problem, to determine the corrective measures necessary and to evaluate their cost. We are told by the Department of the Environment and the Welsh Office that the water authorities have not as yet carried out sufficiently extensive investigations to make a report or any formal recommendations to the two Departments with a view to deciding where remedial costs should lie.

7.19 In the meantime no action is being taken by the water authorities or the N C B in remedying problems or in dealing with past or present owners of abandoned mines. The N C B take the view that effects on water courses are usually localised and unlikely to justify the expense necessary to control any problem. They accept that consents are required where waters are pumped from otherwise unworked mines to protect adjacent workings. There is an urgent need to establish the extent of the pollution and the range of costs of the necessary remedial treatment. The question of responsibility should also be settled. We recommend that discharges from mines abandoned by the predecessors of the N C B, or even the N C B some long time ago, should be considered as a form of dereliction comparable to abandoned pit-heaps. We recommend that provision should be made through central government to meet the costs of any essential work, as is the case with derelict land clearance schemes.

7.20 Throughout this Report we stress the need to avoid the creation of future dereliction by the proper planning and after-care of new developments. The closure of pits, as we also stress, is an integral part of the mining industry. Thus we recommend that when a pit is to be taken out of operation there should be close cooperation with the water authorities to assess and plan for the effects of the cessation of pumping. The N C B should carry the costs of any necessary preventive or remedial measures. Where appropriate, conditions should be imposed as part of a planning consent. We also recommend that Section 46(6) (b) of the Control of Pollution Act 1974 should be amended to ensure that the powers of the water authorities under Section 46(5) to recoup the costs of remedial works should apply in respect of mines being closed by the N C B.

The E E C Groundwater Directive

7.21 A new mechanism for control of water discharges will be introduced at the end of the year when the E E C Groundwater Directive (European Communities 1979) comes into force. This Directive requires control over all activities (including mining) which result in certain listed substances reaching groundwater, including a number of metals (though not iron or aluminium), other toxic and persistent substances, as well as those which may affect the taste or odour of groundwater. However, since the Directive is concerned with the authorisation of "activities", discharges of listed substances into groundwater from mining operations which have already ceased are outside its scope. For existing and future operations the Directive will be implemented through existing legislation. Water authorities are to be made statutory consultees when planning permission is sought for mining and mining waste operations, and they will be able to bring to the attention of planning authorities any requirements needed to protect groundwater.

NOISE FROM COLLIERY OPERATIONS

7.22 Although noise from colliery activities can cause local problems, the evidence did not suggest that noise from deep mining operations is a widespread cause of complaint or concern. The levels of noise from mining operations are well below the level which could cause any damage to the hearing of members of the public. However, any noise such as that from night operations for example, can cause severe distress and annoyance and should be dealt with wherever possible. Under Part III of the Control of Pollution Act 1974, local authorities have powers to take action against the N C B where they are satisfied that noise amounting to a nuisance exists or is likely to occur or recur. They may serve a notice requiring the abatement of the nuisance and the execution of works necessary to achieve this. However, if the noise occurs during the course of trade and business, it is a defence to plead that the best practicable means have been adopted for its control.

7.23 Over the past 15 years the N C B have become more experienced in the use of techniques for measuring and reducing noise levels both underground and on the surface at collieries. There are no precise standards relating to noise nuisance which the Board are in general required to apply. British Standard 4142 on "methods of relating industrial noise affecting mixed residential and industrial areas" gives guidance on the conditions which may cause complaint in different kinds of environment, for example, urban, suburban, rural and at various times of day. When planning the development of new mines the N C B spend a considerable amount of effort to keep down the noise levels at nearby residential developments. The Board start with the basic criterion that sites ought to be at least 1 kilometre from existing villages. Considerable care is then taken with the assistance of noise contours, to plan the use of the site to keep noise levels acceptable. The Board discuss with the local authority both the siting and type of equipment to be used at the mine. At Selby, specific agreements under Section 52 of the Town and Country Planning Act 1971 have been reached which, taking account of the area in which the mine is located, specify appropriate levels for operations at various times of the day and night. (Section 52 of the 1971 Act enables local authorities to enter into agreements with any person interested in land, for the purpose of restricting and regulating the development or use of the land).

7.24 During the construction period of a new mine, which is usually the noisy part of the colliery operation, the Board use established techniques such as the construction of baffle mounds, the location of equipment and the timing and duration of particularly noisy activities to keep noise levels down to agreed limits and in particular to meet lower levels allowed at night times.

7.25 In so far as the operations of existing collieries are concerned the N C B have also developed monitoring programmes and techniques of analysis of the effects of the production phase of colliery operations. The effects of earth moving operations at the edge of the tip are probably the most continuously noisy of operations and do give rise to complaints from time to time which are investigated by the Board. However, phased work on tips helps restrict the maximum noise for any location to a limited period of time. With many older collieries, development of residential and other properties has followed the colliery and buildings can therefore be much closer than the 1 kilometre the N C B try to plan for in the construction of new mines. Even so there seems little evidence of widespread disturbance or complaint.

Conclusions

7.26 The Commission have reached the view that noise from deep mining operations is not a major environmental problem and is comparable to that likely to arise from other kinds of industrial plant. The N C B are fully aware of potential problems in the operation of existing collieries and of the need to prevent unnecessary disturbance from noise from new collieries. We do not therefore envisage any major new problems arising.

DUST

7.27 Again the Commission received no evidence to suggest that dust from colliery operations is a serious problem although on occasions there can be localised difficulties. The two major sources of dust are tipping operations and coal stocking and handling.

Tipping operations

7.28 Operational plans for tipping provide for dust suppression measures and aim to minimise the amount of dry dirt exposed to prevailing winds in dry seasons of the year. Where possible, and particularly with new tips, the N C B aim to provide an earthwork screen at the edge of the proposed tipping area early on in the operations. This is grassed and planted with trees where possible. Retreat systems of tipping which enable tips to be seeded at an early stage, also help dust control. The use of heavy vehicles can create dust problems but spraying is used where appropriate. The N C B suggest that some conveyor systems may be more environmentally acceptable although the Commission have no direct experience of these.

7.29 On our visit to the Yorkshire Coalfield we saw examples of both good and bad tipping practice. At Kinsley Drift, spoil was being blown away by high winds in a great cloud of dust from the machine depositing it. It is the responsibility of the N C B to ensure that what might be termed "good housekeeping" practices are adopted as a matter of course at all their operations. We return to this point in paragraph 7.32 below.

Coal stocks

7.30 Coal stocks provide the second potential source of dust. Usually these are kept close to the pit-head, sometimes stored in permanent bunkerage, sometimes in an open yard. In certain areas, where the demand for a particular kind of coal has decreased, stockpiles reach the size of small tips and can be kept for a period of years. We had evidence in Wales of one stockpile that had been grassed over. We also received complaints from local authorities about the use of G D O sites for stockpiling. The General Development Order gives the N C B the right to carry out ancillary developments in connection with coal industry activities within immediate vicinity of the pithead. These provisions are simply intended to allow the N C B, in common with all other industries, a degree of flexibility in carrying out work on sites in connection with the use for which they have planning permission, including deemed planning permission. It does not appear to us that they should, in fact, encompass the transport of coal for stocking purposes from one colliery to another. Stockpiling on a large scale for a number of years can have adverse consequences for the immediate neighbourhood. The local planning authority should be in a position to impose conditions on stockpiling, or put forward alternative proposals, designed to protect those neighbourhoods. We recommend therefore an amendment to the G D O to make it clear that any permitted ancillary developments relate only to the activity of that particular colliery and cannot be used for the ancillary activities arising from the operations of any other colliery.

IMPACT OF TIPPING AND COLLIERY OPERATIONS

7.31 The total impact of mining operations on a locality is to a very large extent dictated by the nature of the tipping operations that accompany the coal extraction. However, the condition of the colliery buildings and the general appearance of all on-site activities are also important. During our visit to Yorkshire we were very impressed by the substantial improvements that could be achieved by maintaining high standards of tidiness, applying a coat of paint and carrying out a little judicious planting of flowers and shrubs. At Maltby, Kiveton Park, and Thurcroft we saw examples for ourselves of what could be achieved for comparatively little expenditure. In another part of the Yorkshire Coalfield and in South Wales we saw rusted barbed wire around tips, decaying derelict buildings and general on-site sloppiness of a quite different order. Even in the good examples it was not generally evident that adequate architectural and landscape design skills were being employed.

7.32 These contrasts illustrate the importance of "good housekeeping". We have discovered that it is the variations in local standards of good housekeeping that often give rise to complaints. They arise, in part at least, from the absence of sustained pressure from Central and Area headquarters to achieve higher standards. In the past this has been due to the inability of the N C B's internal organisation to deal with such issues, but lately things have been changing. We examine these changes more fully in Chapter 12.

RELATIONSHIPS WITH LOCAL AUTHORITIES

7.33 The N C B suggest, and our observations confirm, that better results are achieved the greater the degree of co-operation between the local authorities and the N C B. One example is provided by the scheme whereby South Yorkshire County Council provide advice and pound for pound cost sharing of amenity improvements such as modest shrub and trees planting and grassing of colliery entrances. Co-operation between authorities and the Board can have beneficial effects much wider than small amenity schemes: joint planning to mutual benefit can both help the Board's operations in a practical way and work in the wider interest of the area; an example is provided by the quarry reclamation scheme at South Maltby which we discuss in Chapter 9. This relationship, of course, requires a constructive and sympathetic approach on both sides. We saw examples of the best and worst of relations. It is essential that all parties recognise that their mutual interests are best served by a high degree of co-operation.

Chapter 8   Subsidence

8.1 We now turn to what in our view is one of the more significant impacts of deep mining – subsidence. Shifts in the land surface can damage all kinds of buildings and agricultural systems. In particular, subsidence can damage individual homes with all the shock, and emotional stress and strain that this involves. In this Chapter we look at the extent to which subsidence can be predicted, measured, prevented or limited. We recognise that a degree of subsidence is at present an unavoidable consequence of modern deep mining techniques. We examine therefore the adequacy of the provision of compensation for those whose property is damaged by it.

MEASURING AND PREDICTING SUBSIDENCE

8.2 In Chapter 5 on deep mining operations we explained briefly how longwall mining methods lead to immediate subsidence of the surface. This is further illustrated in Figure 8.1 (see page 50) which shows how cumulative displacement may produce subsidence of the surface. The N C B have built up a considerable corpus of knowledge about subsidence from analysis of empirical observations of subsidence cases in different mining and geological conditions, and are now able to predict the effect on the ground surface with a high degree of sophistication. They have produced a Subsidence Engineers Handbook (first published in 1965 and revised in 1975) (N C B 1975) which is the standard reference work for predicting the extent of subsidence: this uses graphical methods which, on the basis of the dimensions of the mine, can determine the extent of subsidence, the degree of strain, and the slope of the surface in the subsided area. Mathematical and computer programmes have also been developed, but these are more complex than the standard graphical method.

8.3 The major factors affecting the extent of subsidence are the depth of the coal seam, and the width and thickness of the coal faces extracted. The thicker the faces, the greater the void which will be left after mining with consequently greater eventual subsidence at the surface. But the deeper the coal seam, the more the effects of subsidence will be attenuated through successive strata, so that the effect at the surface will be less, although it will extend over a wider area (see Figure 8.2). For example, the extraction of a coal face 250 metres wide and 1·5 metres thick at a depth of 300 metres would produce maximum subsidence of 1·2 metres; at a depth of 600 metres it would produce subsidence of 0·5 metres. The subsidence never exceeds 90% of the thickness extracted and may be considerably less.

Figure 8.2   Varying subsidence profile at different levels

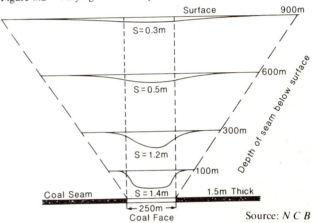

Source: N C B

8.4 The effects of subsidence are not restricted to the area vertically above workings but extend for some distance beyond that area. The angle between the edge of the workings and the outer limit of ground movement is defined as the angle of draw and in most British coalfields it averages about 35 degrees. Thus workings at any one point will have effects on the surface for a distance away from the edge of workings equal to about 0·7 times their depth. Conversely, any point on the surface will be affected by workings within the area subtended in all directions by the angle of draw, that is to say, the "critical area" – the circular cross-section at seam level of a cone of which the apical angle is twice the angle of draw and the diameter of which is about 1·4 times seam depth. In a typical case where the seam is 500 metres deep, the critical area for a given point might be about 700 metres in diameter (see Figure 8.3).

Figure 8.3   Diagram of subsidence

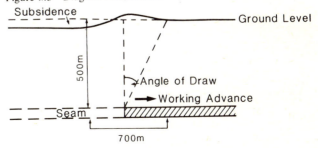

Figure 8.1   Subsidence caused by cumulative displacement

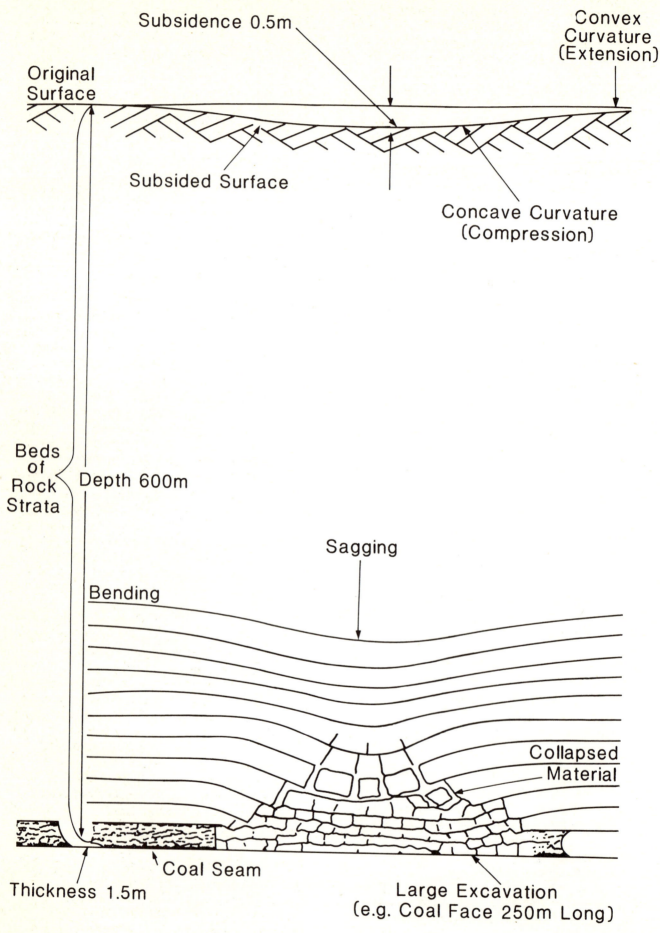

Subsidence 0.5m

Convex
Curvature
(Extension)

Original
Surface

Subsided Surface

Concave Curvature
(Compression)

Beds
of
Rock
Strata

Depth 600m

Sagging

Bending

Collapsed
Material

Coal Seam

Thickness 1.5m

Large Excavation
(e.g. Coal Face 250m Long)

Source: *N C B*

8.5 Subsidence at the surface usually follows the advance of the coal face. Field measurements show that initial subsidence takes place almost instantaneously as coal is extracted. However, any given structure may continue to suffer subsidence while coal is worked within its critical area. Thus for a 500 metre deep seam, any surface structure will be affected by all workings within 700 metres of the point vertically beneath it. With coal faces up to 250 metres long, at a seam depth of 500 metres, at least three adjacent faces would have to be worked to extract the coal within the critical area of 700 metres diameter. At a typical seam thickness of 1·5 metres, these faces might take 4 to 5 years to work and subsidence at the surface point would not be complete until they had all been worked. After the coal face has passed out of the critical area, residual subsidence rarely exceeds 5% of total subsidence at any given point but such movements may continue for up to 2 or so years after mining has ceased.

8.6 Gaps in knowledge still remain. For example, it is less easy to predict what subsidence will occur in ground with geological faults, where the effects of stress and tension in the strata will be more complex; and prediction is less reliable in cases of mining in seams below those where some extraction (and subsidence) has already taken place, especially in old uncharted workings. As much future mining may be in such deeper seams, this is an area in which more research would be valuable. We recommend that the N C B should carry out further research into the effect of subsidence on the mass between the coal seam and the surface especially in geological fault conditions.

EFFECTS OF SUBSIDENCE ON BUILDINGS AND LAND

Buildings

8.7 Figure 8.1 shows how the subsided surface may be compressed or extended (in different parts of the area affected). This process, known as "strain", may cause the buildings on the surface to buckle or stretch. The extent of damage may vary considerably, as is shown by the following Table:

Table 8.1   Classes of subsidence damage

Class of damage	Description of typical damage
1 Very slight or Negligible	Hair cracks in plaster. Perhaps isolated slight fracture in building, not visible on outside.
2 Slight	Several slight fractures showing inside the buildings. Doors and windows may stick slightly. Repairs to decoration probably necessary.
3 Appreciable	Slight fractures showing on outside of building (or one main fracture). Doors and windows sticking, service pipes may fracture.

Table 8.1   Classes of subsidence damage–continued

Class of damage	Description of typical damage
4 Severe	Service pipes disrupted. Open fracture requiring rebonding and allowing weather into the structure. Window and door frames distorted, floors sloping noticeably, walls leaning or bulging noticeably. Some loss of bearing in beams. If compressive damage, overlapping of roof joints and lifting of brickwork with open horizontal fractures.
5 Very severe	As above but worse, and requiring partial or complete rebuilding. Roof and floor beams lose bearing and walls lean badly and need shoring up. Windows broken with distortion. Severe slopes of floors. If compressive damage, severe buckling and bulging of the roofs and walls.

Source: N C B

Damage typical of these categories is also illustrated by the photographs at Plates 8.1, 8.2 and 8.3.

8.8 Some estimate of the degree of damage can be made by plotting the likely degree of strain in the ground against the size of the structure in question. The N C B have devised graphs, based on these two factors, which enable this to be calculated. For example, a degree of strain commonly encountered is around 0·1% (that is, 1 millimetre per metre). This might have the effect of extending or shortening a building 20 metres in length by 20 millimetres, and could be expected to cause only "very slight or negligible" damage. If the building were longer, more serious damage could be expected. But for the 20 metre building, strain of 0·3% (3 millimetres per metre) or 60 millimetres over the whole of the building – might cause "appreciable" damage, and an even greater degree of strain would probably be needed before severe damage was caused.

8.9 But the size of building, and degree of ground strain, are not the only facors affecting subsidence damage although they are the two most important. The N C B's graphs allow some assessment of the possible interruption of use to factories, schools and occupiers, and the possible cost of repair work. But they do not permit accurate predictions of damage to individual homes or buildings because precise effects depend on further factors beyond the N C B's control and probably outside their knowledge such as the shape, the method of construction, and how the building is secured to the ground. A well maintained building stands up to strain better than one in a poor state of repair.

Pipes, sewers and cables

8.10 Pipes and cables which are tightly anchored to the subsoil in subsiding ground may be compressed or extended in part, with the ground. This problem can be allowed for when new pipes are installed as these can be designed to deflect and telescope at the joints. Cables may need expansion joints. Gas pipes, particularly, may need special precautionary measures. Sewage systems may be affected where subsidence produces a slope in the ground greater than, and in an opposite direction to, the existing sewage gradient, and a backfall in the system may result. Effects can be serious: for example, in the Lower Dearne valley the Water Authority considered that a comprehensive replacement scheme, involving two separate sewage disposal systems, was the only practicable way of coping with the constraints on development of the existing system imposed by subsidence damage.

Land Drainage

8.11 Subsidence raises the level of the water table relative to the surface although the exact effect this has depends upon the internal drainage of the land. Some precautionary measures can be taken: for example, flooding of existing rivers may be averted by building embankments or raising existing ones, and field drains may be relaid to cope with internal field drainage. But in low-lying areas or areas of poor drainage, land may be left permanently flooded or water-logged. This may represent a permanent loss of agricultural land. Changes in drainage may also upset ecosystems or alter wildlife habitats (see Plate 8.4).

Other effects

8.12 Land subsidence and the risk of flooding can upset communications. Thus the N C B had to pay to divert the British Rail main line around the Selby coalfield, since much of the area was already subject to flooding even before mining began. The Sheffield–Barnsley–Leeds main line is subject to speed restrictions following mining activity. Highways may also suffer from subsidence damage, leading to extra maintenance costs.

EXTENT OF SUBSIDENCE DAMAGE

8.13 N C B figures indicate that the number of cases in which severe damage to dwellings occurs is proportionately small. An N C B investigation of the effects over a 10 year period in the East Midlands (covering the Nottinghamshire, Derbyshire, and Leicestershire coalfields) showed that two out of every three houses in the area affected suffered no subsidence damage. Two thirds of those damaged were in the "very slight or negligible" category. Houses severely damaged represented less than 0·5% of the total numbers damaged. In absolute terms, however, the figures appear more significant. Altogether 62,500 houses were damaged to some degree: although in 40,000 of these cases damage was only very slight there were nevertheless 2,000 cases of appreciable damage and 250 cases of severe damage. Averaged evenly over the period this would represent approximately one case of severe damage per fortnight over the areas affected and nearly 4 cases of appreciable damage every week.

8.14 In the same area over the 10 year period covered, the Board dealt with claims in respect of agricultural land covering 728 hectares – which represented about 1·4% of all the agricultural land within the areas of mining influence. The N F U consider, however, that the effects on agricultural production may be greater than these figures suggest. Subsidence damage to farm land does not become immediately apparent. As we stated above, drainage systems can be rendered useless and differences in the extent of subsidence in adjoining fields can hinder practical farming: wet or cold patches in the soil may hinder crop growth; sloping ground surfaces may rule out the use of machinery; and the surface may become subject to soil erosion and cracking. As more mining is likely to take place in rural and agricultural areas hitherto unaffected by mining, more good agricultural land may be affected by subsidence in the future. This may mean some shift in the general effects of subsidence damage from buildings to land and communications, particularly to increased risk of flooding. Traditional mining areas have been densely populated and most of the major conurbations, apart from London, are in or near them. Such built up areas will continue to suffer from subsidence to the extent that mining continues in traditional mining areas, particularly those with old housing stock.

8.15 Subsidence, as explained above, affects sewerage and drainage systems also. The water authorities' problems arise mostly through damage to water mains and sewers; occasionally damage is also caused to the structures of service reservoirs or sewage treatment plants. We understand the water authorities by and large inherited a tradition of good relations and co-operation with the N C B. Generally there are local technical groups which monitor operations, assess damage and advise on preventive and repair measures. The main complaint from the water authorities is that compensation does not cover costs of preventive reinforcement of new structures or their costs of survey and inspection. In relation to new developments by the N C B, the water authorities have shown themselves well able to negotiate better preventive terms by agreement, often including contributions from the N C B towards the costs of preventive works, the setting up of formal technical liaison and consultative teams and the agreement of the N C B to provide more than the formal period of notice of mining works specified under existing legislation.

8.16 There may be cases where severe subsidence sterilises the land and effectively prevents future development. We consider that in such cases thought should be given to putting that land to constructive use. The Nature Conservancy Council (N C C) have suggested that land flooded through subsidence could become a valuable habitat for wildlife as has already happened at several sites, such as Stodmarsh National Nature Reserve in Kent and at Gresford Flash in Clwyd which has a full range of hydroseral plant communities. This struck us as a useful suggestion and there may well be other possibilities. We recommend therefore that where such sterilisation takes place, the N C B should discuss with local interests and with environmental bodies what alternative uses might be made of the site. Such land may also be eligible for derelict land grant.

ATTITUDES TO SUBSIDENCE

8.17 Whatever the proportion of people affected by subsidence, it is a major problem for those whose property or land is damaged. Farmers or industrialists may face inconvenience or economic loss. However, those who suffer most are the individuals whose homes are damaged. They must first undergo the stress caused by the uncertainty as to whether their home will be damaged, when this will happen and how severe the damage will be. They must then suffer both the inconvenience of negotiating with the N C B the nature of the repairs, and the disturbance whilst those repairs are carried out. The situation is even worse where temporary repairs are necessary because further subsidence may take place later. In addition subsidence generates a sense of unease even among neighbouring families not immediately affected.

8.18 Anxiety can be partly allayed by adequate information and we consider it important that possible subsidence effects should be discussed and publicised as widely as possible with local residents and other interests before mining starts. We understand that, as general practice, the Board are willing to answer enquiries from anyone about mining activities and their likely subsidence effects, although they do not volunteer information unasked. In recent years they have organised local Subsidence Liaison Committees which can act as a link with local interests. Under Section 2 of the Coal Industry Act 1975 the Board must also publish notices (in the London Gazette and appropriate local newspapers) before it exercises its general right to withdraw support; and the Board are, increasingly, publishing such notices even in cases where because of rights under previous legislation they are not required to do so. In the case of Selby and Belvoir, the Board went further and distributed information to individual householders about N C B liabilities for subsidence.

8.19 We endorse this effort. We do not consider it appropriate to lay on the N C B a legal obligation of advance notification to individual householders. But we recommend that they should make every effort to ensure that individual owners and business interests are aware of likely subsidence – even when publication of notices is not obligatory. We recommend that suitable notices be posted in streets when mining is about to take place within the critical area. We also recommend that the degree of publicity at Selby and Belvoir should become standard practice for all new mining projects.

8.20 Subsidence from the breaking and collapse of pillars of coal left for support in old workings, may still occur in older mining areas, particularly where earlier mining was uncharted. For example, the recent cracks which have appeared in the M1 motorway may well be due to mining subsidence. We are satisfied that the N C B's modern methods and expertise have removed this danger from new mining areas although this will be little consolation to those living in older mining areas, who are already subject to an accumulation of environmental stresses. But the improved prediction techniques of present and future mining, the detailed documentation of mine workings, and the deeper seams being worked reduce the likelihood of severe subsidence and make it easier to control the effects than in the past. The next section discusses possible control measures in more detail.

POSSIBLE METHODS OF MINIMISING SUBSIDENCE

8.21 Subsidence effects may be reduced
(a) through mining techniques; or
(b) through preventive measures to existing structures or introduced in the design of new ones.
These approaches are briefly discussed below.

Through mining techniques

8.22 The extreme option is to avoid mining the area concerned. Any mining proposal involves a judgement of likely costs against gains of coal, and the Board must weigh the balance in each case. Although no formal cost-benefit analysis is carried out, the Board's 5 year plans are extensively discussed among all the different departments at Area level, and the likely effects on property and communities will be among the factors taken into account. Subsidence is likely to be a major factor raised during planning inquiries, as at Selby and Belvoir. But, unless the area affected is heavily built up, or the coal reserves seem too small to justify the likely consequent structural damage, we do not consider it realistic to avoid mining because it may cause subsidence.

8.23 Alternatively, it may be possible to design the mining layout so as to reduce the risk of subsidence. One method is partial extraction, leaving wider pillars of coal between the faces and working smaller faces than usual. This is inevitably a compromise solution as part of the coal will be sterilised, but it can permit typically around 50% and sometimes 70% of coal to be extracted in heavily built-up areas with reduced risk of subsidence. The N C B quoted several recent cases where it had been used successfully. At Coventry Colliery, for example, 45% of a 2 metre seam of coal was extracted at a depth of 530–720 metres. Although the workings lay under some large industrial undertakings with sensitive machinery, few claims for damage were received. Examples in residential areas were also quoted: at Sherwood Colliery in

Nottinghamshire, 50% of a 2·3 metre seam, 450 metres deep, was extracted with only very slight damage resulting to 6% of the houses in the working area. At Horden Colliery, Durham, 60% of a seam over 1 metre thick and 268 metres deep was worked under residential property without any claims resulting for damage. At Selby the planning permission stipulated that subsidence should not exceed 0·99 metres.

8.24 Alternatively, where particular land or buildings need complete support, large pillars of coal may be left in place. The N C B consider this to be appropriate only in a very small number of cases, where complete stability is essential. While any surface developer can approach the N C B to negotiate for a pillar of coal, he will not generally have the right to demand this, and N C B agreement might involve substantial expense to the owner. A planning permission for a new development as at Selby, may require the N C B to leave a pillar without compensation. Where buildings of architectural or historical importance are involved the N C B may offer to leave such pillars. In these cases, the pillar is extended beyond the edge of the structure to be supported, by a circle with radius equal to half the depth of the coal workings. At Selby Abbey, the pillar was extended further with a radius 0·7 times the depth of the coal workings, for additional security.

8.25 This method is different in scale from the older practice of room and pillar workings described above: the surface above these pillars is completely protected. But it can disrupt the continuity of the underground mining layout and make it difficult to mine adjacent areas. Nevertheless, pillars and partial extraction have a valuable part to play in enabling mining to take place in areas where the risk of subsidence might otherwise make it unacceptable, that is, under heavily built up areas or under buildings of major architectural or historic significance. The partial sterilisation of coal entailed may be a necessary price to pay for the coal extracted.

8.26 Backstowing (see Chapter 9 on Spoil Disposal) may also reduce subsidence – by up to 50% in the most effective cases, where the spoil is packed tightly into the void. This would reduce the worst effects on the ground, although it would not necessarily halve the numbers of dwellings affected. While it would not be realistic to use backstowing solely as a method of reducing subsidence, the mitigation of the effects it can offer might be an additional reason for using it when appropriate for spoil disposal purposes.

8.27 Finally, the possible impact of technological developments on subsidence should be noted. Developments permitting deeper mining than at present would increase the area at risk although reducing the severity of the impact. On the other hand, in situ gasification might reduce subsidence since only the coal would be extracted. At present, however all these developments are in their infancy.

PREVENTIVE MEASURES ON THE SURFACE

8.28 Preventive measures can be taken on the surface either through the adaptation of existing structures, or by precautionary measures taken in the design of new structures. We examine these separately below.

(a) Existing structures

8.29 The N C B have powers, but are not obliged, to carry out preventive works to property where they consider that such works would prevent or reduce subsidence (Section 4 of the Coal-Mining (Subsidence) Act 1957). Such works may aim to make the structure more flexible or to reduce the amount of ground movement affecting it. Compression can be eased by removing part of the connections between existing buildings, or by cutting "slots" in the structure; bulging can be prevented by inserting tie rods, or by supporting window openings or large archways. Such works do not eliminate the risk of subsidence after mining, but they may reduce significantly its incidence and its severity. In most cases subsequent damage will be not more than "slight".

8.30 The N C B may either carry out these works themselves or reimburse the owner of the property for their cost. But if the owner unreasonably refuses to consent to the work being carried out, he loses his rights in respect of subsequent subsidence damage which might have been prevented. The evidence suggests that generally the Board prefer to cope with the damage after it has occurred, rather than undertake preventive measures on any large scale. It may well be cheaper for the Board to do this, especially as preventive works do not eliminate the risk of later subsidence and since most dwellings will, in the event, probably not be damaged. We recommend, however, that the possibility of such measures should at least be a matter for full public discussion in areas to be affected by mining.

(b) Precautionary measures in designing new structures

8.31 Precautionary measures taken in the design of new buildings can effectively reduce potential damage. These include building in separate smaller units; using special frames allowing ground movement to take place without damage (such as the C L A S P) system of flexible building, first used in Nottinghamshire and now used extensively in coal areas in the U K and abroad); and placing foundation slabs on smooth beds of granular material (the slabs can slide along the bed when ground movement occurs).

8.32 Under current procedure (Article 15(1)(d) of the General Development Order 1977), local planning authorities must consult the N C B before granting planning permission for buildings in a notified area of coal working. The Board then make recommendations where appropriate for precautionary measures, or on the siting or phasing of buildings – for example, to avoid building on or near a known geological fault where subsidence effects may be worse. However it is the developer's responsibility to take such precautionary measures. The N C B suggest that local authorities differ in their attitude towards the Board's recommendations: some make specific precautionary measures a condition of planning consent, others require the developer to discuss appropriate measures with the Board and others may take little notice of the Board's observations. The N C B also suggest that where possible the surface development should be phased with the mining layout. This would imply that developers should postpone building work until the ground settles after mining. However, this may involve unacceptable delays. Barnsley M B C for example quoted a strategic Industrial Estate in their Borough where three-quarters of the development has been subject to several years delay before mining ceases.

8.33 The developer now bears the cost of any precautionary measures himself because the N C B do not actually require them to be taken. Some evidence, notably from Barnsley M B C, recommended that the cost of such precautions to new buildings should be borne by the N C B. It was urged that developers would have a greater incentive to incorporate precautions if the cost were met by the N C B, which in turn might benefit the locality by attracting new development. It would also bring the N C B's responsibilities into line with their existing responsibility for actual subsidence damage and powers of preventive works on existing dwellings.

8.34 It is essential that the N C B should keep the local planning authority informed of the programme of work in mining areas. If the N C B recommend that precautionary measures be adapted in new structures, after full consultation with the local planning authority and the developer, then we recommend that the local planning authority should include the appropriate planning conditions in any planning permission. This is in the mutual interest of the developer and the N C B. It should serve to limit potential damage to the new building and avoid unnecessary inconvenience or economic loss caused through disruption of its use. It should also mean, and this seems reasonable, that the N C B do not face additional costs for repair to damage which could have been avoided.

COMPENSATION

8.35 We are satisfied that there are severe practicable limits to minimising the incidence of subsidence. For the foreseeable future therefore the Board will have to operate on the present basis – the payment of compensation for the restitution of the damage caused after it has taken place. In this section therefore we look at the basis for and the operation of the compensation code.

Rights to work coal

8.36 The principle of the payment of compensation arises as a corollary of the Board's legal right to work coal if planning persmission has been granted. The N C B has a freehold interest in unworked coal (under the terms of the Coal Industry Nationalisation Act 1946). It also has the right to withdraw support from land on the surface if that land is owned by another party. The origins of these rights are historically complex: in some cases, for example, the right to withdraw support may have been negotiated with the surface owners. Under Section 2 of the Coal Industry Act 1975 the N C B now have a general right to withdraw support where this is requisite to enable coal to be worked, subject to payment of compensation for subsidence damage or making the damage good and subject to any restriction contained in a planning consent. In addition, the Coal-Mining (Subsidence) Act 1957, which we discuss in further detail below, imposes obligations for repair. Special arrangements apply to protect railways and certain other installations from the effects of mining operations: these are known as Mining Codes. The general principle behind these is that the undertaking concerned does not initially own the minerals beneath the land it uses but that it has the right to request that particular areas of minerals should be sterilised, if it thinks fit, subject to payment of compensation. This clearly can help to protect railways from the possible disruption and damage of subsidence. But if the undertaking does not require such sterilisation then the mine owner may work the minerals without liability for damage, provided that he works them in a proper manner.

8.37 Under the Coal-Mining (Subsidence) Act 1957, the N C B have a duty to carry out remedial works to buildings, services or land which have been damaged through mining subsidence, to make them "reasonably fit". Alternatively, the Board may pay for the work to be done. But the Act does not impose any duty to make

Table 8.2 Subsidence costs at current and constant prices

	N C B Payments – current prices	N C B Payments – Constant prices (1)
	£million	£million
1969/70	5·1	9·2
1970/71	5·4	8·7
1971/72	6·3	9·1
1972/73	6·2	8·2
1973/74	7·9	9·6
1974/75	10·1	10·1
1975/76	14·4	11·6
1976/77	17·0	12·1
1977/78	26·0	16·6
1978/79	30·9	17·8
1979/80	42·6	21·3
1980/81	54·7	23·3

Notes   (1) At 1974/75 prices
        G D P Deflator June 1981 Economic Trends
        Source (for Current Price Figures): N C B

the property better than it was before the damage took place. The Coal Industry Act 1975 also requires the Board to pay proper compensation. We understand that on average, the Board receive something like 14,000 new claims each year, most of which relate to private dwellings. The cost of subsidence compensation payments has increased considerably in recent years, as the N C B figures in Table 8.2 for the period 1969/70 to 1978/79 show. This increase is partly due to the general increase in prices, affecting land, property and building work, and partly due to N C B's own acceptance in recent years of liabilities beyond their strict legal obligations.

8.38 A voluntary Code of Practice was adopted in 1976 following discussions by a Government Working Party. This provided that compensation should be paid for all physical damage caused by subsidence to household goods, furniture and fittings, stock in trade, plant and machinery; that farmers should be compensated for loss of crops and stock; and that arrangements should be made to avoid or alleviate individual hardship. And in 1978, under an agreement with the Royal Institute of Chartered Surveyors (R I C S), the Board agreed to contribute towards surveyors' fees incurred by claimants, in cases where compensation or the cost of repairs exceeds £250. In practice this "contribution" often represents the whole of the Surveyor's costs.

8.39 The Table below, supplied by the N C B, gives a breakdown of claims dealt with in 1977/78 for two areas affected by mining subsidence. This shows that over half the claims settled in Western Areas and over 70% of those settled in the Barnsley Area amounted to over £250.

Table 8.3 Subsidence compensation claims in the Western and Barnsley areas

Repairs and/or compensation	Western Area	%	Barnsley Area	%
Up to and including £250	517	49	576	28
From £251 to £500	253	24	638	32
From £501 to £1000	166	16	501	25
From £1001 to £2500	78	7	195	10
From £2501 to £5000	25	2	51	3
Exceeding £5000	18	2	41	2
Total number of claims for Dwellings	1057	100	2002	100
Houses purchased and demolished	8		16	
Houses purchased and repaired for resale	7		3	
Total number of houses purchased	15		19	

Source: *N C B*

8.40 However, there has been continuing pressure to alter the Code still further. Mr Jack Ashley M P introduced a Bill in 1979 designed to widen the N C B's liabilities: particularly, to require payment for loss of value of property due to subsidence, and compensation for stress and loss of earnings from time off work resulting from the subsidence. This Bill also proposed that an independent commission be set up to decide on the Board's liability and to administer the scheme. Currently the N C B decide in the first instance whether the damage was caused by mining subsidence. If the claimant disagrees, he may take the dispute to the Lands Tribunal or the County Court. The onus is then on the Board to prove that it is not mining subsidence that has caused the damage. The Board have recently accepted that questions and disputes should be settled by an arbitration if the claimant so wishes.

8.41 This Bill fell, but discussion of the proposals in it has continued. Several bodies giving evidence to us gave a number of recommendations for altering the compensation code, many on similar lines to Mr Ashley's proposals. The Department of Energy and the N C B have also been reviewing the arrangements.

8.42 Most complaints about the present arrangements related to the operation of the Code of Practice rather than the scope of the general provisions. Different bodies instanced delays in carrying out work: verbal promises of repairs given but not fulfilled; thoughtless repairs such as the use of new bricks which did not match the existing ones; or inadequate supervision of work. Another common complaint concerned delays in carrying out final repairs; delays of up to 2 years were quoted. We recommend that all work should be carried out as speedily, sensitively and efficiently as possible. We accept that where further ground movement is expected it would be wasteful to attempt final repairs. But this in our view increases rather than reduces the need for rapid and unobtrusive interim repairs.

8.43 The evidence suggests that there is room for tightening up on local operation of the Code; we have already suggested this to the N C B, who have agreed to check on local operation. We also discussed possible widening of the scope of the Code. The main suggestions put to us in evidence were for additional payment for stress and inconvenience suffered by the householder, and compensation for diminished property values.

8.44 Stress and inconvenience are probably best coped with by imaginative and compassionate operation of the existing Code of Practice. We did not think it appropriate to suggest additional measures or financial compensation here. Loss of property value may occur as a result of anticipated or actual subsidence. The expectation of subsidence occurring may make a house less saleable, because even if a potential buyer knows that the Board will repair the damage, the prospect of inconvenience acts as a deterrent. We do not, however, recommend any statutory provision to deal with loss of value arising from the prospect of subsidence. A person whose house is liable to be compulsorily purchased and cannot as a result be sold at a reasonable price may, under the blight provisions of the Town and Country Planning Act 1971, require the authority concerned to acquire it forthwith. He cannot, however, require the authority to buy land which is not liable to be compulsorily purchased, even if the prospect of public works on neighbouring land reduces the present market value of his property. Nor is there provision for compensation for such a loss of value. We did not consider coal mining sufficiently different from other forms of public development to justify legislating for cases where the value of property which is not liable to be compulsorily acquired by the Board is reduced by the expectation of subsidence.

8.45 Loss of value caused by actual subsidence is a different matter. It seemed to us that there was a case on grounds of equity for compensating the owner of property which suffers a residual loss in value as a result of actual subsidence in spite of repairs carried out by the Board under the 1957 Act, for example, if a house is left with a slight tilt, or obvious repairs to an unsightly crack which make it less attractive than its neighbours. The N C B implicitly recognise there is a case for additional compensation here as we understand they make discretionary cash payments in certain cases in addition to carrying out repairs. We believe there should be a right to compensation in such cases, rather than payment being left to the discretion of the Board. There is a long-standing provision, currently in section 10 of the Compulsory Purchase Act 1965, for compensation where land not compulsorily acquired is nevertheless injuriously affected by the execution of works. We understand that over time judicial interpretation has limited the right to compensation under this provision to cases where the following four rules are satisfied:
(1) the injurious affection must be the consequence of the lawful exercise of statutory powers, otherwise the remedy is by action;
(2) the injurious affection must arise from that which, if done without statutory authority, would give rise to a cause of action;
(3) the value of the land or interest must be directly affected by interference with some legal right, public or private, which the claimant is entitled to make use of in connection with his property;
(4) the damage must arise from the execution of the works and not from their authorised use.

These rules constitute a stringent test. It nevertheless seemed to us that the existing provision for compensation for injurious affection of land was an apt precedent for the liability which in our view should be placed upon the Board to compensate for any residual loss of value resulting from subsidence. We recommend therefore that, for the future, there should be additional provision in the 1957 Act for compensation for such residual loss of value, taking into account but not superseding the duties already placed on the Board by the Coal-Mining (Subsidence) Act 1957 to make reasonable repair.

GENERAL CONCLUSIONS

8.46 In the present state of mining knowledge, some degree of subsidence is an unavoidable risk of deep coal mining. Further research may improve prediction techniques but is unlikely to remove the problem for the foreseeable future. Each development proposal requires a balance between possible subsidence, with its human and monetary costs, and the gains of coal. Although subsidence will be only one of the factors to be considered, it may in extreme cases rule coal mining out of the question or render the land above completely unsuitable for future use. Local opinion will want to be satisfied that the effects will not be unduly deleterious if coal mining proposals are to gain public acceptance. It is thus in the nation's interest as well as the N C B's own interest to ensure that preventive and precautionary measures, and compensation, are planned and undertaken as effectively and as flexibly as possible, in order to reduce the stress and inconvenience for those directly concerned.

Chapter 9 Spoil

INTRODUCTION

9.1 In Chapter 5 we gave a brief description of deep mining activities and the type of environmental impacts which they cause. We identified two as being the most significant and warranting special concern: subsidence and spoil disposal. The purpose of this Chapter is to consider in detail the problem of spoil disposal. We begin by examining the past and future scale of spoil production and its geographical distribution. We then consider the means the N C B can use to dispose of the spoil in future.

MINING SPOIL

The production of colliery spoil

9.2 We described in Chapter 5 how new spoil is produced in underground mining and that it consists of a variety of materials in sizes ranging from boulders to cobbles and clay particles. When coal was worked by hand much of the associated spoil remained underground. The miner could avoid extracting dirt by adjusting the action of his pick to the dimensions of the seam; and much of the spoil produced in making roadways and other works was stored in the void created by the extraction of the coal. The introduction of modern mechanised techniques greatly increased the proportion of spoil produced with the coal, and brought about changes in methods, which for the first time made it generally cheaper and more convenient to dispose of waste above ground. Coal cutting machines extract dirt along with the coal which are then conveyed to the surface before the dirt and coal are separated. Roadways and other underground works are constructed on a much more substantial scale than before, thus increasing the production of spoil. The roof behind the advancing face is now allowed to collapse and a permanent system to support it is no longer required. As a result the quantity of spoil extracted with a given amount of coal has increased dramatically. Figure 9.1 illustrates the increase in the amount of colliery spoil produced and the worsening of the coal to spoil ratio, particularly during the 1950s and 1960s, in relation to the progressive introduction of power loading at the coalface and the increase in the proportion of coal which is mechanically cleaned.

Figure 9.1 Coal and waste production 1920–1980

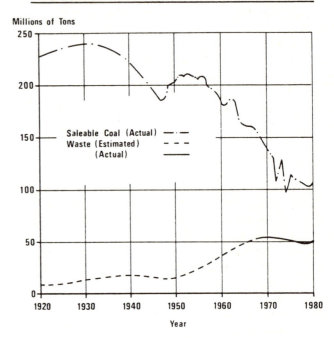

ANNUAL SALEABLE COAL OUTPUT AND WASTE DEPOSITED ON LAND

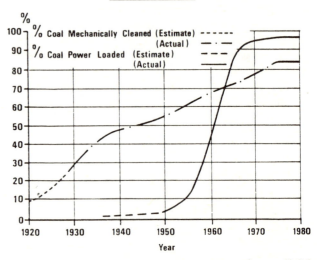

PERCENTAGES OF SALEABLE COAL MECHANICALLY CLEANED AND POWER LOADED

Source: *N C B*

9.3 The present national ratio of coal to spoil production is 2:1. The N C B have stated that they expect no significant change in this ratio as the changes in mining practices which caused the ratio to worsen in recent years are now complete. We see no reason to disagree with this, although as we discuss in the next paragraph, regional variations on the coal to spoil ratio are important. However, in order to arrive at an estimate of spoil production over the country as a whole over the next 20 years, we have applied this ratio of 2:1 to the higher and lower production figures in the Department of Energy's "Energy Projections 1979". Making allowance for opencast production which does not cause spoil problems, this indicates that 1,130–1,230 million tonnes of spoil could be produced between 1981 and 2000. The equivalent annual rate is 56–62 million tonnes. If all this new spoil were to be tipped, over the next 20 years about half as much spoil again would be tipped as during the history of the industry to date. Although new techniques, such as in situ gasification and microbiological degradation, may offer the prospect of spoil free energy production, they seem very far from practical application and they could themselves have adverse environmental implications. We have concluded that over the next 20 years spoil will be one of the major environmental problems arising from deep mining.

Spoil disposal

9.4 By far the greatest proportion of spoil is tipped on land. In 1976 this amounted to about 50 million tonnes (88% of the spoil produced in that year) of which 4 million tonnes was used in local land reclamation projects, with the rest going to conventional spoil tips or, to a lesser extent, lagoons. About 7 million tonnes (12%) was disposed of using marine disposal methods, the majority being tipped on the foreshore rather than into deep water, although this figure has since decreased (see paragraph 9.34). About 0.5 million tonnes was removed from existing tips for commercial use. These proportions do not vary much from year to year. The N C B's Minestone Executive has the responsibility for promoting the commercial use of minestones (unburnt colliery spoil), burnt shales and all surface minerals other than coal. The disposal of spoil by any other means is the responsibility of the producing colliery, from whose budget it is financed. Apart from the activities of the Minestone Executive we were surprised to find that there is no central planning and coordination of waste disposal.

9.5 The continued tipping of spoil will have major implications for land use and the environment. The N C B estimate that every 50 million tonnes require approximately 200 hectares of new land, and, if this ratio were to remain constant, then 4,500–4,900 hectares would be needed to accommodate the waste production we have estimated will arise to the year 2000. The middle of this range would be the area of a town the size of Middlesborough. If 80% of the spoil were to be deposited in very large heaps of 18 million tonnes each, then around 50–55 tips would be needed with an annual land requirement of 226–246 hectares each. A tip this size is very large. It would be required for a colliery producing 2 million tonnes of coal a year over a 20 year period, although it is true that the technique of progressive restoration enables land to be restored on one part of the site as tipping proceeds elsewhere. There will also be a continuing need to treat the liquid waste from coal preparation plants known as "tailings". Conventionally this is treated by settlement in lagoons which have a much greater land requirement than tips for dry waste.

The geographical concentration of tipping

9.6 The problem is exacerbated by the current geographical concentration of both derelict spoil heaps and active spoil tips, plant, buildings and lagoons. Table 9.1 shows that in 1978/79 the Yorkshire/Derbyshire/

Table 9.1 Colliery waste disposal 1978–79

N C B area	Coal output (million tonnes)	Waste output (million tonnes)	Coal/ waste ratio	Disposal (million tonnes) Land	Marine	No of Tips (a) Active	(b) Closed	(c) Disused	Total	No of collieries
Scotland	8·1	2·4	3·4:1	2·2	0·15	30	19	206	255	16
North East	12·9	6·1	2·1:1	1·5	4·6	33	26	43	102	28
North Yorks	8·2	6·2	1·3:1	6·2		44	9	9	62	18
Doncaster	7·0	2·9	2·4:1	2·9		11	11	—	22	10
Barnsley	7·4	4·5	1·6:1	4·5		31	34	6	71	18
South Yorks	7·6	5·7	1·3:1	5·7		30	10	6	46	18
North Derby	7·4	3·8	1·9:1	3·8		22	22	16	60	11
North Notts	11·0	6·0	1·8:1	6·0		31	8	—	39	15
South Notts	8·6	6·5	1·3:1	6·5		12	4	3	19	12
South Midlands (incl Kent)	8·5	2·7	3·1:1	2·7		28	22	4	54	22
Western Staffs, Lancs	10·9	3·7	2·9:1	3·5	0·25	53	21	37	111	
South Wales	7·6	6·0	1·3:1	6·0		50	78	163	291	37
Total	105·2	56·5	1·86:1	51·5	5·0(d)	375	264	493	1102	223

Source: N C B

Notes: (a) In use at collieries;
(b) originating colliery still in operation;
(c) originating colliery not in existence;
(d) 3·5 million tonnes on foreshore, 1·5 million tonnes into sea.

Table 9.2   Land justifying restoration attributable to spoil heaps, England 1974

| County (grouped coalfield) | Area covered by permissions (Hectares) | | | | Derelict land: spoil heaps (Hectares) |
| | Affected by: | | | Not yet affected by: | |
	(a) Spoil heaps and tips, plant, buildings and lagoons	(b) Excavations and pits	Total	(a) and (b)	
Northumberland	365	280	645	635	386
Tyne & Wear	55	—	55	—	375
Durham	409	336	745	331	710
Northumberland & Durham	829	616	1445	966	1471
North Yorkshire	138	—	138	—	—
West Yorkshire	1475	320	1795	348	598
South Yorkshire	2040	96	2136	199	572
Derbyshire	318	193	511	105	716
Nottinghamshire	953	104	1057	314	213
Yorks/Derby/Notts	4924	713	5637	966	2099
Leicestershire	241	19	260	15	78
Warwickshire	122	—	122	—	72
Kent	113	—	113	28	—
Cumbria	2	120	122	—	605
Cheshire	18	—	18	—	50
Lancashire	19	—	19	1	468
Greater Manchester	83	45	128	28	1237
Merseyside	179	—	179	7	203
Lancs Coalfield	299	45	344	36	1958
Staffordshire	613	211	824	70	470
West Midlands	4	—	4	—	358
North and South Staffs Coalfield	617	211	828	70	828
Salop	33	—	33	30	403
Gloucestershire	4	14	18	128	2
Total	7184	1738	8922	2239	7516

Source: *Department of Environment, 1975*

Nottinghamshire Coalfield accounted for 62% of waste output. Over one third of the total national volume of spoil currently arises within the N C B's Barnsley, Doncaster, North Yorkshire and South Yorkshire areas alone. South Yorkshire, North Yorkshire and Barnsley have coal:waste ratios of 1·3:1, 1·3:1 and 1·6:1 respectively, considerably worse than the national average of 1·9:1, that is, producing more waste per tonne of coal. Table 9.2 shows how spoil tipping has been concentrated in the same areas since the traditional method of spoil disposal has been local tipping. In 1974 the Yorkshire/Derbyshire/

Nottinghamshire coalfield contained 69% of the area of England affected by active spoil tips, plant, buildings and lagoons connected with coal; nearly half of this area was within the administrative counties of South and West Yorkshire; and the same area also possessed 28% of the area of derelict spoil heaps (including those not associated with coal mining). Other regions have a greater concentration of inactive and disused tips. For example, although South Wales has 26% and Scotland 23% of all spoil tips in the United Kingdom, over 80% of them are inactive or disused. In comparison only 40% of tips in Yorkshire are inactive or disused.

9.7 In Chapters 3 and 6 we discussed the prospects for the future development of the deep mining industry. Our main conclusions were that the main bulk of the N C B's deep mining activity up to the end of the century will continue to be concentrated in the main Yorkshire and East Midlands coalfield. In addition, it seems possible that substantial new mines could be opened up in the South and West Midlands and even further south in Oxfordshire; those are all areas with no tradition of mining. Thus, there appears to us to be two distinct elements in the question of spoil disposal in the future. The first involves the environmental impact of spoil tips in greenfield sites such as the Vale of Belvoir: spoil disposal was one of the most important questions discussed in the public inquiry into the N C B's proposals for mining in the area. Of course, the environmental impact of spoil tips in traditional mining areas is just as important. However, there is another problem which we foresee arising in the Yorkshire and Midlands coalfield. This is simply that given the amount of spoil which has already been tipped there, and the amount which is likely to be produced over the next few decades, there may not be sufficient land to accommodate the new spoil in an acceptable manner. This point was made most forcibly by the Strategic Conference of County Councils in Yorkshire and Humberside (S C O C C): they fear that "over the next twenty years the communities in the Yorkshire coalfield face having as much waste dropped on their doorsteps in the next two decades as has been tipped there over the last 200 years". Clearly then the practice of tipping spoil locally is central to the problem of assessing the environmental effects of coal mining and we now discuss this in more detail. There is an allied problem of the reclamation of derelict tips which is also very important particularly in areas such as South Wales and Scotland which have a very high proportion of inactive and disused tips.

Spoil tipping operations

9.8 The principal environmental effects of operational unreclaimed tips are visual intrusion, noise and dust from vehicle movements, loss of land and potential water pollution. After reclamation, the main impacts are the changed appearance and ecology of the land, and possible modifications in the pattern of land use from say some reduction in agriculture production or the creation of a new park. The impacts of the operational and pre-reclamation phases will be experienced most acutely by those in the immediate vicinity of the tip, and may affect their daily lives persistently over a long period. They may also constitute one of the greatest single effects of coal mining on the character of the area, for instance influencing its attractiveness to new development and employment.

9.9 There is, however, an important distinction to be drawn between older tips and new tips. Broadly speaking, the extent of the adverse impacts while a tip is being used, and before reclamation, will be less for more recent tips and greater for older ones because of the improvements in tipping practices which the N C B have recently introduced. Older spoil tips are generally conical or irregular in shape and have high profiles. Because of this they can appear considerably more prominent than new tips, seeming to neighbouring residents to dominate the view of the surrounding area (see Plate 9.1). These tips can only be reclaimed by major regrading sometime after operations have ceased when, with sensitive treatment, they can be integrated into the landscape.

9.10 In recent years the N C B have changed substantially their spoil tipping practices. Safety has been a prime consideration in doing so. After the Aberfan disaster the N C B established common constructional standards and management practices for safe tipping which ensure that there is little possibility of tips becoming unstable. Burning on new tips has also been eliminated. Since the passing of the Mines and Quarries (Tips) Act 1969 only minor incidents have occurred: 27 were reported by H M Chief Inspector of Mines and Quarries in 1969–70 but between then and 1977 only 22 have been reported. The N C B's Technical Handbook on tipping (N C B 1970) was the first to be published in the world. Environmentally the main effect of these safety measures has been that tips are now constructed with lower profiles which means that while they are generally less intrusive they also take up more land – yet another example of the balance that has to be struck between concern for human safety and for the environment.

9.11 However, the N C B have also begun to place more emphasis on minimising the impact of tipping operations, on avoiding pollution, and on the landscaping and restoration of tips. Before the 1950s, spoil was transported to tips by inflexible systems such as aerial ropeways, conveyors, railways and tracked inclines. Tipping took place at as high a point as possible and the material was simply allowed to accumulate below in a loose state. As the spoil was dumped the tips tended to be high, steep sided and conical. Mobile earthmoving plant was introduced into the industry in the late 1950s and this, combined with the better environmental standards which came to be expected and enforced through the development control system, led to the construction of lower and less intrusive tips. The newer flatter sites have a proportionately greater land take and are often larger in absolute terms because of the N C B's recent policy of concentrating coal preparation and waste tipping in some areas at selected collieries. A particular example is the Barnsley area where the N C B intend to link most of the area's collieries into three groups – South Kirby, Grimethorpe and Woolley. At Woolley for example the spoil heaps covered 50 hectares in 1956 and permissions since that time have extended this to

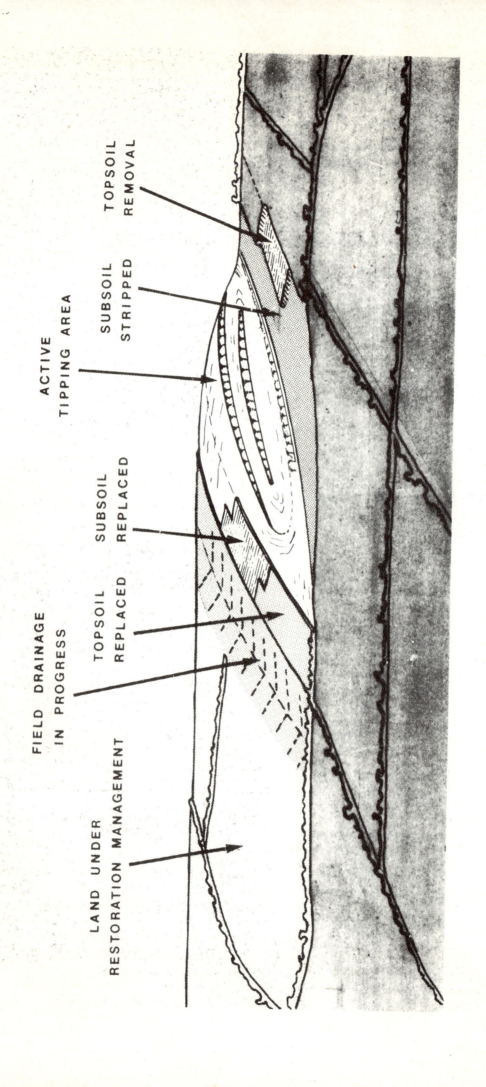

TOPSOIL
REMOVAL

SUBSOIL
STRIPPED

ACTIVE
TIPPING AREA

SUBSOIL
REPLACED

TOPSOIL
REPLACED

FIELD DRAINAGE
IN PROGRESS

LAND UNDER
RESTORATION MANAGEMENT

Source: N C B

Figure 9.2 Progressive restoration of colliery spoil tip

Plate 10.2a   see below.

Plate 10.2b   Thornley Colliery, Durham before and after reclamation. Durham County
Council has run one of the most successful derelict land reclamation schemes in the country
with financial assistance from the Department of the Environment in the form of Derelict
Land Grant.
(*Photograph by courtesy of Durham County Council*)

Plate 9.1 Spoil tip at North Gawber Colliery, Yorkshire. The estimated dirt output is 400,000 tonnes per year which is deposited on a site for which planning permission was given in 1966. Some progressive restoration has been undertaken but the site overlooks housing. (*Photograph by courtesy of South Yorkshire County Council*)

Plate 10.1 Derelict colliery buildings at the former New Monckton Colliery, Yorkshire. Coaling ceased in 1966 but this part of the site remains as shown in 1981. (*Photograph by courtesy of the Strategic Conference of County Council in Yorkshire and Humberside*).

Plate 8.4 Flooding in Barnsley. Flooding caused by subsidence of the river bed at Grange Lane thus preventing adequate discharge of flood water.
(*Photograph by courtesy of Barnsley Metropolitan Borough Council*)

Plate 8.2 (*above*) Severe subsidence damage to housing. This category is typified by cracks in the masonry of external and internal walls and the separation of ceilings from walls.
(*Photograph by courtesy of the National Coal Board*)

Plate 8.1 (*left*) Slight subsidence damage to housing. This category includes plaster cracks in walls and ceilings.
(*Photograph by courtesy of the National Coal Board*)

Plate 8.3 Subsidence damage to houses in Barnsley. The worst cases of subsidence damage can make houses uninhabitable. They may also result in ugly supporting timbers.
(*Photograph by courtesy of Barnsley Metropolitan Borough Council*)

cover 93 hectares. A planning application was made in 1980 to increase the capacity of the tip and planning permission was given for an additional 2·75 hectares. At Gwrhay in South Wales the spoil from 3 collieries, as well as redistributed material from an existing tip, will eventually cover an area of about 200 hectares.

9.12 The N C B intend to reduce the environmental impact of modern tipping sites by increasing the use of progressive restoration techniques and by improvements in tip design. In a scheme of progressive restoration spoil is deposited on part of a site as another part is restored. Thus the amount of land taken out of use for tipping at any one stage is minimised and the visual impact of the operation is reduced. Figure 9.2 illustrates diagrammatically how such a scheme might operate in practice. The design and construction of specific tips will vary depending on the geological and landscape characteristics of the locality where the waste is to be deposited and on the type of waste which is being produced. However, the N C B's objectives in designing tips is to maximise the volume of spoil that can be tipped on a minimum acreage of land consistent with landscaping and contours which would allow the restored tip to blend into the surrounding landscape. Other design and operational measures can be taken to limit the environmental impact of a tip. The boundary of a tipping area may be used and treated first to provide a screen for subsequent operations which may also work towards less densely populated areas of the surrounding locality thus reducing the visual impact of the work as well as the nuisances from dust and noise. Other control measures such as spraying the surface with water and silencing mobile plant, can also be applied. The most recent example of this approach to tip design is the N C B's proposals for tipping in the Vale of Belvoir. We understand that the tipping operations would involve progressive restoration to landforms which the N C B consider would blend into the surrounding landscape. Visual and acoustic screens would be built at each of the proposed sites and operations would work towards less densely populated parts of the surrounding area.

9.13 In general we have been impressed by the N C B's attempts to improve spoil disposal techniques and practices. During the course of our Study the N C B have decided that they should update the Codes and Rules and Technical Handbook to include all necessary guidance on the safe construction of tips using environmentally acceptable operating techniques, and their satisfactory reclamation and restoration in accordance with a pre-determined plan. In the past the N C B have employed consultant landscape architects for advice on particular tips, but this is by no means general practice. We are pleased therefore that, also during the course of our Study, the N C B have decided to recruit a landscape architect to advise on tip design. If these proposals had not been made by the N C B we would have made recommendations to that effect. We

recommend also that landscape design proposals should be incorporated in tip reclamation schemes from the outset – indeed incorporated in drawings as part of the planning application. The bland silhouette is not the only solution; leisure uses are more important than agriculture in some contexts; and planting for the early establishment of a varied and self-sustaining ecology demands a degree of imagination and subsequent after-care which is not as often evident as we would wish. We also consider that the N C B should be allowed to use other supervisors apart from the Agricultural Development and Advisory Service (A D A S) of the Ministry of Agriculture, Fisheries and Food (M A F F) and in Scotland, the Department of Agriculture and Fisheries for Scotland (D A F S) to oversee the restoration of land and should seek advice from the N C C on wildlife interests.

9.14 The evidence suggests that where new local tipping is to take place such arrangements can if properly implemented limit the environmental impact of tipping. However much will depend on how successful the N C B are in implementing their new methods of tipping and restoration in practice and there is not yet sufficient evidence about this for us to reach a firm conclusion. Some of the claims for the benefits of modern tipping may be overstated. We do not know as much as we would like about the levels of agricultural productivity which can be achieved on restored land. It also has to be acknowledged that however well spoil tipping is carried out it will cause disturbance. We have referred to the size of the spoil tips which the N C B intend to use in the future: they are by any standards substantial man-made developments with small possibility of their resulting in any substantial environmental gain. At Belvoir, operations would carry on for 50 years at least. We recognise that there may be occasions where the impact of a tip even designed to the N C B's highest standards may have such a heavy and adverse local environmental impact that it might still be considered unacceptable.

Spoil tipping controls

9.15 The distinction between environmental effects of older and newer tips is roughly paralleled by the planning regimes under which they were established. We have already described how the town and country planning system has been instrumental in enabling local planning authorities to ensure that new developments are planned, constructed, and operated to adequate environmental standards. This has also applied to spoil tipping. At first, consents were simply approvals to use land for tipping, but over the years planning authorities have come to include conditions governing such factors as height, side slope, soil stripping and replacement, planting and landscaping and the discharge of drainage. Clearly it is essential that there should be an adequate framework of control if the impact of local tipping is to be minimised. We believe the development control

63

system is adequate to deal with new tips requiring specific planning permission: in particular, modern permissions require restoration and thus avoid the problem of dereliction (see Chapter 10). There are, however, two other matters which concern us. Older planning permissions tend not to incorporate these higher environmental standards. Secondly, the N C B do not require planning permission to tip on land which was being used for that purpose on 1 July 1948 because of the provisions of the General Development Order which we described in Chapter 4.

9.16 The fact that older planning permissions tend not to incorporate the conditions which would now be included is particularly important, since many sites approved as much as 30 years ago are still used for tipping or are occupied by unreclaimed tips. If a local planning authority wish to modify a permission by imposing additional conditions, their actions are subject to liability to compensate the developer for consequential loss or damage. A further problem with some older planning permissions and with G D O tips is that permission is not required for reworking the tips for coal, burnt shale or minerals which involves installing a washery and lagoons at the tip or transporting spoil from it. In some cases, reworking takes place on tips which have already been reclaimed or where some vegetation cover has been established naturally. The Government has recently introduced the Town and Country Planning (Minerals) Bill which includes clauses affecting planning permissions for waste tip reworking. Local planning authorities would be required to review all sites within their areas where mineral workings are or have been taking place. Where appropriate, they would be able to modify the requirements of the planning permission without any liability to pay compensation unless the costs of implementing the changes were substantial in relation to the value of the right to work the mineral. These arrangements would not apply to any proposal to alter the fundamental nature of a planning permission or to a planning permission less than 5 years old, or if a similar order had been made within the preceding 5 years. In addition, the legislation would require planning permission to be sought for the reworking of waste tips including G D O tips. Planning authorities would also be able to impose planning conditions to provide for a 5 year period of after-care to be supervised in England and Wales by A D A S or the Forestry Commission, in conjunction with the planning authority. In Scotland D A F S performs the same functions as A D A S. Certification of satisfactory restoration would be required.

9.17 In view of the adverse environmental effects of both tipping to outmoded standards and of reworking old tips, we welcome the Government's proposals. We recommend that the schemes for the tipping and after-treatment of waste should include proposals for landscape design and that the length of the supervisory period should be kept under review. The experience of opencast restoration suggests that a 5 year supervisory period may not always be adequate, given the timescales of bio-geochemical and hydrological cycles.

9.18 The Town and Country Planning General Development Orders give planning permission for the deposit of colliery spoil on land used for that purpose on 1 July 1948. In some parts of the country the N C B still own many G D O sites and the highest incidence of them appears to be in West Yorkshire where there are no fewer than 66. We have received a great deal of evidence from local authorities that the N C B 's G D O rights in respect of spoil tips should be abolished. In our visit to Yorkshire the local authorities left us in no doubt that in their view the N C B were able to evade normal planning considerations and ignore their views. We have already explained that G D O rights are not unique to the N C B. However, it appears to us that the N C B 's G D O rights in respect of spoil tipping are too permissive and make it difficult for local planning authorities to control the environmental impact of a substantial amount of tipping. No requirements for waste management or after-care can be imposed on the N C B for spoil tipped before 1974, nor are they under any obligation to follow particular practices for waste deposited since then, unless expressly required to do so by the relevant local planning authority at individual sites. Tipping may be, and in some areas not infrequently is, recommenced at a disused tip. Finally planning permission is not required for reworking the tips for coal, burnt shale or other minerals.

9.19 We recognise that there are strong arguments for abolishing entirely the N C B 's G D O rights in respect of spoil tipping. However, in addition to the Town and Country Planning (Minerals) Bill, the Government are also proposing to revise the G D O. They have outlined their proposals in a consultation document (Department of the Environment, 1979a). We have considered the net effect of these proposals and how far they will go to meet our concerns about this aspect of the operation of the G D O. First, in our view planning permission should be required for the reworking of G D O tips and this would be provided for in the Minerals Bill. Secondly, we believe that the N C B should not be able to retain derelict G D O sites indefinitely in order to rework tips or to deposit spoil arising from operations at another colliery. Local authorities should also be able to require the restoration of non-operational sites. The Town and Country Planning (Minerals) Bill would introduce procedures for determining the cessation, prohibition or suspension of operations. The prohibition and suspension order procedures would not apply to tips as such, but the amended version of the G D O would provide the right to deposit waste on a site for which "mining operations are permitted and lawfully used for that purpose". Our understanding is therefore that, where a prohibition order has been made and confirmed in relation to a site, it would be unlawful to deposit waste there without a fresh grant of planning permission and local authorities would be able to impose restoration conditions through the prohibition order. We consider this would be an adequate safeguard and we recommend that the Government ensure that the G D O is revised accordingly. Finally, we believe local authorities should have adequate control powers over operational G D O tips. They already have the power to require the N C B to submit a scheme for their approval for the deposition and treatment of waste deposited after 1 April 1974. We welcome the proposed amendment to the G D O which would enable local authorities to apply tighter restoration conditions in such schemes.

9.20 Some evidence suggested that local authorities in some areas have been reluctant to use their existing powers to require the N C B to prepare schemes for the future tipping of waste. In our view, this reluctance reflects a conflict of interest, particularly in areas where the community and the industry are interdependent. Such conflicts can arise inevitably, particularly in the present economic climate. We have been impressed by the profound change in the climate of local opinion as a result of the deterioration in the employment situation which has come about in the course of our Study. For example, during our visit to Wales we were left in no doubt that employment was paramount in current circumstances. Nevertheless the local authorities must accept that it is not possible to maximise employment without incurring environmental disadvantages. They cannot expect central government or central agencies to bear the cost of remedial environmental measures if they themselves are not willing to share this responsibility by making use of their existing powers.

Lagoons

9.21 Lagooning takes place both at sites operated under the G D O and at those operated under specific planning permission. Environmentally, it is much more intrusive than dry tipping because for the same volume of solids the land take is very much greater, and it may take more than a decade before the settlement process is completed and the lagoon can be capped with dry waste as a prelude to restoration. When seen from high ground lagoons appear very obtrusive. There is the major hazard of drowning particularly where children overcome barriers and trespass or stray onto the site. The liquid tailings deposited in lagoons are generated in the largest quantities at collieries producing coking coal, where they may form 15–20% of the total waste. The N C B consider that at least as a proportion of total waste the creation of tailings will not increase, though a continuing trend towards cleaner power station coals would alter the position. At present, the main alternative to lagooning is the dewatering of tailings by means of pressure filtration, which costs about £5·30

per tonne of dry solids compared with £2·50 for settlement in lagoons. Research is being carried out at the Board's Coal Research Establishment into the fluidised bed combustion of tailings with a view to using the coal content as a source of heat to evaporate the water and produce ash which is safe for tipping. We recommend that wherever possible the use of lagoons for the treatment of tailings should be avoided because of their environmental impact. We accept that the presently available alternatives are more expensive, but this does not mean that they should be ruled out as a treatment method in every case. We have also noted that in the N C B's proposals at Belvoir the coal preparation process has been designed in such a way that liquid effluent would not be discharged. We recommend that new deep mines should be designed in a similar way wherever possible.

Spoil tipping: conclusions

9.22 In summary, it is mainly the older spoil heaps which have given rise in the public mind to the "slag heap" image of coalmining which is still widespread today. Methods have improved in recent years and N C B current best practice is such that the environmental impact of tipping is substantially less than in the past. Nevertheless, a degree of disturbance will occur particularly in greenfield sites, even where these new methods are used. Furthermore, in traditional mining areas such as Yorkshire, the sheer quantity of spoil to be disposed of will still be a problem, given the amount that has already been tipped there. We have therefore thought it necessary to examine next the alternatives to local tipping which might significantly reduce the amount of spoil to be deposited on land.

ALTERNATIVES TO TIPPING

9.23 In principle, there are several methods of reducing the need for the conventional local tipping of mine waste. Dirt may be stored underground or backstowed; left in the coal which is sold; used for local or remote land reclamation; or sold for commercial use. Some of these alternatives could also cut down the requirement for lagoons. These methods are considered in turn below.

Backstowing

9.24 Backstowing involves the permanent placing of mine waste underground. It has two environmental advantages: it reduces the requirement for surface disposal; and it also limits subsidence. Backstowing was widely practised in Britain before the Second World War when the void created behind the coal face by extraction of the seam was used to store waste and support the roof. In this way, the costs of transporting waste to the surface and disposing of it were avoided and the roof over the working areas was preserved. With the introduction of modern mining techniques incorporating powered supports and shields, additional support was no longer needed and the natural rapid collapse of the roof into the void behind the advancing face was allowed to occur unhindered. The rate at which the mechanised face progressed now exceeded the rate at which dirt could be fully stowed, so that mining economics no longer favour backstowing. The technique is currently employed only exceptionally at a single British colliery. The N C B have presented forcefully to the Commission their argument that backstowing is uneconomic, unsafe and dangerous to the health of men working underground. The N C B also claim that its practice would be of little help in reducing subsidence in modern circumstances nor would it be likely to change significantly the scale of demand for surface spoil disposal.

9.25 In essence there are three stowing methods: gravity, hydraulic and pneumatic. Gravity stowing uses conveyors to transport waste material produced on the surface back to the longwall. Generally, it is only feasible where the workings are at a gradient away from the face, which is extremely rare in Britain. In hydraulic stowing, the waste is mixed with large quantities of sand and water and pumped to a dam behind the face where the water is allowed to drain away. This form of stowing has been practised fairly widely in Poland and the N C B have said that it is believed that the Poles absorb 20% of their mining waste in this way. In the view of the N C B, this method cannot be applied here primarily because of different geological conditions. They have told us that "the effects of large quantities of water upon British mine roofs and floors would be disastrous". We accept the N C B's general view that present methods of hydraulic stowing are not viable in the U K. The general principle of pneumatic stowing is that suitably graded waste is discharged at pressure into the goaf from a pipeline traversing the face. It is this system which has been investigated most extensively. The N C B have pointed to severe potential incompatibilities between pneumatic stowing operations and modern highly productive longwall faces because of economic and operating considerations and on health and safety grounds. They consider that, if the rate of mining coal were to be determined by the rate of stowing by fully packing the spoil, then output could be halved in the case of the type of high production face currently being introduced. Output would be cut by two-thirds if the experience in the Saar Coalfield in Germany was repeated in this country. There would also be a considerable increase in airborne dust levels which would be injurious to the health of miners.

9.26 We accept that backstowing is unlikely to change the need for local tipping significantly in the immediate future. Nevertheless we consider that the N C B's view on the feasibility of backstowing is not necessarily conclusive in the long term. As we have briefly explained, the term embraces a number of different technical options and the differing layout and geology of individual collieries might make some options technically feasible. A system of partial backstowing in association with long wall retreat mining has been in operation successfully at Park Mill Colliery in the Barnsley area of the N C B since 1976. In our view the qualitative and quantitative character of the spoil disposal problem is such, particularly in terms of public acceptability of new major projects, that the N C B should accord to backstowing a far higher priority in its research programme than appears to have been the case hitherto. There is a real danger here that inherited technical wisdom can obscure the potential of what must appear a radical solution. We recommend therefore that the N C B, in conjunction with the Research Councils, should greatly strengthen its research into backstowing, even if that eventually requires substantial modification to the longwall process.

Leaving dirt in the coal

9.27 Another obvious possibility is to leave in the coal the dirt which is removed by coal cutting machines. Currently it would only be advantageous in the case of the power station market where there are already arrangements for the commercial utilisation of pulverised fuel ash on a large scale. For example the Longannet power station in Scotland uses run-of-mine coal direct from a mine on the site although higher grade coals have to be added as a sweetener. Nevertheless even in the power station market there are many potential and serious drawbacks. Power station plant has to be specifically designed to handle dirtier coal; increased air pollution might result; there could be adverse effects on boiler plant and the disposal of pulverised fuel ash is not itself without environmental problems although of a lesser scale than the disposal of colliery spoil. Leaving more dirt in coal is not therefore an acceptable environmental alternative at present although the development of fluidised bed combustion may in the future enable coal with a higher ash content to be burned in an environmentally acceptable way.

Local land reclamation

9.28 Local land reclamation is an alternative to local tipping which is generally no more costly and can result in environmental gain. It takes place in relatively small scale projects within a few miles of the originating collieries. These involve the filling of quarries, disused mineshafts and opencast workings and the covering of municipal refuse tips, often in conjunction with restoration conditions attached to minerals planning permissions and opencast authorisations. A comprehensive inventory of available landfill sites is likely to result from the plans that waste disposal authorities are required to prepare under section 2(1) of the Control of Pollution Act 1974.

9.29 However, the scale of waste production in major coal mining areas (including industrial, trade and domestic wastes) is such that it seems to us to be unlikely that there will be enough holes to accept all the waste. In any case the holes available near collieries tend to be relatively small. The N C B have explained that in their North Yorkshire Area they currently dispose of 36% of the total spoil produced into new land formations created by opencast coal workings and other quarries. The N C B plan to increase this to 66% in two or three years time and to maintain this level for ten years and possibly much longer.

9.30 We have two comments about this scheme. Firstly, it has arisen from taking advantage of special local circumstances as the N C B themselves recognise. It is a very welcome example of what can be achieved by taking advantage of such circumstances. It does not mean necessarily that local land reclamation offers a solution to spoil disposal in other regions, nor even within South and West Yorkshire. Secondly, the N C B have emphasised that the effectiveness of much of this policy depends upon their receiving authorisations to work opencast sites and receiving them in time. We discuss opencast mining in Chapter 11. We fully support all attempts to co-ordinate deep mining with other activities, including opencast mining, to minimise its environmental impact. For example, at Maltby, the N C B have co-operated in a scheme co-ordinated by South Yorkshire County Council to fill a void created by the quarrying of limestone with spoil from the nearby colliery. We recommend that the possibility of disposing of colliery spoil should be taken into account when considering both individual opencast proposals or the phasing of a regional programme. Nonetheless we do not consider that the need to dispose of colliery spoil from deep mining can be used as a general justification for opencast mining. In conclusion, local land reclamation can be important in particular localities and we recommend that every opportunity should be taken to make use of it. However in total it seems likely to make only a very modest contribution to spoil disposal in the future.

Remote land reclamation

9.31 Remote land reclamation is much more costly than local land reclamation or disposal and it would have to be on a much larger scale than local projects to be viable. Preliminary investigations have been made by the N C B and local authorities of two possible schemes, one transporting 125 million tonnes of spoil from the North East Leicestershire Prospect to worked out brickpits in Bedfordshire over a 50 year period; the other taking 80 million tonnes from the Yorkshire coalfield to the mudflats at Pyewipe on the Humber estuary over 20 years. The main drawback to the N C B of such schemes is the very considerable increase in cost over local tipping. The N C B have estimated that the cost of tipping in Bedfordshire would be about £1·38

per tonne of coal more than tipping locally. It should also be remembered that production costs at Belvoir would be much lower than the national average and the benefit of the reclaimed land would also have to be taken into account. The N C B's policy towards remote disposal might be compared with that of the C E G B who transport pulverised fuel ash from several Midland sites to Peterborough and accept the extra costs. These are estimated to be a revenue cost of £5 per tonne compared to a national average of £1–2 (costs for 1979/80). In view of the complex factors involved when comparing the circumstances of the N C B and C E G B, it would be invidious to draw firm conclusions based on the C E G B's Peterborough scheme. The reasons for this apparent difference of approach might be partly historical in that the N C B have long accepted spoil heaps as the "best practicable means" of disposing of spoil. The N C B also confront a much wider range of production costs and have to handle relatively more spoil in relation to the amount of coal produced than the amount of pulverised fuel ash handled by the C E G B in relation to the amount of coal consumed.

9.32 There is another important consideration to be taken into account in assessing the contribution which remote disposal might make to reducing the amount of local land tipping. This is the number of sites in addition to the Bedfordshire brickfields and Pyewipe which could be used for the long distance disposal of colliery spoil. On the basis of evidence provided by the Department of the Environment, we have concluded that there is a very limited number of potential sites available. Furthermore, many of the sites are located on the coast and there would be objections to using many of them on the grounds of nature conservation. Land reclamation programmes undertaken for spoil disposal or to create new land could not be justified unless the final use of the site was sufficiently in the national interest to take priority over the nature conservation interests likely to be affected. The capacity of sites is also limited. It has been estimated that the Bedfordshire brick pits could accommodate 132 million tonnes of spoil and a further 168 million tonnes if clay working proceeds. It has been estimated that the Belvoir mines alone would produce 125 million tonnes over the next 50 years. Moreover, there will be many competing claims from other sources of waste, such as the domestic waste of London which in 1977/78 amounted to about 2·7 million tonnes.

9.33 There is therefore limited scope for using remote land reclamation to dispose of colliery spoil. It should not, however, be ruled out entirely. Although it might not reduce substantially the amount of spoil to be disposed of locally, there may well be cases where remote land reclamation could be adopted where new mines are to be developed in particularly sensitive areas or in localities which have already been seriously affected by the tipping of spoil. In cases such as these, the possibility of undertaking a scheme of remote land reclamation would in our opinion deserve to be considered very seriously. However, it is clear to us that the number of sites which could be used in this way is limited. As such they represent a national asset which could be used to dispose of other waste material. We believe, therefore, that decisions concerning the disposal of colliery spoil by remote land reclamation will need to be taken in the context of defined national priorities for the use of the few sites that are available.

Marine disposal

9.34 Marine disposal as carried out at present is really a form of local tipping for collieries located close to the sea. In 1978/79, 3·5 million tonnes of spoil were tipped onto beaches and 1·5 million tonnes into inshore waters. Beach tipping, which should be distinguished from coastal land reclamation, takes place mainly in the N C B's North East Area from 3 Durham collieries (tipping from a fourth has recently stopped). Spoil has been washed down the coast causing damage to amenity and the habitats of many marine organisms. This tipping has been a cause of concern to local authorities and central government for some time. In May 1974 the then Minister for Planning and Local Government announced that the Government, the N C B and the local authorities had concluded that the tipping must be brought to an end and the beaches reclaimed. A working party was established to consider how this could be achieved and recommended the use of pipelines to deposit waste 500 yards out to sea. We support attempts to bring beach tipping to an end and restore the areas concerned. We recommend that spoil should not be tipped on any further beaches and that the ecological consequences of pipeline deposition should be fully examined as recommended by the working party before it is widely adopted.

9.35 Only limited dumping by barge to sea is carried out involving 4 collieries on the North East coast. It may be considered an alternative form of remote disposal to large scale reclamation projects, but except at coastal collieries, the transport costs might be expected to be higher because of double handling. Unlike land reclamation, there would be no environmental gain at the disposal site and careful study would be required of the possible effects on marine ecosystems, including fish spawning and nursery grounds, and on the structure of the seabed. We do not believe that marine tipping is an acceptable alternative to local tipping and we strongly recommend that in view of its adverse environmental impact the N C B ought not to use this method of spoil disposal unless and until suitable techniques for safeguarding habitats and amenity are found.

Commercial use

9.36 As we have already described, commercial use of spoil makes virtually no impact on spoil disposal from

current production although considerably more is exploited commercially from existing heaps. The following Table sets out the way in which spoil was used commercially in 1976.

Table 9.3 *Commercial uses of spoil in 1976*

Commercial use	Source of spoil ('000 tonnes)	
	From current coal production	From existing heaps
Fill	225	5500
Brickmaking	150	400
Cement Works	65	65
Light aggregate for concrete	—	1000
Total use	500	7005

Source: *European Communities Commission 1979*

9.37 In 1978 the Minestone Executive established a Technical Services Department to carry out and encourage research and development. The Executive have drawn our attention to advances with cement stabilised minestone material, reinforced minestone material and the manufacture of light and strong aggregates. They are to be commended on the efforts which they have made to develop the commercial use of spoil, but we have concluded that there is limited scope for reducing the scale of the disposal problem by this method. Demand is always likely to depend on the state of the construction industry and there seems little prospect of an immediate improvement. The most significant commercial use of spoil, 77% of the total, was for bulk fill material in civil engineering works, mainly road construction. However, major disadvantages of not being as close to major engineering works as alternative material, and of being highly price sensitive to transport costs, constrain attempts to promote its use. The Department of Transport and the Welsh Office operate a dual tendering system intended to promote the use of waste material in road contracts and the Government has also certified that unburnt spoil is safe for bulk use. The N C B consider that greater use could be made of colliery spoil and wish to see the introduction of an improved dual tendering system covering all appropriate building and civil engineering projects. We recommend that the Government examines the N C B's suggestions.

9.38 The second largest market, 1 million tonnes (13%) is light aggregate for concrete. Burnt spoil from tips rather than currently produced spoil is used, though in the future it may be possible to convert unburnt spoil into a suitable form through fluidised bed combustion. Two plants are at present in satisfactory operation; pollution problems have lead to the closure of another. Research has been carried out into the possible use of unburnt spoil in non-structural concrete, but initial results were unpromising.

9.39 The other existing markets for spoil are brickmaking (8%) and cement manufacture (2%). The variability in the quality of colliery waste, the availability of good quality clays in more convenient locations, and in the case of brickmaking competition by lightweight concrete blocks, limit the use to only 0·75 million tonnes a year in these sectors. Various other uses have been suggested for spoil; these include use in road pavements and blended cement; the recovery of aluminium; and the combustion of tailings for power generation, though none appear to be practicable on a large scale in the foreseeable future.

9.40 The use of waste materials to substitute for aggregate was considered in the Report of the Advisory Committee on Aggregates (1976). In their view the national aim should be to achieve a situation in which all wastes suitable for use as aggregates were so used. While emphasising the benefits of the increased utilisation of waste materials in terms of reducing the land required for extracting natural aggregates and helping conserve these resources, the Committee concluded that in view of the absence of major stocks in the South East where demand for aggregates was strong, and of the high costs of transport, there was no prospect of them becoming the principal source of supply.

GENERAL CONCLUSIONS ON ALTERNATIVES TO LOCAL TIPPING

9.41 The methods discussed above, either singly or in total are unlikely to make a significant reduction in the quantity of spoil that will have to be disposed of by local tipping. This does not mean that apart from marine and beach tipping, they should be ignored. On the contrary, they can be very important in solving the problem of spoil disposal at particular collieries and within particular regions. We recommend that they should be adopted wherever feasible and appropriate. However our conclusion points to continued, large scale local tipping and hence to the priority of seeking to minimise the adverse environmental impact of the local tipping that will have to take place if coal production is to expand and we have shown that considerable improvements have already been made. We recommend that the N C B continue to investigate what other improvements can be made.

9.42 However, the problem of spoil disposal arises not only because of the amount of spoil to be disposed of over the next twenty years but also from the lack of any coherent national or regional disposal policy and indeed from the lack of any machinery for developing a policy. Apart from the activities of the Minestone Executive the responsibility for spoil disposal rests with individual colliery managers. Only exceptionally and in the case of the relatively few large projects is there any co-ordination within or between the N C B, local authorities and central government. We believe it to be

essential that there should be better coordination. We discuss this in the concluding section of this Chapter.

SPOIL DISPOSAL: STRATEGIC PLANNING

9.43 Given the population density and the rich countryside of the U K, we consider it extremely important that there should be a deliberate effort by all parties to develop an orderly plan for the disposal of wastes. Our visits to coal mining areas have demonstrated both the need for and the advantage of effective and willing consultation between the N C B and local authorities. Spoil is likely to be one of the major problems associated with coal mining over the next 20 years. It is clear to us that local authorities and the N C B should be discussing the future scale of the problem within regions and the best way to solve it. Only by doing so will they be able to make the best use of the alternatives to local tipping and to minimise the impacts of local tipping. We have already expressed concern about the character and dimension of the problem of spoil disposal in Yorkshire. The N C B have told us in their evidence that they have plans for spoil disposal in the Yorkshire region using opencast sites and other voids to receive spoil, and for the centralisation of tipping and the further development of progressive restoration techniques. On the other hand, the local authorities have said that the N C B's plans are only partially known to them. Although our impression is that the N C B carry out close day to day discussions with local authorities they have been reluctant to publish detailed overall statistics for spoil production. This does not appear to us to be a sound basis for dealing with a major environmental problem. However, we are encouraged that during the course of our Study the N C B have said that they recognise that information about future spoil production and its effect on land use is of public interest. They have therefore published estimates for the Yorkshire coalfield for the period to 1990. We consider these provide a starting point for the consultations we are envisaging.

9.44 The problem is not confined to coal production. It has to be set in the context of waste disposal from the use of coal, and waste from other industrial and domestic sources. There is as yet no comprehensive information on the total amount of waste being produced from all sources within local authority areas. However, waste disposal authorities are required by the Control of Pollution Act 1974 to prepare waste disposal plans; information on the kinds and quantities of waste is necessary for the production of these plans and is now beginning to fill this gap. However, it is possible to assess the amount of waste that is likely to be produced by coal related activities on the same basis as our estimate of colliery spoil. Pulverised fuel ash from electricity generation amounts to about 14 million tonnes a year, which is about a quarter of the quantity of colliery spoil produced each year, although the C E G B sell about 40% of this total. To estimate the production of residues over the coming years we have taken the middle figures in the ranges for power station coal burn in the Department of Energy's 1979 Energy Projections and assumed that the coal waste ratio remains unchanged. We estimate on this basis that 380 million tonnes of power station waste will be produced between 1976 and 2000. Industrial and commercial coal burn produces about 2·6 million tonnes of ash per year and the N C B have estimated that, between 1976 and 2000, about 100 million tonnes of waste will be produced by this sector. If coal burn in industry increases and F B C is introduced, these quantities would correspondingly increase. S N G production is likely to produce substantial quantities of ash but it is difficult to estimate how much by 2000. A "typical" S N G plant could produce over 660,000 tonnes per year of solid waste.

9.45 Outside the energy industries it is likely that the need to deposit other kinds of waste, in particular domestic refuse, will grow. No detailed figures for waste produced by other industrial sources are available as yet but we have been provided with some tentative estimates. For example, industrial waste in the three Yorkshire counties amounted to 4·6 million tonnes in 1978 compared to a production of colliery spoil in approximately the same area of 17·3 million tonnes. In 1977/78 the amount of domestic and commercial waste handled by waste collection authorities nationally was 16·3 million tonnes. These estimates make clear that in the future the waste arising from coal related activities, and coal mining, will be the major contributor in mining areas.

9.46 In general, there is no effective central or regional coordination within or between the N C B, local authorities, and central government. There have been some attempts to tackle the problem. In Yorkshire a Regional Solid Waste Disposal Working Party was set up by S C O C C and published a waste disposal report in May 1978 after which its activities ceased. First steps are already being taken to provide a framework of planning at local level for the disposal of waste other than colliery spoil by the implementation of Section 2 of the Control of Pollution Act in July 1978. This requires waste disposal authorities to survey waste arisings and disposal facilities in their area, and to draw up and periodically revise a waste disposal plan. Although Central Government has no powers to coordinate these plans to form a national strategy, it can adjudicate in disputes between waste disposal authorities and consider the implications of the plans for future policy. We have concluded that in mining areas, colliery spoil is by far the most significant item of waste to be disposed of. However it is at present specifically excluded from Section 2 of the Control of Pollution Act, in spite of the need for the planning of spoil disposal to be tackled with equal rigour.

9.47 It is important that such planning should not lose touch with local problems and impede the search for solutions by the bodies immediately concerned. We therefore consider that spoil disposal should be assessed at the regional level. The Regional Aggregates Working Parties provide a model which could be adapted to the needs of the coal industry. There are now eight working parties in England and two in Wales. They include members of the aggregates industry and local authority planners with technical advice from M A F F, the Department of the Environment and the Welsh Office. There is a National Co-Ordinating Group run by the Department of the Environment in England. The working parties have analysed the current and forecast supply and demand position in each region and are currently developing options for meeting future demand. The information will provide local authorities with the basis for developing aggregates policies in structure and local plans. It will also provide the means whereby central government can consider structure plans within a national perspective.

9.48 We consider that the Department of the Environment should promote and encourage similar ad hoc arrangements between local authorities, the N C B and other producers of waste in those regions where the problems are most serious. We would give first priority to the establishment of a working group to be set up in the Yorkshire and East Midlands Coalfields. Its first task should be to assess the amount and type of spoil to be disposed of in their region and the disposal sites available within it on the assumption that present policies continue. On the basis of that information the Working Parties might then go on to discuss options for the disposal of waste within the region. The role of the N C B would be crucial. However, the local authorities would also have to approach the task in a constructive and responsible way. Both parties stand to gain from mutually acceptable solutions. We have already referred to the N C B's publication of estimates of spoil production in the Yorkshire Coalfield and its effect on land use for the period to 1990. This is a starting point from which the proposed working party could begin its work.

9.49 We have thus considered planning and consultation at the regional and local level. However, it is clear that waste disposal will have national and inter-regional dimensions which it is the responsibility of central government to address in order to ensure that local decisions are compatible with the national interest, and that different regions are treated fairly. We are not advocating a detailed national plan. We recommend that the Government should establish guidelines on the use of different types of site for remote reclamation; on the availability or otherwise of Governmental financial assistance; on the priorities for different types of waste in remote disposal schemes; and on the priorities between regions in the use of remote disposal sites. Such an enhanced role for central government would facilitate decision making in major planning inquiries.

Chapter 10   Dereliction and pit closures

INTRODUCTION

10.1 In this Chapter we discuss the problem of dereliction which can be a major consequence of pit closures. It is important in its own right, but the adequacy or otherwise of the measures taken to deal with it have important implications for dealing with the far more serious human and employment consequences. This is the more important, as we have commented elsewhere, in the light of changes in the climate of opinion as economic recession and unemployment have intensified over the last two years. Employment considerations are now uppermost in the minds of local government leaders and the people they represent in such badly hit regions as South Wales. Thus we wish to stress that our approach to this problem has not been concerned only with such considerations as restoration of the landscape, but also with the critical linkage between the clearance of dereliction and the generation of alternative employment opportunities.

10.2 We endorse the view expressed in the Fourth Report of the Royal Commission on Environmental Pollution (R C E P 1974) that land dereliction may be regarded as a form of environmental degradation just as unacceptable as other forms of pollution. Plate 10.1 illustrates the visual impact of derelict colliery buildings. Derelict land has no statutory definition but is normally taken to mean land so damaged by industrial or other development that it is incapable of beneficial use without treatment. Our visit to the Barnsley area made a particularly strong impression on us. More than a third of the nation's collieries are concentrated in an area within 15 miles of Grimethorpe and the land which has been left derelict after the end of mining activities, combined with the land which is still operational, left a vivid impression of desolation. Dereliction makes an area less attractive to new industrial development which could offer alternative employment opportunities. In urban and industrialised areas it ties up sites unproductively which could be used as additional recreational open space. These problems are intensified in the mining areas of South Wales, where the topography imposes severe constraints on the amount of land available for development. Land has to be reclaimed to allow much needed factory space to be built; over half of factory space built by the W D A in the last five years has been on reclaimed land.

THE SCALE AND CONCENTRATION OF DERELICTION

10.3 Mining dereliction has undoubtedly been a major problem in the past and continues to be one. There are no precise figures available about the extent of the problem nationally, but the 1974 Derelict Land Survey of England showed 7,404 hectares were occupied by derelict spoil tips justifying restoration including those caused by activities other than coal mining. The comparable figure for Scotland was 2,400 hectares and for Wales 8,227 hectares. However, coal mining spoil tips probably constitute the largest single source. We showed earlier in Table 9.2 the results of the 1974 Survey in England and the concentration of derelict spoil heaps. The Yorkshire/Derbyshire/Nottinghamshire coalfield contained 28% of the total area affected by derelict spoil heaps, Northumberland and Durham 20% and Lancashire 26%. Virtually all the derelict spoil heaps in Scotland are in the Central Belt. In Wales, 72% of the area affected by derelict colliery tips in 1972 was located in the two counties of Gwent and Mid-Glamorgan.

10.4 Although the nature of the problem varies from region to region, local authorities have expressed a general concern about dereliction. In traditional mining areas such as Durham and Wales, where the present reduction in mining activity is likely to continue, the most important problem is to clear up the dereliction of the past and ensure that this is not added to as the industry continues to reduce its activity. In Yorkshire, a traditional mining area where coal mining is likely to increase, there is a twofold problem, namely, to ensure that past dereliction is cleared so as not to prejudice the future development of the industry; and to ensure that future mining activity does not repeat the ravages of the past.

PREVENTING FUTURE DERELICTION

10.5 The problem of ensuring that current and future mining activity does not cause dereliction revolves around the question of the conditions attached to planning permissions. Modern planning permissions normally include conditions which ensure that land is restored in an adequate manner. For example, permissions for tipping now generally include restoration conditions affecting soil stripping and replacement, planting, and landscaping. These powers to impose restoration conditions would be strengthened by the Town and Country Planning (Minerals) Bill. Planning authorities would be able to impose conditions to provide for a five year period of after-care to be supervised by M A F F, the Welsh Office, D A F S or the Forestry Commission, in conjunction with the local planning authority, followed by the certification of satisfactory restoration. In general, we consider that these powers would ensure that in the future land should not become derelict as a result of mining activity.

10.6 There are, of course, many collieries where older, specific planning permissions will not include restoration conditions. The Minerals Bill would, as we have explained, enable local planning authorities to review all mineral sites within their areas and, where appropriate, to modify the requirements of the planning permission without any liability to pay compensation unless the costs of implementing the changes were substantial in relation to the value of the right to work the mineral. Local authorities would also be able to make orders which would prohibit the resumption of operations where mineral working has ceased for 2 years and which would require the restoration of the land. These proposals are welcome and should help to ensure that current activity is properly controlled. However there is still a substantial legacy from the past that needs to be cleared and this will be added to as mining activity comes to an end at older sites. It is essential that there should be adequate machinery for clearing this dereliction.

METHODS

10.7 The N C B have for a long time followed a policy of putting the sites of closed collieries and other operational activities to some other beneficial use. This is done in consultation with local planning authorities since it is necessary for the Board to apply for planning permission so that the land and usable buildings may be sold or leased for some form of industrial or commercial use. Between 1969 and 1979, about 3,030 hectares of land were disposed of by sale or lease for alternative industrial and commercial occupation. This includes a substantial acreage of land and buildings leased to private industrial tenants through a wholly owned subsidiary company of the Board, Coal Industry Estates Ltd, which was formed in 1973 to manage and put to best use the N C B's land and property not required for operational use. We recommend that the N C B should continue to develop this approach. Clearly there is a limit to the amount of land that is suitable for alternative development and the number of industries that can be attracted. It is therefore

important that there should be an adequate mechanism for restoring land to other uses. We now examine this question.

Dereliction Land Grant

10.8 It is this machinery which the local authorities and the N C B have to date used to clear derelict land. Under section 8 of the Local Employment Act 1972, derelict land grant is paid to local authorities in England by the Department of the Environment at 100% of the cost of reclamation for approved projects in assisted areas and derelict land clearance areas. Broadly speaking, the designation of these area includes all major coalfields. Smaller significant pockets of mining dereliction may be excluded or may straddle the boundaries. The extent of assisted areas is being reduced with the adoption of a more selective regional industrial policy, but this is being offset so far as derelict land is concerned by the creation of more clearance areas. Elsewhere in England, grants are made under section 9 of the Local Government Act 1966 at a rate of 50%, with the local authority meeting the remainder of the cost. In Scotland, the Scottish Development Agency provides financial assistance on the same basis as in the English assisted areas. In Wales, the W D A is empowered to make 100% grants available to all local authorities in Wales, regardless of assisted area status.

10.9 In cases where the local planning authority and the Board agree that the site of a closed colliery or spoil tip is unsuitable for any kind of alternative development the N C B will normally convey the land to the county or district council, frequently at a nominal value. The local authority then prepares a derelict land reclamation scheme for inclusion in its rolling programme to be agreed with the Department of the Environment. On a national scale, considerable progress is being made in clearing dereliction. For example, almost a third of the area affected by derelict spoil heaps in England in 1974 was reclaimed in the next four years; this was partially offset by the handover of more G D O tips to the local authorities.

10.10 The success of the derelict land reclamation scheme has varied between regions. Durham has been notably successful. Between 1964 and 1979 the County and District Councils in Durham reclaimed 2,441 hectares of land removing 22 million tonnes of pit waste. Of course, more land has become derelict meanwhile so that 1,913 hectares of spoil heaps remain to be cleared (Durham County Council 1979). Nonetheless, the County Council considers that the worst eyesores have been removed. Plate 10.2 illustrates one of the derelict land clearance schemes undertaken by the County Council with grant assistance from the Department of the Environment. Part of the reason for this success has been the close cooperation between the N C B and the County. As the

Plate 10.3a see below

Plate 10.3b Ty Trist Colliery, South Wales before and after reclamation. In South Wales the shortage of flat land for development in the valleys puts a premium on reclaiming land for both environmental and economic reasons.
(*Photograph by courtesy of the Heads of the Valleys Standing Conference*)

Plate 11.2 (*top*) Coal gathering and transportation at Shipley Lake Opencast Site, Derbyshire.
(*Photograph by courtesy of the National Coal Board*)

Plate 15.1 (*bottom*) Westfield Gasification Plant, Scotland.
The British Gas Corporation's slagging gasifier at Westfield, Fife in Scotland is a full-scale commercial Lurgi gasifier.
(*Photograph by courtesy of the British Gas Corporation*)

largest single owner of derelict land in Durham, the N C B came to an agreement with the County Council that they would endeavour to make available each year enough derelict land to enable the County to meet the target it had set of reclaiming 283 hectares each year. In return the County Council agreed to pay the market value for materials in a heap, where markets for them could be shown to exist, and to make every effort to agree terms before entry onto the land was taken. Considerable progress has been achieved in Wales: since 1967 about 4,777 hectares of derelict land have been cleared although not all of this relates to coal mining. We were particularly impressed by the work of the W D A which was established in 1975 with the duty to prepare and submit schemes for the improvement and redevelopment of the environment. In the past 5 years since its inception, the Agency has dealt with 1,619 hectares of land of which 1,214 hectares had been made derelict by coal mining. Plate 10.3 illustrates one of the Agency's clearance schemes in South Wales.

10.11   Contrasting problems are presented in Yorkshire. Since 1946 the N C B has handed over 854 hectares of land to the local authorities for reclamation, and between 1974 and 1978, 585 hectares were reclaimed. However, the local authorities have pointed out that at this rate all the present statutory derelict land will not be reclaimed until about the year 2000. This makes no allowance for any dereliction which might be caused by current mining activity. S C O C C wish to see an increase in the rate of reclamation which they believe will depend on the availability of manpower and finance as well as a greater willingness by the N C B to release tips for reclamation.

10.12   Despite the success which has already been gained, the momentum which has been built up should be maintained. There still remains a large amount of derelict land to be reclaimed. In 1974 there were over 1,400 hectares of derelict spoil heaps justifying restoration in Durham and over 2,000 hectares in Yorkshire, Derbyshire and Nottinghamshire. Even if some of the worst cases have already been dealt with, there is still a major problem to be solved. Moreover, as we have explained, there have been local variations in how successfully the problem of reclaiming derelict land has been tackled. We have therefore considered whether the present legislative machinery is adequate; whether the N C B and local authorities are making the best use of it; and whether central Government is providing sufficient resources.

Legislative machinery

10.13   The legislative machinery for reclaiming derelict land appears to us to be generally satisfactory from the national standpoint. Changes in the statutory provisions affecting derelict land procedures were introduced recently in the Local Government Planning and Land Act 1980. For the first time derelict land grant is available to the non-local authority sector including the N C B though at a uniform national rate of 50% of any net loss compared with the 100% grant available to local authorities in most coalfield areas. Grants are also available towards the costs of surveys and investigation of derelict sites and in the case of local authorities will not be tied to the carrying out of an approved scheme as at present. Again, for the first time, grant is now available towards the cost of providing development infrastructure such as access roads, water and sewers though this will be restricted to local authorities. These provisions should provide greater flexibility in the working of the derelict land scheme and are to be welcomed.

10.14   There is one further point of concern. In the future, the capital cost of derelict land schemes which receive 100% grant aid will count against the capital expenditure allocations of local authorities. There may be a danger that in times of financial stringency reclamation schemes may be squeezed out of local authority expenditure programmes by other competing demands on local finances. We recommend the Government should exempt this particular item of local authority expenditure from counting against capital expenditure allocations.

The N C B and Local Authorities

10.15   If this machinery is to work effectively the N C B and local authorities must cooperate to make the best use of it. Overall, however, substantial progress has been made in the last decade, although the number of inactive tips still remains a large proportion of the total. Recent trends are shown in the following Table.

Table 10.1   Inactive spoil tips in N C B ownership 1968 and 1979

	Number of tips in N C B ownership	Number of inactive tips	% of all tips
1968	2024	1445	71
1979	1102	757	69
Change	−922	−688	−2

Source: N C B

In the decade up to 1979, 5,500 hectares were sold by the N C B for treatment under the derelict land programme. The scale of these disposals is substantial by comparison with the N C B's total land holdings of 23,000 hectares connected with collieries in 1979, even when allowance has been made for land acquisition of 3,200 hectares over the period. The N C B have been criticised for alleged unwillingness to transfer tips for which they might possibly find a future use, or which might form part of a wide package of land acquisition and disposal in connection with future development. This is sometimes explained by the N C B's wish to keep open the option of reworking disused tips for coal

73

or other materials. The Royal Town Planning Institute (R T P I) in particular drew attention to the problems faced by local authorities in assembling land for reclamation schemes, and recommended that the N C B should be required to offer land surplus to its local requirements to local authorities within a specified time and at existing use values. The N C B consider such criticism to be unjustified and assert that they do not retain such land unnecessarily. We have concluded that nationally the N C B have been releasing land at a reasonable rate.

10.16 There appears, however, to have been a wide variation in how effectively the N C B and local authorities have been able to co-operate in drawing up restoration programmes. We have already referred to the success of reclamation programmes in Durham and South Wales, and to the close co-operation between the N C B and local authorities which made this possible, and to the success of the W D A in Wales. The position in Yorkshire is less satisfactory. S C O C C explained to us the need for the N C B to release more tips for reclamation if the reclamation programmes were to be accelerated. They also argued the need for more manpower and finance. In the specific context of criticisms levelled at them by the Yorkshire authorities the N C B have provided the following Table.

Table 10.2 Land released (hectares) to local authorities by N C B since 1946 and reclamation schemes begun

	Released to local authority	Reclamation schemes begun at 1 Jan 1980
South Yorkshire	589	249
West Yorkshire	275	211

Source: N C B

10.17 These figures suggest that N C B may be releasing land somewhat more quickly than the two local authorities can reclaim it. The N C B liaise directly with officers of South Yorkshire County Council when that authority draws up its reclamation programme but otherwise it appears that there are no other consultations. Indeed there is not even an agreed statistical base as the local authorities have asked for more information about the figures which the N C B have provided to the Commission.

10.18 The country's coal mining is heavily concentrated in the Yorkshire region and this seems likely to increase. We consider it essential, therefore, that more should be done to clear the dereliction of the past. It seems to us that in Scotland and Wales, through the Development Agencies, much more has been achieved both in the clearance of derelict land and in the provision of new industrial sites than has been the case in the Yorkshire coalfield. We recognise that the regions concerned do not have identical problems. For example, the industrial structure of Wales, with its heavy dependence on coal and steel, together with the

exceptionally high rate of unemployment, has resulted in the overriding priority of the W D A to reclaim derelict land in order to attract new industry by providing factory space. Moreover the topography of Wales creates particular problems of land availability for this purpose. Nevertheless, in Yorkshire the local authorities also see a need for further diversification of employment opportunities by a similar process of providing new industrial sites. We have also stressed the need to clear past dereliction if the continued concentration of coal mining in the region is to be acceptable and to make the area attractive to other industries. We realise that any recommendation specific to the Yorkshire Coalfield will have broader implications for regional policy in England. There may well be other areas such as the North East with claims for special treatment. However, there is a serious need for the N C B and the Yorkshire local authorities to cooperate more effectively in drawing up a rolling programme of restoration on a regional basis if the region is to cope with the planned increase in mining activity over the next two decades. To meet this need consultation may not suffice. There is a strong case for an independent regional Development Agency with its own budget to deal with problems such as derelict land clearance. We recommend therefore that the Government examine the feasibility of establishing such a body in the first instance in Yorkshire.

THE LEVEL OF GOVERNMENT FINANCE

10.19 The other basic question concerns the adequacy of central Government finance for derelict land clearance. During the course of our Study the Government announced a major rationalisation of regional policy. We note that this ensured that, broadly speaking, the coverage of derelict land grant in mining areas will remain unchanged. We are concerned that the new local authority block grant system, which will enable local authorities to transfer funds received under one head of expenditure to other heads, might result in a reduction in local authority clearance programmes because of competing demands on funds. However, the present rate of clearance will not enable the present area of derelict land attributable to coal mining to be cleared nationally until the year 2000. We considered recommending that higher priority should be given to clearing derelict land caused by coal mining, but we recognise that nationally there are other claims on the funds available for derelict land clearance, particularly in inner city areas and also for reclaiming land made derelict by chemical contamination. Taking account of these competing claims, we nevertheless recommend that Government assistance for this purpose for coal mining areas should be maintained in real terms. In addition, there are particularly strong arguments for increasing the speed of reclamation in Yorkshire. We therefore recommend that Government should consider providing additional funds for the clearance of coal dereliction in that area in addition to considering the feasibility of establishing a Development Agency as we recommended above.

CONCLUSIONS

10.20 We wish to re-emphasise the important linkage between pit closures, and clearance of derelict land, and the generation of alternative employment opportunities. We have highlighted the seriousness of these highly sensitive issues in the particular circumstances of a region such as South Wales. We therefore attach particular importance to the efforts of the W D A, in co-operation with the local authorities, to grapple with these interrelated problems in the region. However, the Agency has emphasised to us that following a pit closure it can take four years to create new jobs and not all those jobs will be suitable for unemployed miners. The continued success of the Agency's activity hinges critically on forward planning based on early warning and the closest co-operation between all interested parties.

10.21 We are fully conscious of the anxieties generated by this approach among the coalmining unions, since they see a risk of such consultations becoming an instrument for accelerating pit closures. In strongly advocating such wider consultations we are in no sense seeking to modify the industry's Colliery Review Procedure which provides for the essential detailed consultation between management and unions who have rights of appeal to the Board at national level. Nevertheless, we see this sensitive complex of problems as one of the clearest demonstrations of the need for forward consultation and continuous discussion between the interested parties which is one of the central themes to emerge from this Study.

Chapter 11   Opencast mining

INTRODUCTION

11.1 Public concern about opencast mining is much greater than the concern about a similar output of deep mined coal. The output of opencast coal was 15·3 million tonnes in 1980/81, only about 11% of total coal production, but evidence submitted showed great disquiet about both the short-term impact on an area and its residents as well as the long-term effect on the landscape. A question particularly at issue is whether the economic and technical advantages of opencast mining outweigh the undoubted environmental damage and whether this damage is consistently reduced to a minimum.

11.2 The Chapter examines:
(1)   the history and development of opencast coal
(2)   contribution of opencast coal to the coal industry
(3)   location of present and future activities
(4)   exploration
(5)   opencast authorisation procedures
(6)   mining operations on opencast sites
(7)   restoration for agriculture and forestry
(8)   environmental effects of opencast operations
(9)   land restoration
(10) general conclusions.

HISTORY AND DEVELOPMENT OF OPENCAST MINING

11.3 Opencast mining on an organised basis began in 1942 as part of the effort made during the war to produce as much indigenous coal as possible. Responsibility first lay with the Ministry of Works but passed to the Ministry of Fuel and Power in 1945. In 1952 the relevant organisation of that Ministry was incorporated in the National Coal Board and became the Opencast Executive which is still responsible for opencast mining today. The Executive is a management unit within the Board's organisation and its Chairman is a member of the main Board.

11.4 Originally opencast operations were restricted by the size of the excavating plant available. These limited the depth of excavation to about 15 metres and the ratio of "overburden" to coal to about 5:1. Subsequently developments in the size of plant and extraction techniques have enabled the depth of excavations to increase, the deepest reaching to about

200 metres (Westfield, Scotland) and overburden ratios to rise to about 25 or even 30:1

11.5 Technical developments resulted therefore in the extension of the range of economic mining for opencast coal and enabled a continuous expansion of the industry. However, in 1959 and 1968/69 Government decisions were taken to cut back on opencast production as overall demand for coal was falling. At the time of the 1973/74 oil crisis the position was reversed. A rapid and continuing increase in the demand for coal was foreseen. 'Coal for the Future' (Department of Energy 1977a) set a target figure for opencast production of 15 million tonnes per year by 1985. This target was endorsed by the previous Government and has been confirmed by the present Government. The target figure of 15 million tonnes per year remains the Executive's proposed maximum and was reached for the first time in 1980/81. The following Table sets out annual production for ten years from 1970/71 to 1980/81.

Table 11.1   Opencast production 1969–81

Year	Annual tonnage (million tonnes) (excludes coal extracted by private operators under licence to the Board)
1970/71	8·1
1971/72	10·1
1972/73	10·1
1973/74	9·0
1974/75	9·2
1975/76	10·4
1976/77	11·4
1977/78	13·6
1978/79	13·5
1979/80	13·0
1980/81	15·3

Source: N C B

11.6 Beyond 1985 'Coal for the Future' suggested that the coal industry would be capable of raising total coal production from 135 million tonnes to 170 million tonnes per year with an increase in opencast production from 15 million tonnes to 20 million tonnes per year. This figure has never been given any formal endorsement by Governments. In his introductory statement to the Commission in October 1978 the Chairman of the N C B, Sir Derek Ezra, told the Commission "so far as opencast production is concerned there is a policy, agreed with Government, to build up opencast production to 15 million tonnes a year or more. The Board consider that this policy can be sustained in the longer term. However, there are geological and logistic constraints to increasing opencast coal to a greater extent than this''.

THE ROLE OF OPENCAST COAL

11.7 Opencast coal mining is more economical in its use of resources than deep mining. It produces coal of low ash, generally good quality which is needed for special markets (at least 10%), which deep mined coal supplies cannot meet or for blending to up-grade poor quality coal (about 30%). Furthermore, opencast coal is an integral part of the N C B's overall production and makes a vital contribution to the N C B's finances. In 1980/81 the N C B made an operating loss of £134·9 million on deep mined production of 109·6 million tonnes (before grants and interest) and a profit of £156·5 million on 15·3 million tonnes of opencast production. The financial contribution from opencast is, and will continue to be, an important factor in the industry's return to fully economic working and its further development. Meanwhile, it serves to counterbalance the losses resulting from high cost deep mining capacity.

11.8 The N C B and the Department of Energy are fully aware of the impact on the environment that opencast operations, by their very nature, must make. The Department of Energy have stated to us that the target of 15 million tonnes was adopted as the maximum level of opencast coal that could be expected to be produced without unacceptable consequences for amenity or invasion of private rights and that it is the Secretary of State's responsibility when granting individual authorisations to extract opencast coal to maintain a proper balance between the need for the coal and amenity and environmental considerations.

11.9 On the other hand, the Commission received very forceful evidence from many bodies including the local authority associations and individual local authorities, the R T P I, the Committee for Environmental Conservation (Co En Co), the N F U and the Council for the Preservation of Rural England (C P R E) as well as from particular interest groups such as the Opencast Mining Intelligence Group (O M I G) and the Methley Environment Group to suggest that the public does not accept this view of opencast mining. In this evidence, it was argued that opencast mining is one of the most environmentally damaging processes being carried out in the U K. It causes immediate and medium term environmental pollution, social upheaval and disfigurement of the landscape, and in the long-term, a permanent loss of good agricultural land and degradation of the rural landscape. A further argument associated with this view is that as opencast coal is quickly accessible, these reserves should be confined to

supplying occasional peak demands which could not be satisfied within an acceptable timescale by an expansion of deep mine production. This view ignores the practicability of rapid fluctuations in opencast output. Moreover, it would involve significant sterilisation of land. An emergency reserve would inevitably inhibit the building of factories or houses on the land in question.

LOCATION OF OPENCAST MINING

11.10 The map at Figure 11.1 indicates present and potential opencast working and areas of shallow coal. This shows that, although opencast coal is quite widely dispersed, there is some concentration of activity in West and South Yorkshire, Northumberland, Durham and in Scotland and South Wales. The following Table further illustrates this point.

Table 11.2   Number of opencast sites by N C B region and coalfield 1981

Opencast Executive Region	Number of authorised and working opencast sites as at 28.3.81 (excluding sites under contractual restoration)
1 Scottish	10
2 North East	13
3 North West	10
4 Central West	5
5 Central East	15
6 South West	14
Total	67

Source: N C B

There is unlikely to be any significant shift in the geographical location of opencast activity from the current main areas of operation but with some decrease in South Wales and an increase in the Central West Region.

EXPLORATION

11.11 A continuous programme of new sites will be necessary if the Opencast Executive are to maintain production at current levels. Forward programmes are prepared for 5 year periods to include prospecting for, as well as the working of sites, to enable new sites to be brought into operation as old ones are worked out. The Opencast Executive have increased their overall prospecting activity considerably since 1974 and as a result have increased proven opencast reserves to 300 million tonnes. The current rate of proving is running at about 25 million tonnes a year. We were pleased to note that this is about twice the rate of production and should allow some flexibility and choice in the programming of new production and sites.

11.12 Under Article XX(iv) of the G D O 1977, the Executive do not need specific permission to drill boreholes, they must simply notify the county planning authority. Once it is decided to explore a site the

Figure 11.1   Areas of present and potential opencast working

Opencast Executive Region   **3**
Working/Authorised Sites   •
Programmed Sites   *
(Identified to Local Authorities)
Areas of Shallow Coal   ////
(These include sites which have
been worked and areas which
can never be worked due to
geographical features)

Scale
                         miles
0   25   50

N

1 Scottish

2 North East

3 North West

5 Central East

4 Central West

7 South West

Source: N C B March 1981

Executive seek the views of the N C B Deep Mines Area, the Mines and Quarries Inspectorate and the I G S. Rights of entry are then sought. Notification of the proposals is given to all statutory undertakers, local authorities and relevant Government Departments such as M A F F. All owners and occupiers of land within the proposed site boundary are approached and agreements reached on the terms and conditions of entry into the land for the purpose of exploratory drilling.

11.13 After about 3–4 months of site prospecting it is usually possible to detect the pattern of the site. About 2–3% of sites are abandoned at this stage because of geological or other physical constraints or because the ratio of overburden to coal would make extraction uneconomic. After more concentrated drilling a working site proposal is prepared for the remainder at the Region's headquarters. Some of these proposals may be amended later as a result of negotiations with local authorities or other bodies, or dropped because of falling demand for certain kinds of coal. Generally working site proposals are sufficiently detailed to provide the basis not only of the application for authorisation to the Secretary of State but also detailed guidelines for the contractor for the site.

OPENCAST AUTHORISATION PROCEDURES

11.14 Originally, rights of access to prospect for coal and subsequently to work it were granted under Defence (General) Regulations 1939. These remained in force until 1958 when the Opencast Coal Act replaced them. This was subsequently amended by the Coal Industry Act 1975. The legislation requires the N C B to seek an authorisation from the Secretary of State for Energy before working any coal by opencast operations. The authorisation covers both the working of the coal and the restoration of land affected. When the land is in agricultural use at the time of the authorisation, the restoration has to be such as to secure that it is "reasonably fit" for use as agricultural land, except where the Secretary of State is satisfied that the land will not be used for agricultural purposes when the authorised workings have been completed. No separate application for planning permission is necessary as this is, in effect, covered by the grant of authorisation.

11.15 The 1958 Act specifically requires that both the Board in formulating proposals and the Secretary of State either in granting an authorisation or imposing conditions must take into account any effect which the proposals would have on the natural beauty of the countryside or on any flora, fauna, features, buildings or objects.

11.16 If, on publication, any objection is made to a proposal by the local planning authority, any other specified local authority, any owner, occupier or lessee of the relevant land, then the Secretary of State must hold a public local inquiry before determining the application. The Secretary of State may also decide to hold a local inquiry if he thinks fit where no objection has been made.

11.17 The legislation also provides the N C B with powers to make orders granting them compulsory working rights. Such compulsory orders grant the Board rights for the temporary occupation and use of the land. Rights of entry to land both for exploratory purposes and for preparatory working prior to the submission of an application for an authorisation, and compensation payable to those whose land will be affected by the coal working and incidental operations are also provided for.

11.18 The formal application to the Secretary of State for Energy must be advertised and served on all parties with a legal interest in the site, as well as County and District Councils. It includes a full description of the land with its surface features and statutory undertakers' apparatus, a description in detail of how the coal is to be worked and the measures to be taken to protect the environment and a description of how the land is to be restored. If there are no statutory objections within 28 days the Secretary of State normally issues the authorisation which covers the conditions governing working operations and restoration. Before making his decision the Secretary of State consults both the Department of Environment and M A F F (D A F S in Scotland and the Welsh Office in Wales) who advise on the agricultural aspects of applications for authorisation for opencast workings.

11.19 If statutory objections are received and not subsequently withdrawn, the Secretary of State for Energy is obliged to hold a public inquiry. Up to the end of 1973, 19 public inquiries were held out of a total of 215 applications. From 1 January 1974 until 1 May 1981 the figures were as follows:

Table 11.3   Opencast authorisations 1 January 1974–1 May 1981

Authorisations without a Public Inquiry	93
Authorisations after a Public Inquiry	25
Favourable decision after a Public Inquiry	1
Decision awaited after a Public inquiry	1
Applications rejected after a Public Inquiry	4
Applications withdrawn	1
Other applications under consideration with the Department of Energy	15
Total applications	140

Source: N C B

11.20 The Commission received strong representations in the evidence about these procedures. Bodies including the Local Authority Associations, Convention of Scottish Local Authorities,

Environmental Health Officers' Association, R T P I, N F U, O M I G and several individual local authorities, argued that the present procedures were unsatisfactory and that opencast coal working should be brought within the normal minerals planning control machinery. Ministers are aware that there is substantial public concern and asked us to evaluate and comment on these representations. However, it is important, before reaching conclusions, to take into account the full effects of the operations themselves and of the subsequent restoration. We rehearse therefore the arguments for and against a change of procedures in the last part of this Chapter.

Private opencast coal working

11.21 The N C B Opencast Executive have powers to grant licences to private individuals for the winning of coal by opencast methods. Total production from such sites accounts for a small proportion of national opencast coal production, averaging about 1 million tonnes per year. Site licences normally restrict extraction to about 25,000 tonnes and sites are accordingly small in area and often worked within 1–2 years. A licence is only granted once planning permission has been obtained from the local planning authority and if the coal is not within an area of interest to the Opencast Executive. M A F F (D A F S or the Welsh Office) are normally consulted by the local planning authority as to what conditions if any should be imposed in respect of the restoration of the site to agriculture if that is the proposed after-use. The licensee is required by the Opencast Executive to take out a bond in the form of an insurance policy which is intended to cover the cost of restoration in the event of bankruptcy or the failure of the operator to restore the site to the agreed conditions. In addition, the licensee must either pay a royalty, regardless of coal quality, or dig, load and transport the coal to an N C B disposal point for an agreed price per tonne. Many of these private sites produce coal as a secondary mineral with fireclay or brick forming the primary mineral worked and these may have a longer working life. The technical problems relating to private sites are in many ways identical to those encountered on larger sites worked on behalf of the Opencast Executive. In the remainder of the Chapter we concentrate therefore on the Executive's operations.

MINING OPERATIONS ON OPENCAST SITES

11.22 Coal production and coal preparation on opencast sites is carried out by contractors working for the Opencast Executive. The Executive currently have 19 production contractors on their tender list. Contractors are required to nominate a "Manager" under section 98 of the Mines and Quarries Act 1954 and give him the authority to be responsible to Her Majesty's Inspectorate of Mines and Quarries for all operations on the site including health and safety. The Executive also require the contractor to nominate an official to be responsible for dealing with and taking any necessary action on any complaint about working operations made by local residents, other members of the public or local authorities.

11.23 The authorised conditions for site working are written into the production contract between the Executive and the contractor. Over a period of time a comprehensive set of about 65 standard conditions have been evolved, any of which, when included in an authorisation, have the force of ordinary planning conditions. Some 28 of these conditions are designed to deal with the main environmental hazards of opencast working, such as noise and dust control; others deal with restoration practices to be adopted by the contractor. In addition, site engineers from the Executive's Area office have a responsibility to make sure that the controls designed to reduce the environmental impact of the operations are put into practice.

11.24 When a contractor takes over a site, he first completes any necessary preliminary works, such as the diversion of main services, the provision of proper drainage and the construction of oil traps wherever oil is stored or vehicles are serviced. The main excavation begins with the stripping of the topsoil, then the subsoil. These are stacked in separate mounds and are usually used to provide baffle embankments on the site perimeter. The Opencast Executive tell us that the embankments are usually seeded to grass to help provide screening from the operations. All the available topsoil on a site is recovered and sufficient subsoil to provide a layer of about 600 millimetres of subsoil. If there is a potential deficiency in soil thickness for subsequent resoration, additional soil-making material may be recovered from the overburden during later excavation. A total minimum depth of 1 metre of topsoil and subsoil is normally required.

11.25 The initial strip of overburden is excavated next and stored at ground level. Where necessary blasting is carried out to loosen both coal and overburden. Draglines then excavate to the lowest seams in a progressive strip cut, replacing overburden from the next strip in the void produced after each slice of coal is extracted. The material from the initial strip is then replaced in the final void after all the coal is extracted. Over burden tips and soil mounds are shaped and graded to make them as unobtrusive as possible; agreed heights are often specified by local arrangement and care is taken to prevent protrusion above the skyline at critical vantage points. Experiments with grass seeding are currently being carried out to help to make large mounds less obtrusive. Plate 11.1 together with Figure 11.2 illustrate operations at an opencast site.

Plate 18.1a Sheffield city centre in the early 1950s. The picture shows the effect of atmospheric pollution – much of it due to the widespread burning of coal on domestic fires. (*Photograph by courtesy of the National Coal Board*)

Plate 18.1b Sheffield city centre in the 1960s. The introduction of clean air legislation led to dramatic improvements in air quality. (*Photograph by courtesy of the National Coal Board*).

Plate 11.1 (*above*) Shilo South Opencast Site, Derbyshire. Figure
11.2 (*right*) explains the operation of this opencast site.
(*Photograph and diagram by courtesy of the National Coal Board*)

Figure 11.2 Shilo South opencast site Derbyshire (see Plate 11.1)

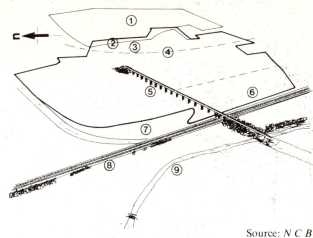

Source: *N C B*

The Shilo South Extension opencast coal site was authorised in June 1978 and commenced coaling in December 1978. This 114 acre site contains about 297,000 tonnes of coal and lies due west of the village of Awsworth (1) on the borders of Nottinghamshire and Derbyshire.

The Eastern boundary of the site coincides with the rear gardens of many of the houses with a baffle bank (2) screening them from the edge of the workings (3) where early restoration is now taking place.

The site is to be crossed by the Awsworth by-pass (A6096) (4) and the replaced soil is therefore being consolidated. The site is also crossed by the disused Nottingham–Derby railway line including the Bennerley Viaduct (5). This is listed as a structure of Special Architectural or Historic Interest and has therefore been protected during the working of the site.

Coaling is currently taking place on the site of a former slag heap (6) with the coal being taken by an internal haulage road to the Board's Bennerley Disposal Point (7). The coal is crushed and then put into railway wagons before being despatched to the C.E.G.B. or industrial customers by British Rail (8).

In the foreground can be seen the Erewash canal (9) which is being used for recreational purposes.

11.26 The coal is removed from the successive coal seam faces and loaded direcly into coal lorries. These may be normal road vehicles which can haul the coal directly on to public roads or may be larger if hauling over an internal road direct to an Executive disposal point (see Plate 11.2).

11.27 Most opencast coal is transported from the site by road. Points of access to the public highway from the site are agreed in advance with the local highway authority and are designed with vision splays approved by the authority. To prevent mud or coal being deposited on the road, washing facilities must be installed on the site before coaling begins.

RESTORATION FOR AGRICULTURE

Contractual restoration

11.28 When the excavation of coal on a site has been completed the contractor carries out the initial restoration of the land under the supervision of A D A S in England and Wales and D A F S in Scotland. First, the overburden is graded out to the required contours. The aim is to ensure that when the subsoil and topsoil are replaced after subsequent settlement, the site will blend into the surrounding contours and can be adequately drained. A meeting is held with the planning authority, landowners or occupiers and M A F F to approve the achieved contours. A landscape architect is occasionally involved, but not as a matter of course. If these are acceptable the subsoil and topsoil are replaced. The subsoil is usually replaced in two separate layers about 0·3 metres deep. At each stage the layers of overburden, subsoil and topsoil are rooted (deep harrowing) to break up the surface and any large stones or other extraneous material are removed. This work should be carried out in dry weather. Care needs to be taken to avoid intermixing of the layers and to ensure soils are evenly spread with the minimum of compaction. If any ponding appears in the surface because of settlement, these areas are stripped and re-graded. At each stage of spreading, M A F F usually inspect and approve the work before the next stage takes place. When the top soiling has been completed satisfactorily, the contractor has met all his obligations and the site is handed over to A D A S who carry out a 5 year period of agricultural rehabilitation at the expense of the Opencast Executive.

Five year restoration period

11.29 When the site is taken over from the contractors a meeting of interested parties is held to agree a basic plan for restoration which sets out the size and shape of enclosures, the position of new ditches and water courses, the types of fence to be erected and the location of new hedges, shelter belts and woodlands. The Executive inform us that the views of the owner, especially if he or she is also the farmer, carry great weight at this stage. If there is any disagreement between the parties about restoration proposals worked under an authorisation, then the Secretary of State would be the arbiter subject to the over-riding decision of the owner in matters not subject to legislation. The land contours and the alignment of new and restored water courses dictate the primary pattern. Other major features such as woodlands, hedges or fences can be established as new fields are laid out. Throughout the process the ultimate drainage needs of the site have to be catered for. We got the impression that this kind of design by committee, rather late in the day, was a poor substitute for adequate forethought and a published plan with adequate landscape design input at the inception of the project.

11.30 In the first stage of agricultural rehabilitation, the whole site is usually sown to grass. This assists the re-formation of the soil, decreases water run-off from the undrained site and prevents soil erosion. It also limits damage to the soil from vehicles crossing the site. During this period, the grass may be used for hay or silage. The cutting of ditches is carried out immediately, where possible along the proposed fence and hedge lines. Hedges are provided whenever possible but the evidence suggests that farmers often prefer to restrict the use of hedges to the major land sub-divisions in order to retain maximum land use and maximum flexibility. Where hedges are not planted, a standard timber fence is used as the main stock-proof fence. Stone walls are occasionally used if this is traditional to the area.

11.31 The aim of the agricultural rehabilitation during the 5 year period should be to achieve good soil structure. Conditions vary from area to area but because the heavier soils are common to most coalfields, and because of high rainfall, laying down to grass is the most common treatment. An appropriate programme of seeding and cropping is followed through the 5 year period involving the use of about twice as much fertiliser as would normally be used on an ordinary site. Until about the fourth and fifth year, drainage continues to be by the initial open ditches and water courses. However, the practice is now increasing of laying a skeleton system of drains at an early stage for completion later. By the spring or early summer of the fourth and fifth year, the under-drainage should be complete, so the land can again be cultivated and sown. Some restoration plans include the provision of woodland destroyed during the working of the site or for new woodland if this is considered advisable. However, most farmers, again, wish to keep as much land for productive use as possible. Thus the percentage of land use for tree planting tends to be small at the end of the day.

Costs of restoration

11.32 The Executive quote the average total cost of rehabilitation in 1980/81 at about £3,635 per hectare. This is broken down as follows:

Table 11.4   *Percentage costs per hectare of agricultural restoration*

Treatment	% of cost
5 years agricultural management	28
Permanent Land Drainage	36
Fencing and Ditching	22
Water Supplies	1
Hedges and Woodlands	7
Farm Roads and Paths	6
	100

Source: *N C B*

A rehabilitation allowance is also paid to the occupier of the land when it is taken over at the end of the 5 year period. This is settled by the District Valuer during the original negotiations for the Executive's rights of occupation of the land. Its purpose is to help re-establish the occupier on the land and to encourage the use of additional fertilisers.

RESTORATION FOR FORESTRY

11.33 In South Wales, where opencast mining can take place on land of poor quality unsuitable for intensive agriculture, more difficult areas have been restored as forest. In these instances the planting, management and maintenance of the restored woodlands is carried out by the Forestry Commission acting as agents for the Executive. The Commission usually then take over the land on completion of satisfactory establishment of the trees.

ENVIRONMENTAL EFFECTS OF OPENCAST MINING OPERATIONS

11.34 Opencast mining, by its very nature, has a major impact on the land taken for the site and ancillary workings and on the surrounding neighbourhood. We have received a substantial amount of evidence on the effects of opencast mining. We have also seen opencast sites in Yorkshire, Durham, Scotland and Wales. The criticisms of opencast mining which are made most forcefully, and which our own experience bears out, focus first on the impact on the surrounding neighbourhoods of the operations themselves and, secondly, on the long-term effects on landscape and on the agricultural quality of restoration. The criticisms need also to be considered against the change in the type of opencast sites which appears to have been taking place over the last few years. In the 1950s and 1960s opencast mining proved to be an excellent and economic method of effecting the restoration of derelict land. Although there are many sites where this dual purpose exists, such as the Meadowgate site we visited in South Yorkshire, the overall proportions have been falling. At the end of February 1980 the Opencast Executive had some 10,500 hectares of land under contract, or authorised but not yet contracted; of this about 1,500 hectares, or about 14·4%, could be attributed to some form of derelict land clearance. In many instances the clearance of dereliction amounted to a very small area of a much larger site. Increasingly as the Executive's programme has expanded, sites have been located in or around the edge of urban areas, notably in the North West (around Wigan) and in Yorkshire (around Leeds and Barnsley) thus affecting larger numbers of residential properties, and in rural areas, such as Cumbria, absorbing better quality agricultural land or affecting areas of higher landscape value. We examine separately below the impacts of opencast operations on those who live near such sites and on landscape and agriculture.

Effects of opencast operations

11.35 We have received a substantial amount of evidence on the discomfort and frustration experienced by those affected by opencast works. The effects on residential areas are exacerbated in the older coalfields where the environment has also been affected by deep mining operations. For example, the projected expansion of opencast mining in West Yorkshire is a cause of considerable concern. In the Metropolitan Borough of Barnsley, an area which already has a substantial concentration of deep and opencast activity within or adjacent to its boundaries, the Opencast Executive have notified the local authority under Article XV of the G D O of their interest in a further 43 potential opencast coal areas. Together these would cover an area of almost 9 square miles (2,400 hectares). A large number would affect residential areas. Others would take up some of the few remaining areas of open land within the Borough (see map at Figure 11.3).

11.36 The environmental impact and resultant loss of amenity in an area can have a number of facets. There is the visual intrusion of the site, the general deterioration in living conditions caused by dust and noise; the anticipation of the shock from blasting noise and vibration; the early awakening by heavy machinery when work starts at 6 am and the increased wear and tear or damage to buildings, roads and services due to increased heavy traffic. A combination of these effects taken together can, for those badly affected, add up to a very severe diminution in the quality of life. In addition, there are the less tangible effects to be taken into account such as the blight on residential properties.

(a) *Noise*

11.37 The major source of disturbance is the noise from the stripping and re-spreading of the top and subsoil and the construction of the baffle mounds on the site perimeter. This period of earth moving usually takes about 4–6 weeks and is carried out by machinery similar to that used for road construction and other earth-moving operations. These scrapers and dozers can produce noise levels of up to 94dB (A) (decibels measured on the A weighted scale) at 10 metres. Subsequently the baffle mounds and the working depth of the excavation can reduce the noise level of the operations outside the site. The remaining noise apart from blasting arises mainly from the dump trucks used to carry overburden from the excavation to storage areas. As these have to climb fully laden from the excavation level to the top of the overburden mound, they are obliged to operate on full power for much of the time. They also operate warning sirens when in reverse and these have caused complaints from local residents at night.

11.38 Conditions are usually laid down in authorisations on timing and frequency to control the noise of blasting. The criterion for control is the "peak particle velocity" (ppv) of the ground motion. The Executive operate well below the level of safe vibration at which structural damage is unlikely to occur (12 ppv compared to 50 ppv). However, there can be some disturbance from vibration and air pressures. The Executive are engaged on research both internally and externally and in conjunction with the Royal School of Mines at Imperial College, University of London, and the Construction Industry Research Association aimed at reducing the nuisance caused by blasting.

11.39 Haul trucks which take the coal from the site through residential streets are often very noisy. Occasionally where rail loading facilities are available, further noise can be generated at night when trains are loaded in storage areas.

11.40 We are satisfied that the Opencast Executive are aware of the importance of noise control although there may be variations in practice from site to site. Noise levels are monitored at site boundaries or the nearest noise-sensitive property depending on where the conditions in the authorisations fix the noise limits. Noise limits are imposed in terms of 'Leq' which the Department of the Environment recommended in 1976 should be adopted for those noise sources where no other measure was in general use. (Leq, in effect, measures the average noise level from the site in energy terms over a specified period.) Coaling operations are generally planned to avoid the noise level at the site boundary exceeding 65 dB(A) Leq by day and 55 dB(A) Leq by night. Noise levels are usually lower than this, but are higher when topsoil is stripped or re-spread, which are activities excluded from the noise limits.

11.41 It has not been standard practice for the Secretary of State's authorisations to specify maximum noise levels. In certain cases, noise limits for site working have been specified. These are sometimes lower than those proposed by the Executive. For example, at Butterwell, Northumberland, a noise-limiting scheme was imposed in 1975 of 57 dB(A) Leq, by day and 50 dB(A) Leq by night. The Executive are not yet satisfied that they can in practice work to such levels at other sites. Noise monitoring research has been carried out at Butterwell involving the local authorities, the Executive and the Department of the Environment's Building Research Establishment (B R E).

11.42 The B R E have been responsible for some general work with the Opencast Executive on "Noise from Opencast Mining Sites" (B R E 1979). The B R E saw no reason why Leq should not be used to express noise for opencast sites, but concluded that insufficient is known about public response to such noise to enable levels of likely acceptability to be recommended. There is general agreement on such levels for road traffic noise and aircraft noise. These are particularly useful, in some cases essential, at the planning stage of new proposals. It would be particularly helpful both to the Executive and to those likely to be affected if it proved possible to draw up recommendations for assessing the noise impact of opencast operations. We recommend that the N C B make funds available to promote further research in this field.

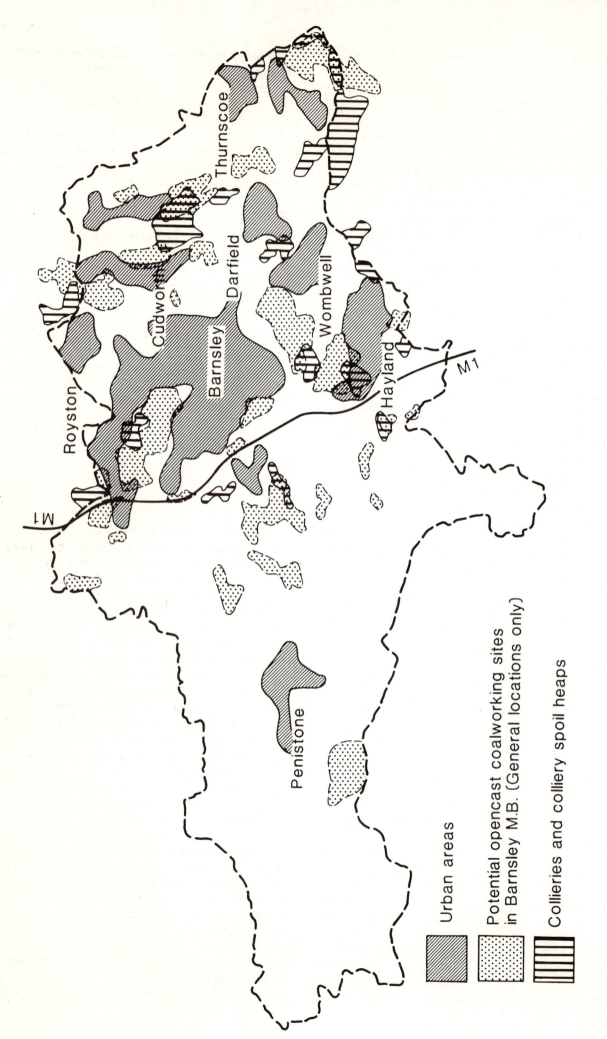

84

Figure 11.3   Potential opencast mining activity in Barnsley M B (sites notified under article 5 of the G D O)

Source: *Barnsley M B C*

Thurnscoe

Cudwo

Darfield

Barnsley

Royston

Wombwell

Hayland

M1

M1

Penistone

Urban areas

Potential opencast coalworking sites
in Barnsley M.B. (General locations only)

Collieries and colliery spoil heaps

11.43 During the course of the Commission's Study, further discussions have taken place between the Department of Energy and West Yorkshire County Council about the conditions to be included in authorisations. The Department of Energy have now informed the County Council that they propose to include in the standard condition provisions to the effect that the Board must use the 'best practicable means' as defined in Section 72 of the Control of Pollution Act 1974 to control noise. In the Code of Practice, agreed last year (see paragraph 11.68) the Executive will discuss noise levels with local authorities and length of working hours, before an application for authorisation is made. Details will then be included in the application and could also be included in the authorisation.

(b) *Dust*

11.44 The second major source of irritation is caused by dust from the site. In dry and windy conditions dust comes from the dirt haul roads within the excavation areas and from haul trucks moving on to residential streets and from the mounds of topsoil, subsoil or overburden before seeding has taken place. We had evidence of families unable to use their gardens, hang out washing or open windows in the worst cases. On the site, contractors are required to keep haul roads damp at all times during dry weather. Plant and vehicles must have engine exhausts which point in an upward direction and blasthole drills are required to be fitted with dust-collection equipment. Special arrangements are usually made for dumptruck tipping on overburden tips or for dragline casting when high winds are blowing. The Executive have also engaged the Royal School of Mines to devise a sample designed to help measure the degree of nuisance likely to be caused by airborne dust at particular times.

11.45 We are satisfied that the Executive take steps to ensure dust control insofar as possible at present although conditions will vary from site to site. It seems to us, however, that this is one of those aspects of operations which is only controllable within certain limits, as is to some extent the case with noise.

11.46 We are satisfied that the Opencast Executive and the Department of Energy are aware of the need to minimise the impact of opencast operations on the immediate neighbourhood insofar as they are able. Before a formal application for authorisation is submitted to the Secretary of State, informal consultations usually take place with those having an interest in the land, with County, District and Parish Councils and with Statutory Undertakers, for example, water authorities, M A F F and interested local bodies such as amenity societies. Discussions may cover conditions to be imposed on the contracts so as to limit the environmental impact of the proposed site. In many workings, where the restoration has been aimed at the improvement of derelict land, the precise proposals will

have been worked out at this stage between the Executive and the relevant local authorities. In recent years the Executive have also developed a policy of offering to form a site liaison committee, if required by local people, to provide an opportunity for discussion of the effect of the proposals in the locality. Under the Code of Practice recently agreed with the local authority associations, such committees will be established as a matter of course.

Role of the contractors

11.47 It has been suggested that although the Executive might operate from the best of intentions, there is no guarantee that the contractor on a site will fulfill all the environmental conditions of the grant of authorisation. We have no direct evidence ourselves to this effect. If it is the case in certain instances then it is the responsibility of the Executive's site engineer to ensure that practices are changed. The Executive have a powerful sanction in the ability to exclude a contractor from their tender list. Local planning authorities also have the power to enforce conditions attached to the authorisations granted by the Secretary of State for Energy, as well as those attached to planning permission for private licensed operators. It is their responsibility to ensure that this happens. We do not see this, therefore, as an unresolvable problem.

Proximity of sites to residential property

11.48 We have reached the view that there are limits, even with most careful controls, to which the intrusion, noise and dust can be reduced by pollution control measures. We have therefore considered the case for keeping operations further away from residential areas. Several recommendations to this end (some for a distance as great as 900 metres) have been put to us. The Opencast Executive have told us that, during the Committee stage of the Opencast Coal Bill, assurances were given that no opencast mining would take place within 50 yards of occupied properties. This has been enforced in all authorisations since the Opencast Coal Act, although in many instances the Executive have attempted to keep the excavation line further away. On the Glyn Tai Site in South Wales, for which a public inquiry was held, the Department of Energy have said that the excavation area to be authorised should not approach a line nearer than 82 yards (75 metres) from the nearest occupied dwelling.

11.49 We find it hard to accept that in many cases a distance of 50 yards provides adequate protection from opencast operations. We believe that the Executive should draw up its site boundaries with a view to causing the minimum disturbance possible to residential property, bearing in mind the economics of extraction from the site in question. A minimum safeguard is, however, advisable, so we recommend that the Executive, the local authority associations, the Department of the Environment and the Department of Energy, consult with a view to assessing the feasibility of extending the 50 yards limit on the basis that the revised figures should be clearly understood to be a minimum limit.

Discretionary power to purchase

11.50 It has been suggested to us that, in certain cases, the hardship to individual property owners would be reduced if the Opencast Executive made full use of the discretionary power to purchase properties which are not actually required for the workings and ancillary activities. In some cases the owner of the odd property near the boundary would be only too happy to sell to the Opencast Executive rather than to live adjacent to the operations. In certain instances the use of such powers might also help to speed up the planning process by removing the only objectors. We recommend therefore that the Opencast Executive make full use of the power to make discretionary purchases of property, with the agreement of the owner, in such cases.

"Temporary" nature of opencast operations

11.51 There is some implication in the Opencast Executive's approach that because the operations on any one site are temporary, a greater degree of disturbance may be acceptable than in the case of a deep mine. The argument is contentious but it has been put forcefully to us that in many areas it is completely misleading to describe the Executive's operations as temporary. In many localities the Executive pursue a rolling programme of sites which means that an area can be continuosly affected for many years. The Executive may also come back and re-open old sites where coal can now be extracted at greater depth. We accept this view, especially as temporary effects can include not only the physical impacts of operations as discussed above, but also loss of property values.

Loss of property values

11.52 Those most immediately affected by a reduction in property values are those who are obliged to sell for specific reasons such as change of job or illness. There appeared to be some confusion as to whether the Executive had or had not the power to make discretionary purchases in these instances; the Executive's practice varied. We understand that the Executive are now willing to consider purchasing a property where the owner is forced to move because of a change of job or on medical grounds and has been unable to sell the house at a reasonable price and within a reasonable time. Any sale is to be on terms negotiated by the district valuer as if the opencast operations in question were not taking place.

11.53 We consider, on grounds of equity, that the loss of amenity, disturbance and loss in property values are sufficiently detrimental to those who live close to opencast sites to justify the provision of some form of

compensation. We had in mind the kind of compensation provided in the Land Compensation Act 1973 for loss in property values resulting from the physical effects of works such as noise, vibration, smell, fumes, smoke, artificial lighting and the discharge on to the land in question of any solid or liquid substance. However, the planning compensation code only applies where there are no rights of redress against developers through the courts. The N C B have no immunity from claims of statutory nuisance under the Control of Pollution Act 1974 or from nuisance claims through the civil courts. In statutory nuisance cases the application of best practicable means can constitute a defence. In civil cases best practicable means is not necessarily a defence but a plaintiff must show that enjoyment of his property has been impaired. We conclude that it would be inappropriate to place the N C B at a disadvantage vis-à-vis other developers, although we believe, in practice, most individuals would be deterred from taking legal action by the potential costs and the additional stress likely to be incurred.

11.54 The Executive must take even greater care, therefore, to limit the impact on residential properties by careful programming of sites, a subject to which we return in the concluding part of the Chapter, in setting the boundaries of individual site proposals and by sympathetic exercise of the discretionary power to purchase. The local authorities must also be diligent in the use of their powers to enforce authorisation conditions and to take actions for nuisance where appropriate.

LAND RESTORATION AND RECLAMATION

11.55 Reclamation of land after opencast working is a major activity of the Opencast Executive and about 14% of their costs are devoted to it. In the 1950s and 1960s most opencast mining took place in areas of derelict or poor quality land, and its subsequent reclamation restored it to profitable farmland or for other uses; however, although this still happens, there are now many more sites, as we indicate above, absorbing better quality agricultural land, or affecting areas of higher landscape value.

11.56 For a long time the Opencast Executive has had a justly deserved reputation for the quality of much of its restoration work on derelict sites. There are many examples where derelict land has been restored for recreational use in conjunction with the local authorities concerned, such as Pugneys in West Yorkshire and Shipley Country Park, Derbyshire. We ourselves visited one such site in the process of being worked, the Meadowgate site in South Yorkshire. This provided one of several examples of collaboration between the N C B and the Opencast Executive. The Opencast Workings had been designed to leave a big hole near an adjacent active colliery for the concealed deposition of mine waste while the old tip had been re-graded and landscaped. When the work of restoration is complete, the Opencast Executive and the five local authorities involved will not only have improved the visual amenities of the area but will also have provided a recreational park with excellent sporting facilities, and a site for the unobtrusive disposal of dirt from the nearby deep mine.

Long-term effects of restoration

11.57 Recently doubts have begun to emerge about the long-term effects of restoration. The criticisms made to us concentrate on the diminution in landscape quality, the loss of soil fertility and the loss of agricultural productivity. Before we examine these in more detail we should like to draw attention to the report completed in 1980 by Co En Co "Scar on the Landscape?". This and subsequent evidence presented to the Commission by Co En Co, the local authority associations, the N F U, O M I G and other bodies, and followed up at first hand in Durham and Scotland, support the conclusions that land restored to date shows evidence of continued low agricultural yields. It also points to a lasting effect on the ecology of areas with the loss of hedgerows and trees and associated plant and animal life and a substantial reduction in the quality and variety of the rural landscape.

11.58 The problems seem to start with the quality of the restoration work carried out by contractors before M A F F begin their 5 year period of aftercare of a site. Although the planning conditions are designed to prevent shoddy restoration work, we had evidence to suggest that contractors can ignore such conditions by, for example, replacing sub and topsoil under wet conditions. Bad work at this stage can result in types of rock, for example blue clay, coming to the surface, compaction of the top and subsoil and poor drainage resulting in ponding in the surface. We have already referred to the responsibilities of the Executive and the local authority in relation to the enforcement of conditions attached to a grant of authorisation. M A F F also have a role to play here.

11.59 M A F F's 5 year after-care service should serve to ensure that the best possible care is taken to bring land up to maximum quality before it is finally handed over to the farmer. However, in practice, the service is thin on the ground (38 man years in 1979 for all the A D A S services involved, that is, Land Service, Agricultural Science Service, Land Drainage Service and Agricultural Service) and there is some evidence to suggest that many farmers continue to need advice after the initial 5 year period. We believe there is scope for considerable expansion of this work by M A F F. As there will be further tasks for the Service, if the after-care provisions for other kinds of mineral extraction, including deep mined coal, proposed in the Town and Country Planning (Minerals) Bill, become law, we hope the Ministry will be able to take the opportunity to strengthen the Service to take account of this work and to increase the effort on opencast

restoration. In addition we see no reason why the Opencast Executive should not employ private contractors to supervise and carry out restoration work if they think this appropriate.

11.60 The Commission found that although a large research programme is now in hand, there was until recently, because of the decline of operations in the 1960s, little research being undertaken into the operational elements of opencast working, for example, into the conditions under which soils are stripped, stockpiled and reinstated and the effect on the composition of soil. Although the Executive and M A F F are now beginning to build up a programme of research, the results are not likely to be available in the short-term. M A F F have now set up a Land Restoration Research and Development Liaison Committee which includes the Opencast Executive as well as other extractive industries and local authorities.

Loss of use of agricultural land

11.61 Whilst a site is being worked for coal, agricultural production ceases and during the subsequent 5 year restoration period emphasis is placed on restoring soil structure, thus restraining the full exploitation of the grass crops usually produced. The Opencast Executive state that in order to maintain an output of 15 million tonnes of coal a year, some 2,000 hectares must be brought into production every year. (An average site is about 14 hectares.) To sustain a continuous programme about 8,000 hectares will be worked at any one time. A further 1,000 hectares will be simultaneously undergoing restoration and, although agriculturally productive, will be being farmed below potential. Thus they suggest about 10,250 hectares are lost to agriculture at any one time, but this calculation does not take into account subsequent lower yields on fully restored land.

11.62 The N F U have suggested that, where the N C B put forward new proposals for any kind of mining, these should be accompanied by some form of agricultural impact statement which would include an analysis of the effect of the proposals on the food production of the area and the proportion of agricultural land to be taken out of use. We agree with this. Although M A F F formally advise the responsible Minister on the agricultural impact of proposals, we believe that other bodies should also have the opportunity to provide views. Consultations between the Opencast Executive and the local authorisation on 5 year programmes under the Code of Practice should include consultation with agricultural and other scientific interests.

Loss of landscape values

11.63 The criticism was made by bodies such as Co En Co, N C C and local authorities, that although the restoration achieved by the Executive was in scale and keeping with the surrounding countryside, it was basically open and featureless. The loss of mature woodlands, hedgerows and other small scale landscape features not only removes the variety which contributes so greatly to the quality of the British landscape, but can also have an ecological effect by removing the habitat of various kinds of flora and fauna. We believe these criticisms are valid and well illustrated. We have seen examples in the Wear Valley District of County Durham of a contrast of scale in landscapes on different sides of a single lane road and have learnt from the N C C of interruptions to migratory bird habitats in the Lower Aire Valley.

11.64 The problem does not lie solely with the Executive or M A F F who are not completely free agents. It often reflects the wish of the farmer to seize the opportunity to develop more modern methods of farming. We consider that a substantial improvement in practice could be achieved if details of the restoration programme were submitted as part of the application for authorisation and were more fully covered in the conditions attached to the grant of authorisation. This would bring opencast applications in line with the requirements to be imposed on applications for new deep mine proposals embodied in the Town and Country Planning (Minerals) Bill. We recommend, therefore, that the appropriate legislative changes to this end be made as soon as possible.

11.65 We also recommend that the Government's statutory advisers, the Countryside Commissions and the N C C, should be consulsted as a matter of couse on opencast applications. We are aware that in many cases such consultations already take place and that there is satisfactory liaison certainly between the N C C, the local authorities and the Executive in many areas. However, there are occasions when proposals become known too late in the day for the statutory bodies to offer the positive advice they are well equipped to provide. Finally, we believe that adequate landscape design skills should be involved at an appropriate level of authority from the early planning days of the site. We recommend therefore that the application to develop should include a survey of the existing agricultural, landscape and ecological characteristics of the site, a clear visual and verbal description of the forecast condition of the site after mining and restoration, and an assessment of the impact of the operation on the existing state of affairs. Reclamation after opencast working too often exhibits the featureless, bland and boring attributes which characterise the typical pit heap after restoration to agriculture. The British landscape is full of variety and interest. Discontinuities like scarps provide contrasts and sites for wilderness rich in balanced populations of wildlife. Imaginative restoration could seize such opportunities and provide a balance to the excessively powerful demands of agricultural productivity.

88

GENERAL CONCLUSIONS

11.66 Our analysis has led us to the view that even if the greatest possible care is taken in both the extraction of opencast coal and the subsequent restoration of the land, and while acknowledging that in some cases amenities have been enhanced by opencast restoration, opencast mining has a severe impact on the environment in both the short and long-term. In spite of the recommendations made for particular improvements, in practice an over-riding question still remains. Are current levels of opencast production successful in maintaining the balance between environment, amenity and private rights and the nation's need to exploit coal by opencast methods? This is a particularly difficult question in view of the vital role of opencast coal in the N C B's financial structure. The answer in the last resort must be a matter of judgement. After careful consideration of the interests involved we concluded that the balance is moving against environment and amenity. We now examine, therefore, whether this balance might be better maintained by either a change in the authorisation procedures or by a change in the volume of opencast coal production.

Opencast authorisation procedures

11.67 Before we examine in detail the case for a change in procedures, we should like to draw attention to a new arrangement reached between the Opencast Executive and the local authority associations during the course of our Study.

Opencast code of practice

11.68 In the initial evidence to the Commission, complaints were expressed by local authorities at County and District level that in some cases they had no prior knowledge of the Opencast Executive's forward programme of development in their area. They were limited, therefore, in their opportunity both to plan for the effects of opencast mining in their areas and to influence the choice of sites. During the course of our work, the local authority associations in England and Wales reached agreement with the Opencast Executive on a Code of Practice "Opencast Coal Mining Operations – July 1980", designed to achieve a much closer degree of co-operation. Under the Code there will be regular informal discussions between local authorities and the Opencast Executive before the Opencast Executive draw up firm programmes. These will cover sub-regional opencast coal targets relevant to the local authority area, their implications for structure and local plans and the extent and location of derelict land in the local authority area. The Opencast Executive will provide consultative maps to local authorities to help identify areas where opencast coaling is or is not acceptable to the local authority. The aim will be to enable individual prospects to be examined within a total framework. After the exploratory prospecting has taken place the Opencast Executive will provide the local authorities with their preferred site priorities for the area. There is to be no formal procedure for the discussion of individual site proposals in detail but sufficient exchange of information to enable local authorities to examine the impact of these proposals. The Opencast Executive might in turn expect a clear and committing statement from the local authority. On the basis of all these discussions the Opencast Executive will draw up a 5 year programme to be updated at 12-monthly intervals. Local authorities will thereby be able to assess the general and cumulative impact of the proposed sites in their areas.

The case for a change of authorisation procedures

11.69 As explained earlier, Ministers have particularly asked us to evaluate the arguments for a change in authorisation procedures. Besides giving full consideration to the views expressed in the written evidence, we have discussed this subject with the Departments of Energy and the Environment, the local authority associations and individual authorities, the Opencast Executive, Co En Co, C P R E and O M I G. The arguments for and against a change can be summarised as below.

11.70 Those arguments in favour of a change rest largely on the need to incorporate opencast coaling into the normal system of minerals planning. Since 1947, the planning system has demonstrated itself to be just as satisfactory a means of controlling mineral operations, including the deep mining of coal, as it is for non-mineral development. The national need for mineral working has to be taken into account in development plans and the Secretary of State for the Environment, or for Scotland or Wales, as appropriate, has to satisfy himself that this need is being met when approving structure plans. Questions of national and regional need also have to be considered in most planning inquiries relating to mineral working and this was, for example, a major element in the recent Vale of Belvoir Inquiry. Local planning authorities have demonstrated their ability to balance questions of need against environmental considerations and have developed the expertise necessary to impose appropriate conditions to minimise the environmental effect of mineral working and to ensure proper restoration. They are assisted in this by technical advice from the Department of the Environment, M A F F and the Scottish and Welsh Offices.

11.71 As with other minerals, opencast coal working needs to be integrated with the overall planning of the area in which it is to take place. However, this can be difficult to achieve when the planning authority's statutory role is limited to objecting to an application. In particular, planning authorities have sometimes complained that they have been given little advance knowledge of the N C B 's plans, so that their own

development plans have not been able to take into account the need for or effects of opencast working. The Board's proposals have therefore often been unrelated to other planning policies, although there has recently been more consultation by the N C B with county planning authorities over their forward programmes. In addition, the planning authority has no statutory role in relation to the imposition of conditions to control the environmental effects of working, in spite of the fact that they ultimately may be responsible for enforcing them.

11.72 Evidence given to the Commission also claims that the present arrangements do not reflect the accepted role of the public in development control. Under the planning system, most decisions are taken by a local authority where members need to explain , and if necessary justify, decisions to their electorate. When decisions are taken by Central Government, this element is lacking and local residents are deprived of the normal means of ensuring that their views are taken into consideration.

11.73 There has been an increased emphasis on environmental considerations since 1958, as the creation of the Department of the Environment itself demonstrates. Some of those giving evidence suggest that they would have greater confidence in the system if decisions were given by a Secretary of State who is accustomed to balancing conflicting considerations in planning cases, rather than by the Secretary of State who is also responsible for coal policy and for setting the target for opencast coal.

11.74 Arguments in favour of retention of the present system rest on the ability of the Secretary of State for Energy, operating within the doctrine of collective responsibility, to adjudicate on the balance between the national energy requirements and the local planning and environmental issues in the applications put before him, particularly in view of the responsibilities in relation to the environment laid on him under the Opencast Coal Act 1958, the Countryside (Scotland) Act 1961 and the Countryside Act 1968.

11.75 The present system provides the Secretary of State for Energy with the flexibility to regulate the production of opencast coal when fluctuations in supply and demand are anticipated. This flexibility has been used in the past to refuse an authorisation which an Inspector had recommended after a public inquiry and to grant one when an Inspector had recommended refusal. The decisions are still decisions of the Government as a whole, taken only after consultation, during which the interests of agriculture, of amenity and of conflicting land uses are adequately represented by the Departments concerned. The planning conditions governing the working and restoration of the site are agreed with the Department of the Environment, and M A F F, D A F S or the Scottish or

Welsh Offices. Any planning conditions agreed between the N C B and local planning authorities, as is often the case, are also included.

11.76 In every application, it is argued that the need for the coal is weighed against the potential damage to the environment and the views of the Department of the Environment, the Welsh Office, M A F F and D A F S are taken fully into consideration as are those of every person or body which makes representations to the Department of Energy in response to the invitation in the advertisement of the application. Thus local residents have every opportunity to have their views taken into account. In cases which do not evoke statutory objections, the local planning authority would, if it had the jurisdiction, be likely to grant planning permission. But where the local planning authority would refuse planning permission, the N C B would normally appeal. The case would go to the Secretary of State for the Environment, who would institute a public inquiry under similar rules and a similar Inspector, as at present happens when the local planning authority objects to an application. Practically, any change could lead to extra work in local planning authorities and at the Department of the Environment, Scottish Office or Welsh Office in the formal processing of applications and might lead to some duplication of effort as the Department of Energy would still have a continuing need to approve applications on financial and coal policy grounds.

11.77 In our view, one of the most important reasons for bringing opencast coal development within the normal planning procedure would be to provide local authorities with an increased opportunity to influence the choice of sites in the forward programme of the Opencast Executive both for environmental and amenity purposes, and to ensure account can be taken of other planning considerations relevant to their area. The Opencast Executive have acknowledged that they now have sufficient geological information on potential sites to allow for a degree of flexibility in their 5 year programme. They have already begun to provide local authorities with more information in their forward programmes and have now agreed a comprehensive Code of Practice to this end. If the grant of planning permission is within the power of the local authority there would be greater impetus to co-operation and genuine consultation. If local authorities influence phasing and timing of programmes and impose planning conditions, a more mutually acceptable programme should be achieved. The local authorities under the normal planning process will have to explain any refusals of permissions, or conditions attached to a permission. On appeal, the onus of proof would shift to the Opencast Executive to show why the local authority's reasons were invalid.

11.78 Secondly, although we in no way wish to cast any reflection on the individual decisions made under

the current system by the Secretary of State for Energy, nor would we wish to argue that individual results would necessarily have been different if the system had been different, we have no doubt that at present there is a lack of public confidence in the authorisation machinery. We believe this could be largely dispelled if proposals were to be decided within the normal planning framework. We do not suggest that opposition to individual proposals will cease but rather that the interested public will more readily accept that their case has had a fair hearing.

11.79 We do not believe that the proposed change would create much duplication of work. There might be a little more effort required in dealing with applications by local planning authorities. In central government the three main Departments with an interest, Energy, Environment and M A F F as well as the Scottish and Welsh Offices, are already involved. In particular, the Opencast Executive have told us that it would make very little difference to the Executive in practice whether decisions are taken by the Secretary of State for Energy or the local planning authority and the Secretary of State for Environment on appeal (or the Secretaries of State for Scotland or Wales). We even understand that the change in authorisation procedures could be achieved without legislation if the Secretary of State for Energy were willing to discontinue the exercise of his power to give deemed planning permission for opencast working whilst retaining an interest in applications on financial and coal policy grounds.

Recommendation

11.80 We recommend therefore that applications for opencast coal working should be dealt with under the normal minerals planning machinery as for all other new mineral developments and not directly by the Secretary of State for Energy, as at present.

11.81 Before leaving the planning and programming of opencast mining, we wish to suggest that there is much more scope for coordination between the N C B's deep mine and opencast operations. This could be particularly useful, as we state in Chapter 9, in ensuring that appropriate opencast schemes incorporate provision for spoil tipping from deep mines. It would also help the N C B to develop an awareness of the combined impact of deep and opencast mining on any one locality.

The volume of opencast production

11.82 It was suggested in the evidence that the environmental consequences of opencast mining are so serious that the only way to prevent irreparable damage is to reduce substantially annual levels of production. We do not believe it would be right to cut the volume arbitrarily in view of the severe implications for the financial viability of the N C B's overall activities. The profits from opencast production can help finance the costs involved in the sensitive and complex consequences of the run-down of obsolescent deep mined capacity. However, neither do we accept that opencast profits should continue indefinitely to disguise and cushion deep mine loses. We repeat the N C B make a profit of £156·5 million on 15·3 million tonnes of opencast production and a loss of £134·9 million on 109·6 million tonnes of deep mined coal (1980/81 figures). We strongly recommend that as older, more unprofitable and less environmentally acceptable deep mines are closed and more efficient and profitable operations take their place, the volume of opencast mining should be allowed to decline. In the meantime, there should be no increase in the present target of 15 million tonnes per year. The uniquely sensitive character of the British countryside and the high population density in much of the country would not be able to accommodate, without unwarrantable damage, a target in excess of that level.

11.83 We also believe that in view of the scope for greater flexibility in the programming of opencast sites, it would be possible to define more strictly the areas where opencast coal might be mined. This would limit the effects of the operations on people and restrict the longer term environmental damage. At the same time efforts to improve restoration techniques should be maintained. We are attracted by the idea of a set of guidelines on acceptable sites as suggested by the R T P I in their evidence. We have had discussions with the Opencast Executive in order to ensure that such a proposition could be effective in practice. We are satisfied that a series of guidelines could be an effective tool in the planning and programming of opencast mining.

11.84 We recommend therefore the adoption of a series of Guidelines which confine opencast extraction to sites:
  (i) where coal would otherwise be sterilised by built development including major roads;
  (ii) where part of the site (a substantial area) can be environmentally improved by coaling or subsequent restoration or where other environmental benefits can be obtained including the disposal of some domestic refuse or deep mine waste;
  (iii) where there is a demonstrable need for a certain grade of coal for blending or specialised needs which cannot be met from deep mines;
  (iv) to supply demonstrable short-term increases on demand which deep mine production is too inflexible to satisfy.

11.85 We recommend also that such Guidelines should be brought within the recently agreed Code of Practice adopted by the Opencast Executive and the local authority associations and rigorously applied. The existence of the Guidelines would not be inconsistent with the normal planning process but would act as a broad sieve on applications from the Opencast Executive to the local planning authority. A case for a particular site would be stronger for being within the Guidelines and weaker for being outside them.

11.86 The implementation of these guidelines would need to take account of the overall economic benefit to the nation of opencast production. We recognise that the industry is in a state of transition. Profits from opencast will serve in the medium term to counterbalance the losses resulting from high-cost, obsolescent, deep mine capacity. As more efficient deep mined capacity comes onstream, the trade-off between the economic benefits of opencast production and its environmental costs can be weighted to give progressively more consideration to the latter. Imports could provide, together with new indigenous deep mined capacity, an alternative to reliance on currently envisaged targets for U K opencast production. However, the trade off between opencast production and imports needs to be considered not merely in terms of comparative environmental advantages and disadvantages, but also in terms of the balance of payments implications and the degree of risk to the security of national energy supply.

CONCLUSION

11.87 Finally, we have been struck by the extent to which, when considering opencast and deep mine production, we are dealing with two totally different operations involving different management, different unions, and different plant. There is a striking analogy between the position in the coal industry and the former position of the Post Office regarding telecommunications and the postal services. We considered whether to recommend a hiving off of the Opencast Executive as a separate Corporation. We decided against this in view of the complex common marketing and logistic infrastructure shared by two greatly contrasting activities and the possibilities for collaboration between them (see paragraph 11.81). However, we consider that future judgements about the scale and role of opencast production would be greatly facilitated if the Opencast Executive were treated in the N C B 's Annual Report and Accounts as a separate accounting unit and we recommend accordingly. This would permit a much more informed assessment of the appropriate balance between the economic benefits of opencast production and its environmental costs.

Chapter 12   Coal production: General conclusions

12.1 In the preceding Chapters we have examined the environmental effects of the extraction of coal by both deep mine and opencast methods. We have looked at current mining practices with an eye to their immediate impact and to their long term implications. We have also kept in mind that over the next 20 years or so the major part of the mining industry is likely to be progressively more concentrated in the Yorkshire–Derbyshire–Nottinghamshire coalfields, supplemented by further developments in new coalfields such as Belvoir and South Warwickshire and perhaps Oxfordshire and with some continuing decline in the older traditional areas of South Wales and the North East.

12.2 We have worked from certain basic premises. We believe that the N C B like any other developer has a duty to take into account at every stage of planning and operation the environmental effects of a development. We also believe that coal mining should be covered by the same planning and pollution control machinery that applies to other forms of mineral extraction. Conversely, we do not believe the N C B should be unduly penalised by that machinery simply because of the scale of its activities. We have, therefore, when making recommendations in the preceding Chapters, tried to demand no more of the N C B than we believe it right to expect of any other developer.

12.3 We have been impressed by the major improvements in many practices that have taken place over the last few years. Indeed, there has been continuing improvement during the couse of, and to some extent prompted by, our Study. As the longer term prospects for the industry have improved, the N C B have become increasingly aware of the need to relate their activities to the broader community in which they are operating. This is not to say there is no variation in practice or no room for improvements. However, modern methods of mining when properly applied, can substantially limit the impact of those adverse environmental features which are an inherent part of the industry. A well run modern mine such as the Daw Mill Colliery, near Coventry, is a far cry from the traditional image of coal mining which persists even today. There are certain problem areas which we discuss fully above, in particular the incidence of subsidence, the volume of spoil for disposal, the overall effects of opencast mining, and the quality of land restoration and reclamation.

12.4 With present mining methods, some degree of subsidence is an unavoidable consequence of deep mining. As the N U M stated in their evidence "in an extractive industry the total elimination of subsidence is not possible and public attitudes to coal expansion may depend in large measure on the speed and effectiveness with which the industry responds to the damage caused to individuals by its operations". It is thus in the N C B's own interest, as well as essential for those whose dwellings are likely to be damaged by subsidence, to ensure that preventive and precautionary measures as well as restitution of damage and the provision of compensation, are approached as sympathetically, flexibly and speedily as possible.

12.5 We believe the volume of spoil that will have to be disposed of could constitute one of the major problems the N C B will have to face over the next 20 years. The local tipping of spoil can be carried out in ways which will considerably reduce its impact both during tipping operations and on restoration. However, even with the application of best practice, environmental intrusion will be substantial. This will influence the acceptability of the development of major new coalfields. It is already causing problems in some of the traditional mining areas, for example in West Yorkshire, where the extent of past tipping has resulted in considerable environmental stress. We have argued in Chapter 9 that N C B should pursue all realistic alternatives to local tipping, including local and long distance disposal schemes and additional research into mining techniques including backstowing. There will still remain, however, a considerable volume of spoil to be disposed of locally. It is imperative that provision is made to accommodate this spoil by coherent planning on a regional basis involving the industry and the local authorities and with the guidance of central government. This is necessary to ensure that the strains placed on any one locality are not too great and to enable the disposal of spoil to be planned in relation to the disposal of the growing volume of other kinds of waste.

12.6 We have concluded that there is limited scope for restricting the detrimental environmental effects of opencast mining both in terms of their impact on the surrounding neighbourhood and of the longer term implications for the quality of agricultural land and rural landscapes. We were particularly concerned that for historical reasons the Opencast Executive have only recently begun to carry out and sponsor research into the ecological impact of opencast extraction and the effect of the operations themselves on, for example, soil structures and some biological processes. Much of the evidence centred on the justification for the 15 million tonnes target for opencast production. We are not convinced that this represents the considered outcome of a delicate balancing of energy and environmental interests; it seems rather to reflect the importance of opencast coal in the financial accounting of the N C B. We have therefore concentrated on the need to plan opencast coal development to minimise the environmental effect. Building on the Code of Practice agreed during the course of our Study between the Opencast Executive and the local authority associations for joint discussion of 5 year rolling programmes, we have recommended that a series of guidelines should be incorporated in the Code of Practice to help identify sites where opencast extraction might be acceptable. We have also recommended that opencast operations be brought within the normal minerals planning machinery to ensure that local authorities are fully able to influence the choice and programme of opencast sites taking into account other planning considerations relevant to their area. We believe this package will encourage local planning authorities and the Executive to work even more closely together to limit the impact of opencast operations. Even more might be achieved if there were joint programmes between the N C B deep mines organisation and the Opencast Executive at regional level to ensure no one locality suffers a disproportionate concentration of mining activity.

12.7 However, the worst effects of opencast mining can be so severe that, although from the N C B's point of view opencast coal is cheap and easy, the volume should not be maintained at present levels longer than necessary. As more modern and profitable deep mines are brought into operation, and older uneconomic pits phased out, the total volume of opencast production should be allowed to decline. In the meantime, it should not exceed 15 million tonnes per annum.

12.8 The quality of land restoration can affect both agricultural productivity, wildlife and landscape for decades ahead. We are conscious of the vast improvement in design and restoration techniques that the N C B have been developing in respect of both spoil tip restoration and opencast works. Nevertheless, we believe there is scope for a more adventurous and imaginative approach as well as the need for more basic research as described above. To date the major influence on design within the N C B has been through the mining engineers. This was quite right and proper as the first emphasis must always be on safety below ground and above. It was also the need to improve safety after Aberfan that resulted in the development

of modern practices of continuous restoration. However, we believe that there is now greater scope for the Board to give more weight to the services of other professions such as architects, scientists and landscape architects not only in the design of large new projects but also in improving ongoing operations.

12.9 We have pointed out above issues of major concern. We must repeat, however, that we have in many instances been greatly impressed by the extent to which modern coal mining need not be unacceptably intrusive although individual cases must always be judged on their merits. We should particularly mention the N C B's sensitive approach to the major new development at Selby. Such doubts as remain relate to the earlier stages of the N C B's forward programming which we discuss in full in Chapter 21.

12.10 On the other hand the opening of new mines is only one part of the story. A large number of the deep mines in the country represents a legacy from the past when the environmental standards expected of a developer were much lower than today and consequently their impact on the environment is much greater. The results of a concentration of the older collieries in certain areas means that the combined effect is so much more serious. Older mining areas such as certain parts of the Yorkshire coalfield have a heavy concentration of current mining activity, both deep and opencast operations, combined with a disproportionate concentration of dereliction and old spoil heaps. The authorities in the areas concerned took care to stress to us their recognition of the benefits that accrue to the area through the activities of the N C B, especially in terms of employment. However it is clear that the disbenefits in terms of loss of amenity, environmental stress, and inability to attract other industry in the area are suffered by the local community in the wider national interest.

12.11 Although we have some sympathy with this viewpoint, we did not accept all the recommendations that followed for alleviation of the situation. We consider, however, that it is essential to maintain and improve the momentum of derelict land clearance in these areas if any general improvement in environmental quality is to be achieved. Local authorities and the N C B with the new powers available under the Town and County Planning (Minerals) Bill and the Local Government (Planning and Land) Act 1980 should have the tools available to do so. The most important factor is the availability of money. We urge the Government, even in the present economic climate, to maintain the levels of grant for derelict land clearance and to improve these wherever possible. In the case of Yorkshire the scale of the problem is such that we recommend that the Government examine the feasibility of establishing in the first instance an independent regional development agency for this area.

12.12 It is also important to ensure that the highest standards appropriate in the circumstances are applied to new developments in the older areas as well as to new developments on greenfield sites. Thus, in all cases the N C B should employ the current best practice available to limit adverse environmental effects and should adopt the same sympathetic approach to the local communities in terms of compensation or other voluntary provisions (for example the N C B have paid out funds for sound insulation on an ad hoc basis at Selby). We have described potential methods of ameliorating environmental impacts in a variety of circumstances. We also believe the application of the current best practice should ensure that, as changes in operating practice become technically feasible and financially acceptable, they should be introduced throughout the N C B's activities. This would not mean uniform standards throughout the industry, but the achievement of current best practice in the light of varying local circumstances.

12.13 We have also considered the arguments for the introduction of additional forms of compensation both for individuals and for communities in respect of the disadvantages they might be thought to bear in the broader national interest. Although we have some sympathy with these views, we could not accept the arguments totally because it seems to us there are many respects in which the adverse impact of coal on communities is similar to the adverse effects of other industrial concentrations, for example, steel and shipbuilding. We have in Chapter 8 made a specific recommendation on the provision of compensation to individuals for the residual loss of property values after the damage from subsidence has been repaired. We do believe it is possible to sustain a strong case that individuals should not be treated unfairly by industry or by government, and when the property of individuals is purposefully but necessarily devalued by industrial development then the individuals should be compensated by the developer when there is no right of redress through the civil courts. However, in no circumstances would we wish to rely on the provision of compensation to individuals instead of the prevention of unnecessary damage by the adequate planning of new projects.

12.14 We looked separately at the case for compensation for communities. There are already examples of this kind of provision. Under the Shetland Isles Act the local authority is able to use a proportion of the oil companies' profits in the interests of the whole community. In the North Wales Hydro-Electric Power Act, the C E G B is enabled to "collaborate in the carrying out of any informal measures for the economic and social improvement of the administrative areas which by virtue of the Local Government Act 1972 form the new county of Gwynedd". The C E G B are as a result providing financial backing for the establishment of an Environmental Centre in the Snowdonia National Park. During the day to day

operation of the development control system there is also scope for the occurrence of what is generally called 'planning gain'. Sometimes a local planning authority is able to reach an agreement with a developer under Section 52 of the Town and Country Planning Act 1971 requiring the developer to make additional provisions towards, for example, the local infrastructure, where the need arises out of the proposed development but could not be encompassed by planning conditions. There is another sense in which planning gain may be said to operate and that is where a developer offers to include some facility within the proposed development which will be available for the amenity of the community as a whole.

12.15 We conclude that, although particular communities might be expected to bear a disproportionate cost for the benefit of the wider community, it is virtually impossible to devise a fair system of formal compensation or one which could distinguish coal from many other kinds of industrial development. There is scope within the day to day operation of the planning machinery for local authorities and other bodies where relevant to obtain some form of planning gain from a major development. It is preferable therefore to urge the N C B and, indeed, other major developers, to look favourably on such practices rather than try and devise some form of statutory right to compensation. Finally, we consider it important that the N C B, and in particular the Opencast Executive, make their full contribution to the costs of the local community through the rates they pay. Given the breadth of our Study we did not feel able to examine in detail the operation of the rating system. We recommend, therefore, that the Department of the Environment review the existing provisions to ensure that the rates paid by the N C B are appropriate to the environmental stress their operations can cause, as well as the benefits they can provide.

12.16 Against the background of the major environmental issues inherently associated with the production of coal we attached importance early in the Study to a critical scrutiny of the N C B's environmental organisation both at headquarters and in the regions to see how it dealt with the environmental implications of its operations. While we soon became aware of the thoroughness with which the N C B tackles the environmental impact of major new projects such as Selby, we were not at all impressed by the provision made for dealing with environmental policy on a systematic and continuous basis. This was particularly true in respect of refurbishment of existing pits whose life was being extended by new investment. Architectural quality here is often abysmally low. Standards of good housekeeping are also very variable and perhaps too susceptible to the attitude of mind of the colliery manager and his Area Director. This was the more surprising in view of the critical importance of the Board's handling of these matters in terms of public acceptability of its operations. However, during the course of the Study a number of significant changes have been introduced.

12.17 At Board level responsibility for environmental issues is shared between the Member for Science and the Deputy Chairman who also has Mining Policy, Personnel Policy, Organisation, Staff, Statistics, Operational Research Executive and Northern and Midland Area Directors Groups on his plate. In 1978 the Board set up at headqauarters a Working Party on Energy and the Environment chaired by the Board Member for Science to co-ordinate the worl necessary for the Board to contribute to our Study. This Group concentrates on scientific issues but now has a co-ordinating role on environmental matters generally and can call for research support when they need it. More detailed matters of environmental policy relating to collieries are dealt with by the Headquarters Environmental Planning Committee chaired by the Deputy Director General of Mining. This reports as necessary to the Headquarters Mining Committee, which is chaired by the Deputy Chairman and provides advice to the Areas. At Area level, the Area Director, who is personally accountable to the Board for all the Area activities, is supported on environmental matters by the Deputy Director (Administration). A new system of Area Surface Environment Committees, chaired by the Deputy Directors (Administration) has just been established. The Opencast Executive will be represented on both these and the Headquarters committees. The Board believe that the new system will make it easier for the inter-relationship between various environmental policies to be identified and for a co-ordinated and balanced strategy to be worked out and put into effect.

12.18 New policies or standards are usually put to the Board through the Headquarters environmental committees. If the Board approve, any instructions are issued to Headquarters and Area Management supported as necessary by more detailed technical instructions. Nevertheless, the fundamental structure of the industry places responsibility for the day to day expenditure on improvements on the individual colliery manager who has to find the funds involved from within his own budget and at his own inclination. It is reasonable to assume that a manager of a less profitable colliery will put a lower priority on such expenditure than his colleague at a more profitable one. It is likely also that such pits will be in the older coalfields. We believe the Board has a responsibility to ensure that the best environmental practices available are introduced as appropriate in the day-to-day management of its operations and that continual encouragement from the Headquarters through the Area is necessary to achieve this. At Area level therefore, the Deputy Director (Administration) should be responsible for seeking out those collieries where environmental standards fall below an acceptable level and, with the help of the

Area Surface Environment Committees, for providing the necessary advice and technical assistance to effect an improvement.

12.19 Nevertheless our study of the Board's organisation has led us to appreciate the good sense and value of delegating executive responsibility downwards through a very clear line from the Board to the Area Directorate and the Colliery Manager. This derives from the early perils of the industry and the imperative need to identify accountability for health and safety. It also exposes precious veins of local initiative and enthusiasm which can be seen to be yielding rich rewards in, for example, the achievement at Selby. It can, however, expose the whole system to the vagaries of the individual. Thus, because, for historical reasons, concern about environmental issues does not come naturally to the industry, this system is not always adequate.

12.20 We discern a great deal of concern about environmental impact at all levels of the N C B but we doubt whether, at present, there is sufficient manpower with the relevant skills and experience to translate this concern into action. This is apparent not so much in those areas of the problem accessible to engineering sciences, as in those where biology, ecology, landscape design and architecture embody the relevant expertise. This is not to say that we would like to see more centralised control of the Areas from Headquarters in these fields but rather that we would wish to see a strengthening of the relevant resources at the centre in the interest of better Board policy making and more effective advice to the Areas.

12.21 We also consider it essential that someone in a senior position should have a clear responsibility for environmental matters. This could be a Board member if it was compatible with his other duties or it could be a part-time Board member as is the case with the C E G B. In either case he should be able to influence the flow of resources available for environmental matters. However, it is for the N C B to determine the precise organisation and responsibilities of its senior management.

12.22 Further, in the Headquarters Directorates the Chief Architect reports to the Chief Civil Engineer who reports to the Director of Engineering who reports to the Director General of Mining who reports to the Deputy Chairman of the Board. We understand that the Landscape Architect to the Board is being recruited at the same level as the Deputy Chief Architect. We would like to see these skills represented at a higher level with easier access to the Board, and with a Board member as the direct recipient of their views and advice. We understand that it is a part of the "Coal on the Switchback" syndrome that this has not been done, and we acknowledge the good intentions of the Board in advertising for a Chief Landscape Architect at a time of economic stringency and uncertainty. Nevertheless we would press for these reforms regardless of short term financial strictures. The effects of bad decisions in the short term interest can be catastrophic in the long term where the natural environment is concerned. It goes without saying that a strengthened Headquarters resource of architectural and landscape design would naturally take on some of the Board's most difficult and sensitive projects. Equally this must be seen to make them better fitted to select and adequately commission private consultants to cope with the peaks in the work load which lie ahead.

12.23 We consider on balance that, with the above provisos, implementation of the changes introduced during the Study should result in a far greater awareness at all levels that it is in the interest of the industry to adopt a much more positive approach to environmental concerns. Much will hinge on the extent to which the new Headquarters organisation will stimulate the appropriate level of awareness in the regions but without stifling local initiative. The degree to which it is in the N C B 's own interest to sustain and consolidate the changes is reflected in the sensitive issues we have discussed in the preceding Chapters on the environmental effects of coal production. The N C B 's handling of these key issues will be a major factor in securing public acceptance of the enhanced role for coal embodied in its plans.

PART III

Coal use

Chapter 13   Coal markets

INTRODUCTION

13.1 We have looked at the prospects for markets for coal not in order to determine the most likely scale and composition of demand for coal, but in order to forecast whether any major environmental problems were likely to occur given the range of projected demand. Our primary concern has been to assess the environmental implications of the higher end of the range for each sector, since this provides the basis for the "worst case" environmental assessment. If we can be confident that with suitable safeguards and anticipation, these higher levels of coal burn can be achieved without unacceptable environmental drawbacks, then any lesser levels which may in practice eventuate will most certainly be acceptable.

CURRENT PATTERNS OF COAL BURN

13.2 Coal accounted for 37% of U K energy consumption in 1979 on a primary fuel input basis, as shown in Table 13.1. The scale and composition of the coal market have changed significantly since 1950. Figure 2.2 demonstrates the subtantial increase in power station coal burn, and the significant contraction of other markets in response to competition from oil and gas. Coke ovens provide the second largest current market, although demand in this sector has fallen to only half the peak level reached in 1960. Past trends in general industry and the domestic market have followed a similar pattern, with total consumption and market share declining at about the same rate. The loss of the industrial market for coal has been particularly significant; coal now takes only about 15% of the industrial fuel market compared with 78% in the early 1950s.

Table 13.1   Total U K inland consumption of primary fuels for energy use

(Percentage Share)

Primary fuel	1955	1960	1965	1970	1975	1980
Coal	85·4	73·7	61·8	46·6	36·9	36·9
Petroleum*	14·2	25·3	35·0	44·6	42·0	37·0
Natural gas	—	—	0·5	5·3	17·1	21·4
Nuclear/hydro	0·4	0·4	2·7	3·5	4·0	4·7
Total energy (mtce)	253·9	269·4	303·3	336·7	324·8	328·0

*Excludes non-energy use of petroleum products and bunkers.

Source: *Department of Energy*

98

13.3 In the early 1950s and 60s gas works also formed a substantial market for coal, which has subsequently declined to extinction. However, in the longer term, there is a possibility of a substantial market developing for Substitute Natural Gas (S N G) produced from coal and possibly for substitute liquid fuels. The larger scale conversion of coal to other more convenient fuels could become one of the single most important markets beyond the turn of the century, and could result in major changes in the pattern of coal use. It is, however, a long-term possibility whose timing will in part depend on how well supplies of natural gas and oil hold up. Because this would be a development of a totally new market, individual issues arise in it and these are considered separately in Chapter 15.

POWER STATIONS

The current pattern of coal use

13.4 In 1980, 78·1% of the fuel used to generate electricity was coal (see Figure 13.1). The size of individual power stations has increased very considerably over the last 20 years and this has been perhaps the most significant single development of the power industry during this period. Large modern stations operate with greater efficiency than the older

Figure 13.1   Fuel used for electricity generation 1980: U K public supply

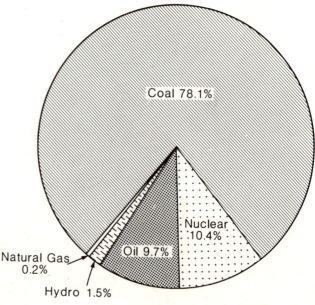

Source: *Department of Energy 1980*

Figure 13.2   Major power stations complete and under construction 1980

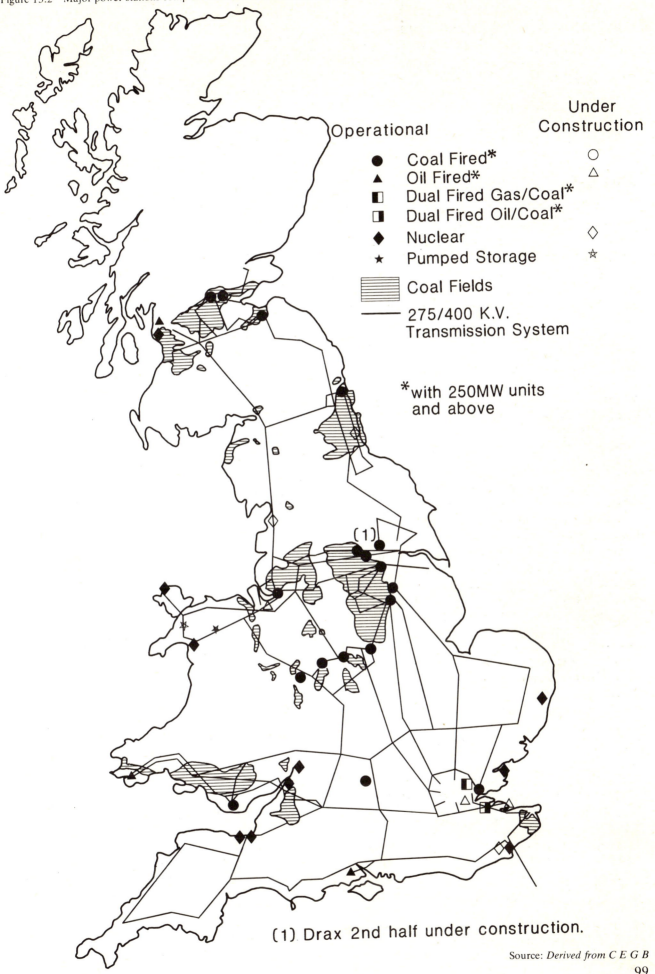

Operational

- ● Coal Fired*
- ▲ Oil Fired*
- ◧ Dual Fired Gas/Coal*
- ◨ Dual Fired Oil/Coal*
- ◆ Nuclear
- ★ Pumped Storage

▤ Coal Fields

— 275/400 K.V. Transmission System

Under Construction

- ○
- △
- ◇
- ☆

*with 250MW units and above

(1)

(1) Drax 2nd half under construction.

Source: *Derived from C E G B*

ones, and so are used more intensively. Thus three-quarters of the total coal burn takes place in only a quarter of the total number of coal-fired stations.

13.5 Figure 13.2 shows the distribution of major power stations. It demonstrates clearly the concentration of the larger coal-fired power stations in the Midlands and in Yorkshire. In fact, over 88% of the coal consumed by power stations in the U K is consumed in or near the coalfields from which it is produced. The principal exceptions are Didcot power station near Oxford, and those power stations situated on or near the Thames Estuary.

Future expectations

13.6 The scale of future coal burn in power stations is likely to depend upon first, the size of demand for electricity; secondly, the scale of the nuclear contribution; and thirdly, relative fuel prices. Oil-fired capacity will be declining by the end of the century. No new oil-fired capacity is being planned. The longer term prospect is that fuel oil will be required increasingly for conversion at refineries into higher grade petroleum products.

13.7 The Department of Energy's 1979 Projections imply that demand for electricity will grow less rapidly than G D P – the reverse of post-war experience, and over a period when indigenous natural gas supplies are expected to begin to decline.

13.8 On purely cost grounds, the C E G B would prefer the majority of their generating capacity to come from nuclear power stations. They estimate the ratio of cost of efficient generation of electricity from nuclear and coal sources to be 2:3 respectively. Experience both in the U K and elsewhere suggests that it will take time to build up the nuclear power contribution. We consider that the 40 GW nuclear contribution assumed in the Department's Projections by 2000 to be unrealistically high. After the difficult period experienced during the 1970s, even the achievement of the more modest nuclear contribution of 22 GW by 2000 will present a formidable challenge to the construction and nuclear power plant industries.

13.9 Against this background, coal will continue to be the main source of fuel for electricity generation in the U K at least until the turn of the century. Table 13.2 indicates the range of estimates of power station coal burn used in various projections.

Table 13.2   *Possible coal burn in power stations in 2000*

million tonnes

Origin of estimates	Coal burn in 2000
Green Paper 1978 (1)	70
Energy Projections 1979:	
(a) 1% Economic growth assumption	65*
(b) 2% Economic growth assumption	66*
(c) 2·7% Economic growth assumption	78
Coal for the future (2)	75–95
World Coal Study (3)	85–95

*If the nuclear capability by 2000 is no more than 22 GW, compared with a 40 GW capability assumed in the projections, these figures could be increased by 10 million tonnes per annum in the 1% G D P growth case and by 33 million tonnes per annum in the 2% case.

Source: (1) *Department of Energy 1978*
(2) *Department of Energy 1977a*
(3) *Coal Bridge to the Future 1980*

13.10 Within this timescale, and given this scale of burn, it seems unlikely that major changes will occur in the pattern of location of coal-fired power stations. The C E G B have estimated for us that no more than two greenfield coal burn sites would be needed to cope with a level of electricity demand which would entail 90 million tonnes per annum coal burn by the year 2000. This limited requirement would result from adopting a policy of expansion at existing sites, and, where possible, complete redevelopment of existing sites as plant reaches the end of its economic life.

13.11 So far as re-development of existing power station sites is concerned, those sites that have become available up to the present have not been suitable for re-development for major stations. However, in the next decade, the possibilities for re-development will become greater as power stations built in the immediate post-war period are taken out of service. Although few, if any, of these sites will be able to accommodate a station with modern 660 MW units, smaller developments with 300/350 MW units may be feasible if such units are technically attractive or economically viable.

13.12 In the longer term, most modern stations have the potential for complete re-development when the existing plant is retired. Modern power stations have a nominal 30 year design life, although with extensive repair and maintenance, it may be possible to extend the life to about 40 years. This means that these sites will become available for re-development towards the end of the century.

INDUSTRIAL USE OF COAL

The current pattern of coal use

13.13 Coal supplied the bulk of the energy needs of the industrial market in the 1950s but was pushed into second place by fuel oil in the late 1960s and into third place with the expansion of non-premium industrial gas sales in the early 1970s. Sales of coal in the general industrial market have stabilised during recent years at around 10 million tonnes per annum.

Future expectations

13.14 A range of estimates of the scale of coal burn in general industry by 2000 is presented in Table 13.3. The Department of Energy's 1979 Energy Projections are built up on the basis that demand in this sector will expand relatively rapidly in the 1990s as industrial consumers move away from higher priced oil products. Indeed, in the 2% G D P growth case, these projections provide for a quadrupling of current levels of industrial coal burn.

Table 13.3   Possible coal burn in general industry in 2000

million tonnes

Origin of estimate	Coal burn in 2000
Green Paper 1978 (1)	41
Energy Projections 1979:	
(a) 1% Economic Growth Assumption	32
(b) 2% Economic Growth Assumption	39
(c) 2·7% Economic Growth Assumption	45
Coal for the future (2)	30–50
World Coal Study (3)	30–50

Source: (1)   *Department of Energy 1978*
         (2)   *Department of Energy 1977a*
         (3)   *Coal Bridge to the Future 1980*

13.15 Oil substitution in industry can make a significant contribution to reduced dependence on oil which is now recognised by all major industrialised nations as the priority objective of energy policy. The most important short to medium term opportunities for oil substitution in electricity generation have already been taken up. Immediate scope in the transport and petrochemicals sectors, where the special characteristics of oil are at a premium, is limited. It is, therefore, in the area of boiler use in industry, commerce and public administration that the next significant opportunities for oil substitution in the U K are to be found. As U K North Sea oil production declines in the 1990s, we shall again become increasingly vulnerable to the full impact of price increases and supply interruptions. The 1980s are, however, likely to provide the U K with a period of comparatively stable supplies and it is during this time that we can best achieve a gradual and long lasting transition away from a pattern of energy use over-dependent upon oil.

13.16 The overall scope for oil substitution in industry-related sectors is substantial – up to about 15 million tonnes of annual fuel oil consumption with additional opportunities for displacing some gas oil consumption. This might be equivalent to about 30 million tonnes of extra coal burn annually. Against this background, we have sought to identify key factors governing the prospects for renewed coal sales to general industry.

Type of customer

13.17 The option of switching to coal is most likely to be attractive initially to large scale consumers where the size of operations and of fuel bills enables the cost of conversion and installing the necessary facilities to be spread more widely. It is already evident that some of the most energy intensive sectors of industry, for example, brick making, are switching back to coal as fuel oil prices rise. These pressures can be expected to spread at higher oil prices to other large users, notably the metal and chemical industries. Of the 30 million tonnes increase in industrial coal sales projected for 2000, some 15–20 million tonnes would represent demand by larger users and the remainder an emerging demand from medium sized and smaller consumers.

13.18 Medium and large oil-fired boilers in industry (with an equivalent consumption rating above one tonne of coal an hour) represent only about 20% of the total number of boilers, but comprise about 75% of the installed boiler capacity. It is in this area that efforts to switch fuel consumption from oil to coal could have the most substantial results. However, the majority of these boilers were either purpose designed for oil-firing and are not at present convertible to coal, or are old coal-fired boilers converted to oil and having a limited remaining operating life. Consequently, fuel switching from oil to coal, based on the technology available in the market today, will involve the installation of new coal-fired boilers and in most cases new storage, handling and other ancillary equipment.

13.19 Of the remaining 80% of the boiler population, over half are small boilers burning less than the equivalent of half a tonne of coal an hour, but they account for only about 10% of installed capacity. This segment of the market may represent about 3 million tonnes of oil burn a year, that is, a potential coal market of 5–6 million tonnes a year. But the obstacles to winning new customers to coal in this size of operation will be much more considerable and only a proportion of the potential is likely to be realised even in the longer term. Similar considerations will apply to fuel switching in smaller boilers used in the commercial and public administration sector which is discussed below.

Price
13.20 As supplies of oil and natural gas become scarcer and more expensive, industrial fuel options can

be expected to narrow progressively towards a choice between coal, electricity and S N G. In the timescale of our Study, considerable uncertainty surrounds what market share might fall to S N G manufactured from coal. Electricity will almost certainly be in extensive demand by the later part of this century and the early part of the next for process and other uses and for its convenience. But it is likely to remain an expensive fuel for lower grade non-premium uses, such as bulk heating and steam raising in industry. In the longer term coal will be the principal option for meeting such demands.

13.21 The pace of the transition will turn critically on coal and oil price relativities. The assumptions in the Department of Energy's projections imply a substantial advantage for coal during the 1990s. As discussed in Chapter 3, these assumptions have been challenged by Robinson and Marshall (1981). However, it is extremely difficult to predict with any certainty the likely pattern of price relativities that will develop during the 1980s, and whether any particular price advantage of coal will be sufficient to promote its wider use in the industrial market. A coal and oil price relativity of advantage to U K coal will be determined largely by the extent to which the best of U K coal reserves are progressively developed at rising levels of productivity.

Costs

13.22 The installed capital cost (including all ancillaries such as boiler house, fuel storage and handling facilities) of conventional coal-fired shell boilers (up to 1·5 tonnes an hour coal consumption) is on average some 2 to 3 times the cost of equivalent output oil-fired boilers. By comparison, coal-fired water-tube boilers (the majority of which burn more than 1·5 tonnes an hour of coal and operate at high pressure and with a high utilisation rate) have a capital cost of about three times the cost of equivalent oil-fired boilers.

13.23 Maintenance and operational labour costs for small scale conventional coal-fired installations are about twice those of equivalent oil-fired equipment but this cost difference progressively decreases with increasing size of installation. It would appear that labour and maintenance costs are currently seen as constraints on oil substitution by some consumers. There are, however, advances in automated coal handling techniques that can reduce these costs, even with conventional coal-fired equipment, and there would be advantages in demonstration and publicity so that decision makers in industry are made more aware of those possibilities.

Rate of penetration

13.24 Energy Projections 1979 would give coal a share of the industrial market of about 50% in 2000

compared with about 15% today. Oil's share of the industrial market between 1960 and 1970 increased from 20% to over 50% – in ten years rather than 20 years.

Availablity of north sea gas

13.25 Gas is a premium fuel with properties more akin to lighter oil products than to coal or refinery residues such as fuel oil. Gas has been sold into non-premium industrial markets mainly for load balancing reasons. But the load balancing justification for "interruptible sales" is likely to diminish over the period to 2000. If larger quantities of natural gas become available, it is unlikely that it would make sense in the probable circumstances of the 1990s to increase non-premium gas sales to industry further, particularly if coal was a viable fuel option for these same consumers.

Technological advance

13.26 Coal is recognised to be less convenient to use than oil or gas. Handling is more difficult; there is dust from coal stocks and ash to be disposed of; and it takes up more room for equipment and storage. The capital cost of coal-fired equipment is substantially higher than for oil or gas-fired competitors. In addition, the need to design to a specific customer's requirements and for particular coals adds to the lead times for installation of coal-fired equipment, while oil and gas-fired equipment can, to a much greater extent, be bought "off the peg".

13.27 Such disadvantages, if not mitigated, must limit the market expansion that coal expects from a price advantage over oil and the withdrawal of gas from non-premium markets. In the shorter term, oil substitution will depend on existing conventional coal burning equipment which is already available with automated coal handling. The N C B is, moreover, currently embarked on a major effort to develop a range of improved coal and ash handling techniques, and there are continuing improvements in the design of conventional coal-fired equipment to increase the degree of automation.

13.28 However, while these are important developments, fluidised bed combustion (F B C) technology could help to overcome further obstacles to the adoption of coal as the leading alternative to fuel oil but only if it is successfully developed and commercialised soon enough. The technology holds out potential advantages over conventional coal-fired equipment. It enables a wider range of coal types and qualities to be used; and it has greater scope to reduce polluting emissions to the atmosphere.

13.29 Work is furthest advanced on fluidised bed systems suitable for the smaller end of the boiler size range and several examples are currently installed in "guinea-pig" firms. Systems suitable for the medium size boiler requirement (1.5–3.0 tonnes of coal an hour) where significant oil savings are available, are being developed by "sizing up" from smaller versions and would, therefore, be likely to take several years longer to bring into mass production. The largest oil savings are likely to arise from replacing boilers at the top end of the range – over three tonnes an hour coal consumption (the water tube boiler market). However, the high cost of these facilities has meant that there has been relatively little demonstration work carried out in this area. Meanwhile, it must be stressed that oil substitution in the interim will depend on existing conventional coal burning equipment.

13.30 Coal-oil mixtures could prove an attractive alternative for some consumers as a transitional step in reducing heavy fuel-oil demand. It offers the prospect of lower conversion costs and fewer problems of adjustment and can use existing oil transport and storage facilities. The oil substitution potential of the technology, on which work is still being carried out, is some 25% of that for full coal conversion. We discuss this more fully in paragraph 16.47.

Use of coal in the Commercial/Public Administration Sector

13.31 The fuel consumption pattern in this sector (Table 13.4) reflects the basic requirements of consumers for cheap, easily stored, handled and distributed, clean burning fuels primarily for space and water heating. Approximately 75% of energy used in 1979 by the commercial and public administration sectors for space and water heating was supplied by petroleum products and 68% (4·2 million tonnes) of this oil demand was in the public sector.

13.32 The average age of boilers in the commercial and public administration sectors is about 15 years, with coal-fired boilers probably being upwards of 20 years old. Therefore, there should be a continuing demand for replacement boilers over the coming decade and fuel choice will have a significant effect on energy demand patterns in these sectors during the 1990s. Given the size of the public sector demand, the overall rate of fuel switching and the longer term market share of each fuel in the public administration and commercial sector will be largely set by government in-house policies.

Table 13.4   Energy consumption 1979 – Commercial/Public Administration Sector

(Delivered energy, million therms)

	Solid fuel	Gas	Electricity	Petroleum products
Pub. Admin/commerce				
Space Heating	515	997	325	2265
Water Heating	139	491	123	390
Lighting/Appliances	—	478	96	20
Total	654	1966	1740	2675
of which Pub. Admin	418	908	539	1818
Pub. Admin/Commerce Space and Water Heating:				
Health				541
Education				495
Govt. Offices				778
Commercial Offices				210
Shops				297
Others				349

Source: *Department of energy*

13.33 From the view point of consumers' convenience and level of capital investment required, both gas and electricity offer considerable attractions as substitutes for oil in these markets and coal will be competing in the longer term with them. The penetration of coal in this market will be influenced markedly by how successfully a range of improved coal and ash handling techniques can be developed together with improvements in the design of coal-fired equipment.

Policy implications

13.34 Coal is the leading alternative to fuel oil, particularly in larger scale industrial boilers and other uses. It is clear, however, that to wean industry away from natural gas and oil could be a slow and costly process. Smaller scale industrial boiler use and heating requirements in the commercial and public administration sectors can expect to face greater competition on grounds of convenience and cleanliness from more expensive gas and electricity. Even in more favourable areas, while relative fuel prices point in favour of switching to coal, there are obstacles to its early adoption, notably: the capital cost of equipment; the relatively short payback period sought by much of private industry; problems of convenience and maintenance; coal's image as the dirty fuel of the past; and, not least, the financial pressures operating during the recession.

13.35 We are concerned above all that any transition from oil to coal should be gradual in view of the unacceptable environmental damage which could result from a crash conversion programme. Such a programme could, for example, exacerbate problems which may face local authorities in their role as regulatory agencies in the area of pollution control. We discuss these problems in Chapter 19.

13.36 A recurring theme throughout our Study has been the need for mechanisms which will translate the implications of national energy policy at the local and regional level. A striking feature of the evidence submitted to us is the general unawareness of local authorities of the possible scale of re-entry of coal into the industrial market. This was most marked during our visit to Manchester. This unawareness is the more serious given the projected quadrupling of industrial coal burn to which we have already referred. We recommend strongly that the Government should take the lead in stimulating a continuing and systematic dialogue with the N C B, representatives of industry and local authorities. The Department of Energy should monitor progress towards achieving meaningful levels of oil substitution in its regular meetings with the N C B.

13.37 We are aware that the N C B have not themselves seriously marketed coal outside the power station sector for many years. Progress in substituting coal for oil in industry will require a major marketing effect by both the N C B and the boiler manufacturers working in close co-ordination.

13.38 We welcome the package of measures announced by the Government in March 1981 whereby £50 million will be committed over 2 years for grants to industry towards the cost of converting boilers from oil to coal. This assistance will be provided in the form of grants up to 25% of capital cost which could lead to an increase in coal burn of about 2 million tonnes a year. This package should help to overcome current problems of capital constraints and should facilitate a gradual rather than a crash transition. We recommend that the Government itself should take the lead in establishing the required momentum in the market. The role of the Property Services Agency (P S A) in the field of energy conservation provides a model. The P S A's annual budget for minor works enables it to determine priorities for energy conservation investment across the whole Civil Estate. The Government should now be seen to give a similar lead in the use of coal in the public administration sector. Where appropriate, specific installations could be used as demonstration facilities with costing and performance data being made available to other interested consumers.

13.39 Some of the constraints on the re-entry of coal into the industrial market may be overcome if F B C technology is successfully developed and commercialised soon enough. There is a real risk that the U K's initial lead in this field is being eroded. Without Government action, systems are unlikely to be available commercially in time to play a significant part

in U K oil substitution during the 1980s or to enable the U K to compete successfully in what should be a growing export business. We recommend vigorous action by Government to promote the commercialisation of fluidised bed technology. One possibility would be the funding of demonstration projects for the most promising systems – as an inducement to speeding up trials, identifying the best systems and then commercialising them. We also recommend an integrated effort by the N C B and boiler manufacturers, with appropriate participation by the Departments of Energy and Industry to formulate an agreed programme for commercialisation and to examine export potential. We discuss in Chapter 18 the need for further research work on the disposal of ash produced by fluidised bed combustion and on the solid residues resulting when limestone is added to the bed. This must be pursued equally vigorously as an integral part of the development programme.

COKE OVENS AND MANUFACTURED FUEL PLANTS

Present pattern of market demand

13.40 This market has been in substantial decline in recent years. In 1980/81 it used 11·3 million tonnes of coal principally for steel-making, compared with 27 million tonnes in 1955.

13.41 There is a concentration of plants in the Northern, Yorkshire, Midlands and South Wales regions, with smaller numbers of plants in the North West and Scotland. All the large British Steel Corporation (B S C) coke ovens are located in the coalfields. The typical location of major steel works on the coast, and of other plants in or near the coalfields, is unlikely to change significantly over the next 20 years.

Future prospects

13.42 The evidence presented to us indicated a limited potential for change. Table 13.5 indicates some recent estimates of the future scale of the market, but this is critically dependent upon the fortunes of the one large customer, B S C. Many factors will operate to limit the expansion of coal in this market, including: the current steel recession which is the deepest for 40 years; the existence of a substantial surplus of steel making capacity; a past and projected decline in the ratio of steel consumption to G D P; the amount of electric arc capacity installed; and the efficiency of the steel-making process.

13.43 The future of solid smokeless fuel plant producing for the domestic market will in part depend on the success with which smoke reducing appliances burning bituminous coal are developed and introduced to that market. We would see this as a possible option in certain circumstances, (see paragraph 20.27) for it would avoid both the local environmental impact of the solid smokeless fuel plant, and the energy inefficiencies of the processes.

Table 13.5   Possible coal use in coke ovens in 2000

million tonnes

Origin of estimate	Coal burn in 2000
Green Paper 1978 (1)	27
Energy Projections 1979:	
(a) 1% economic growth assumptions	12
(b) 2% economic growth assumptions	16
(c) 2.7% economic growth assumptions	19
Coal for the future (2)	20–25
World Coal Study (3)	17·5

Source: (1) *Department of Energy 1978*
(2) *Department of Energy 1977a*
(3) *Coal Bridge to the Future 1980*

DOMESTIC

13.44 The size of the domestic market in 1980/81 was about 8.5 million tonnes. Coal has lost ground over the last 10 years principally in urban areas where low-cost gas is readily available. However, in non-gas connected areas (mainly rural), where the alternative fuels are at a higher cost, coal has continued to retain a major share of the market. More recently, with oil prices rising quickly again, and further concern over supplies and rising gas prices, coal has received renewed interest. Over the next 10–15 years it is possible to foresee a rise of coal demand in rural areas, accompanied by a continued decline in coal demand in urban gas-connected areas. The general expectation for the total market is therefore that it is likely to remain at or around its present level for the forseeable future. We discuss the environmental effects of domestic coal burn in Chapter 19.

Chapter 14   Transport, handling and storage

INTRODUCTION

14.1 We have so far examined in Part III the market prospect for coal as a backdrop to the assessment of the environmental implications of coal use. The first element in that assessment is consideration of the environmental issues associated with the transport of coal from the point of production to the point of use and with the transport requirements of particular markets. We have been aware throughout this area of our Study that the transport of coal is part of the wider question of national transport policy. We have therefore concentrated our attention on those aspects of particular significance for the environmental acceptability of increased coal production and use. We examine the respective roles of road and rail, the potential for increased use of alternative modes and their comparative environmental impacts, taking account of the particular requirements of both bulk users and the smaller industrial and domestic markets. Finally, we examine the prospects for improved handling and storage of coal. This area of our Study highlights the points we stressed at the outset (see paragraph 1.47), namely that we have been concerned not only with wide ranging issues of public policy but also with more detailed matters which may not be of major importance in themselves but which cumulatively give rise to concern. We have therefore felt it necessary to consider the latter in some detail in this Chapter.

TRANSPORT REQUIREMENTS OF PARTICULAR MARKETS – RELATIVE SUITABILITY OF MODES

Power stations and other bulk users

14.2 We noted earlier that coal burning in power stations represents well over half the U K's current coal consumption and will continue to represent over half total coal consumption to the end of the century. The characteristics of this market, which are shared by other bulk users like cement plants and major coke ovens, are demand for large quantities of coal and frequent delivery. This is catered for almost exclusively by rail which can supply the large quantities of coal required and which is flexible enough to permit British Rail to vary the supplying collieries and concentrate deliveries as necessary.

14.3 It is the Merry-Go-Round system (M G R) which

mainly handles these large tonnages of coal. This is the single most important advance in recent years in the capability of rail to handle the mass transport of bulk freight like coal. M G R is a system of permanently-coupled wagons especially designed for automatic rapid loading at collieries (1,000 tonnes in less than half an hour) and rapid discharge at the point of use (Figure 14.1). Both loading and discharge are performed while the train is on the move, drawn by locomotives adapted to travel at very slow speeds during this process. No shunting or marshalling of wagons is required as the train travels to the discharge point on a continuous loop which returns to the main line for the journey back to the colliery. Considerable savings can be achieved in the length of track and sidings required at collieries and discharge points. Although a greater area of land is required to accommodate the rail loop than would suffice for conventional sidings, the area within the loop can be used for coal storage. The system also allows for high wagon use. At present the average cycle time for M G R wagons (this is the average time taken for a wagon to be loaded, delivered, unloaded and returned empty for reloading) is just less than two days compared with the seven or eight days average cycle time for conventional wagons.

14.4 There remains considerable scope for substantial increase in the quantity of coal moved by M G R to power stations. The best M G R supplied power stations currently achieve around seventy per cent availability for supplying power to the grid, though the average performance of such stations is much lower. If the average availability to the grid of *all* M G R supplied stations could be increased to seventy per cent, the amount of coal carried by M G R could be increased from 42 to 60 million tonnes per annum, thus demonstrating the scope for increased movement of electricity coal by M G R even without new M G R supplied stations or further conversion of stations to M G R working. Increases in this decade are, however, likely to arise from this latter source with the completion of Drax B and the probable conversion to M G R of eight other power stations (five have approval and three are currently under review), adding about 14 million tonnes to the present annual M G R tonnage. The tonnage of electricity coal travelling in conventional rail wagons is expected to continue to decrease.

Figure 14.1 Merry-go-round system

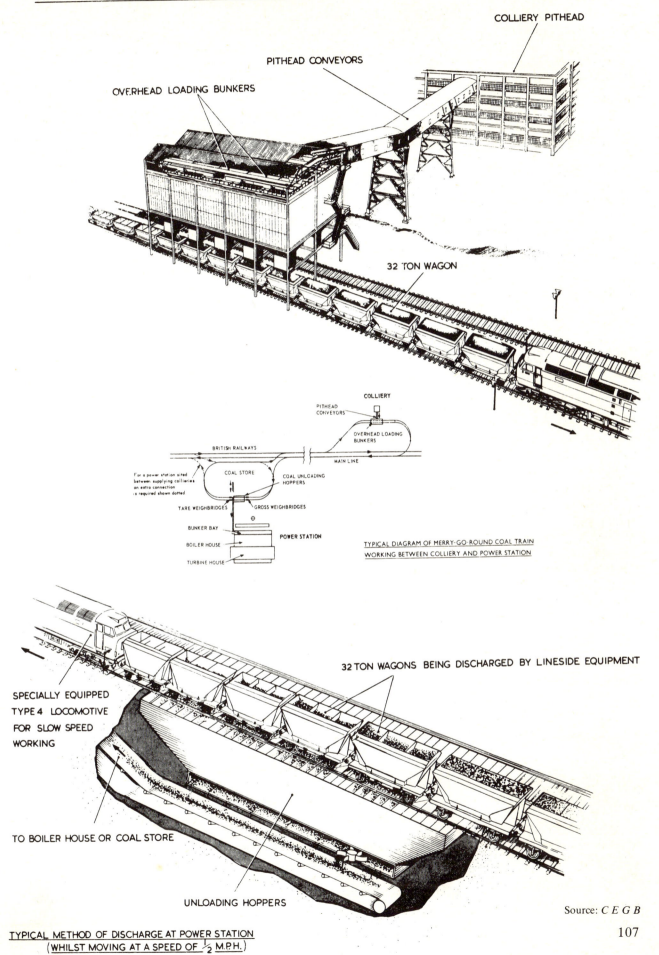

TYPICAL METHOD OF BUNKER LOADING AT A COLLIERY

COLLIERY PITHEAD

PITHEAD CONVEYORS

OVERHEAD LOADING BUNKERS

32 TON WAGON

COLLIERY

PITHEAD CONVEYORS

OVERHEAD LOADING BUNKERS

BRITISH RAILWAYS

MAIN LINE

For a power station sited
between supplying collieries
an extra connection
is required shown dotted

COAL STORE

COAL UNLOADING HOPPERS

TARE WEIGHBRIDGES

GROSS WEIGHBRIDGES

BUNKER BAY

BOILER HOUSE

POWER STATION

TURBINE HOUSE

TYPICAL DIAGRAM OF MERRY-GO-ROUND COAL TRAIN
WORKING BETWEEN COLLIERY AND POWER STATION

32 TON WAGONS BEING DISCHARGED BY LINESIDE EQUIPMENT

SPECIALLY EQUIPPED
TYPE 4 LOCOMOTIVE
FOR SLOW SPEED
WORKING

TO BOILER HOUSE OR COAL STORE

UNLOADING HOPPERS

Source: *C E G B*

TYPICAL METHOD OF DISCHARGE AT POWER STATION
(WHILST MOVING AT A SPEED OF ½ M.P.H.)

107

14.5 The use of M G R for non-electricity coal is currently restricted to certain major coke ovens and cement works and amounts to about 5 million tonnes of coal per annum. This is expected to increase though there are certain constraints. M G R is only feasible for large consumers and for the supply of small sizes of coal where degradation is not a significant factor. Amounts of the order of 1 million tonnes of coal per annum (four 1,000 tonne trains per day) must be involved for full M G R to be economic. Modified M G R (M G R trains loaded conventionally rather than by the expensive overhead rapid load system, thus reducing capital outlay), is viable for 100,000 tonnes of coal per annum (two 1,000 tonne trains per week). There are also problems relating to maximum capacity. The M G R system is necessarily high volume and consequently, once maximum capacity is reached, a further high volume unit (in the form of loading or unloading facilities or an additional M G R train) is required even if the increase in throughput is relatively small. This necessitates much larger additional capital investment than a modest increase in throughput would normally warrant.

Industrial and domestic market

14.6 The (non-bulk) industrial and domestic market is very different in character to that of the bulk user. The transport effort involved is considerably greater than its share of total coal disposals (around 30%) would indicate. Wider geographical dispersal of consumers and relative lack of rationalisation of the transport system for these sectors results in smaller amounts of coal being transported longer distances over a wider area. A major re-entry of coal into the industrial sector will increase the need for coal and ash transport. It will be particularly necessary for local authorities especially, but also industrialists, industrial organisations and the N C B, to make early provision for this changeover. This will involve safeguarding sites for depots which will need to be well located in relation to the major highway network, preserving rail links and making sure that the necessary road infrastructure is available. If action is to be taken to minimise environmental effects, preparations should begin now rather than waiting until much higher levels of coal burn have been attained.

14.7 In addition to the dispersed nature of the industrial and domestic market, there are two further factors which increase requirements for the transport of coal to those markets. Firstly, the more specialised the coal, the longer the distance it may be expected to travel because the supplying pits and regions do not each produce the full market range of coals. Different types of coal vary in thermal properties. chemical constituents, ash content, size, friability and hardness. This is discussed in more detail in Chapter 16. Thus the electricity industry consumes mainly bituminous coal in smaller sized grades; the iron and steel industry requires specialised coking coals; while large or graded lumps, as well as high quality smokeless anthracite, are required by the domestic market.

14.8 The second factor relates to pithead prices, particularly for industrial coal, which can vary considerably. The consumer may find it more economic to buy from a low-cost area further away than from a high-cost area in his locality, even though additional transport costs are incurred. By comparison, power stations tend to take their coal from the nearest collieries because of the need to minimise the average delivered price of coal per therm. Although this need exists for other markets, other factors will more frequently come into play to offset the advantages arising from minimising transport costs in these markets. For example, with regard to siting, factors other than the proximity of coal supplies will be considered and may result in a decision to locate the plant away from the coalfield. Another example is in the manufacture of coke where certain blending requirements relating to coal quality may dictate that no more than a certain portion of the locally available coal be taken, the rest needing to be transported from more distant sources.

14.9 Both rail and road play essential parts in supplying the industrial and domestic sectors. Rail can be used for larger requirements and for supplying coal depots. Lorries, on the other hand, which offer speed (over short distances), convenience and flexibility, can make fast deliveries particularly at points near to a colliery and can deliver direct into the boiler-house or store place. Individual rail-borne delivery for the smaller consumer is not a practicable proposition against the capital outlay involved in trans-shipment and bulk stocking. Such customers are fed by road direct from the point of production, or from Coal Concentration Depots (large mechanised trans-shipment points for coal from rail to road) or smaller coal depots. In addition, road transport is often necessary from opencast sites where, because of their temporary nature, rail links seldom exist. Small amounts of spoil and ash are also transported by road.

(a) Conventional rail transport

14.10 Where rapid discharge and M G R is not economic, or demand is insufficient, coal transport by rail to (non-bulk) industrial and domestic users will be by conventional wagon traffic. Compared with the situation prevailing in the mid-1960s, there are now approximately half the number of conventional wagons moving about one-third the quantity of coal. British Rail have in recent years been turning their attention to improvements in this area. The first of these is the implementation of the T O P S (Total Operations Processing System) computer system for the monitoring and allocation of rolling stock. This has been a major policy decision considered essential for the efficient running of a modern rail network. A major benefit is intended to be an improvement in conventional wagon utilisation. In addition to T O P S, British Rail have been rationalising much of the most uneconomic coal traffic, which mainly consisted of supplying domestic coals to wholesale rail depots. The number of these depots has now been vastly reduced and many of those remaining have large enough capacity to receive full train loads ("block trains") and are fully mechanised for efficient unloading and hence improved turn-around of wagons. Further, the new "Speed link" wagonload service has been introduced whereby overnight transits are available along fixed routes with no intermediate marshalling. British Rail have simultaneously been encouraging expansion of the use of privately owned railway sidings. For the moment, British Rail intend to retain their conventional wagonload services, although doubts about the long-term viability of this service were expressed to us.

(b) *Rail and Road: Coal concentration depots*
14.11 Where coal markets, such as the smaller industrial, commercial and domestic markets, are not rail-linked, much of the coal is transported by rail to coal depots for onward transport by road. Of these depots, there are 58 Coal Concentration Depots and 319 other coal depots. Coal Concentration Depots handle about 2·6 million tonnes of solid fuel per annum; they are generally fully mechanised for wagon discharge, the coal being sent direct by conveyor to bunkers or stocks. Facilities exist for coal to be loaded from bunkers in bulk or from bagging hoppers. The majority of coal depots are, however, smaller, less mechanised depots or station yards which also handle about 2·2 million tonnes of coal per annum. Distribution from coal depots is normally to consumers within a 25 mile radius.

14.12 Coal depots (both Coal Concentration Depots and other depots) are spread widely throughout the country (maps are at Figures 14.2 and 14.3). They would seem to be adequately placed to serve future industrial needs assuming that the present industrial geographical distribution does not change significantly. There would also seem to be adequate provision with regard to capacity. Many Coal Concentration Depots are currently working well below capacity; this is also true of small coal depots. In addition, of the 1,400 British Rail owned coal stocking yards around the country, some 600 could be re-connected to the railway system, according to evidence received from British Rail, though at a cost which in many cases would be significant.

14.13 The trend towards Coal Concentration Depots will probably continue and it is here that the long-term future of a large rail stake in non-electricity coal probably lies. By concentrating supplies into centralised depots, a high enough throughput can be maintained to preserve the economic attractions of continuing bulk supply by rail. However, it is the coal trade and not British Rail which operates these Concentration Depots and many other depots and it is for the trade to make the necessary capital investments to improve facilities, though British Rail can play an important role in initiating developments. As any expansion in production and use of coal places larger demands on the coal trade, so it is likely that the necessary capital investment will be forthcoming to improve facilities of already existing depots.

(c) *Road direct from colliery*
14.14 As well as coal transport by road from concentration and other coal depots discussed above, a substantial portion of coal (19 million tonnes in 1980/81) travels by road direct from N C B points of production. Some of this is destined for power stations but much the larger slice is for industrial and domestic consumption. It is likely that this sector of coal transport will continue to dominate the smaller scale markets which lie within or reasonably close to the coalfield areas.

(d) *Developments in coal transport to industrial and domestic markets*
14.15 We realize that, as road transport will inevitably continue to play a large role in the transport of coal to the smaller consumer, new developments in this field could be very important in contributing to reduced environmental impact. One example of such a development is the coal tanker. This vehicle carries the coal in a totally closed container from which it is pneumatically discharged direct into the stocking area. So far, the high cost of these specialist vehicles has limited their use, except for deliveries to environmentally sensitive users such as hospitals. However, we hope that further development will reduce costs so that use of this method of transport can increase. It is understood from N C B that further work is continuing on the design of specialised road vehicles for coal.

14.16 A further area into which we would recommend research is the possibility of coal transport by container which can be carried by train or lorry and is detachable from both. The back of the container is hinged and the coal can be tipped out in the same manner as from a tipper lorry. Such containers could be used to deliver coal and remove ash. In addition, if there is to be an increased world trade in coal, containerisation could be used to transfer coal from ship to train or lorry. Containerised coal transport is already used on a small scale in the U K to take domestic coal from Midlands collieries to domestic depots in London and by road to port for shipment to Northern Ireland. There is also some small scale containerised transport by road and rail in the North East of England. A further contribution to reduced environmental impact could be at industrial sites particularly if the container carrying the coal could be slotted directly into an automatic boiler feed system.

Figure 14.2   Fully mechanised coal concentration depots

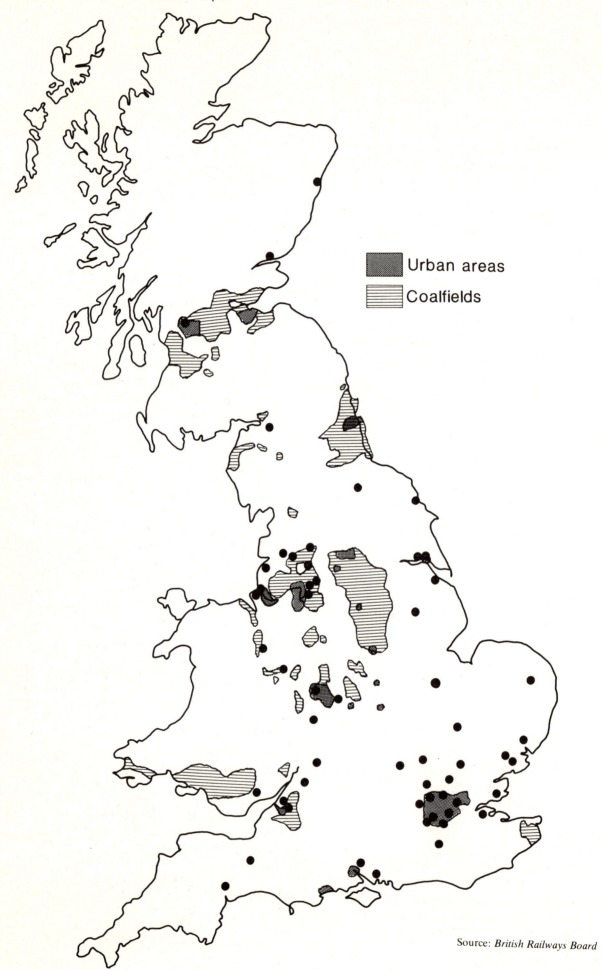

Source: *British Railways Board*

110

Figure 14.3   Partly mechanised coal concentration depots

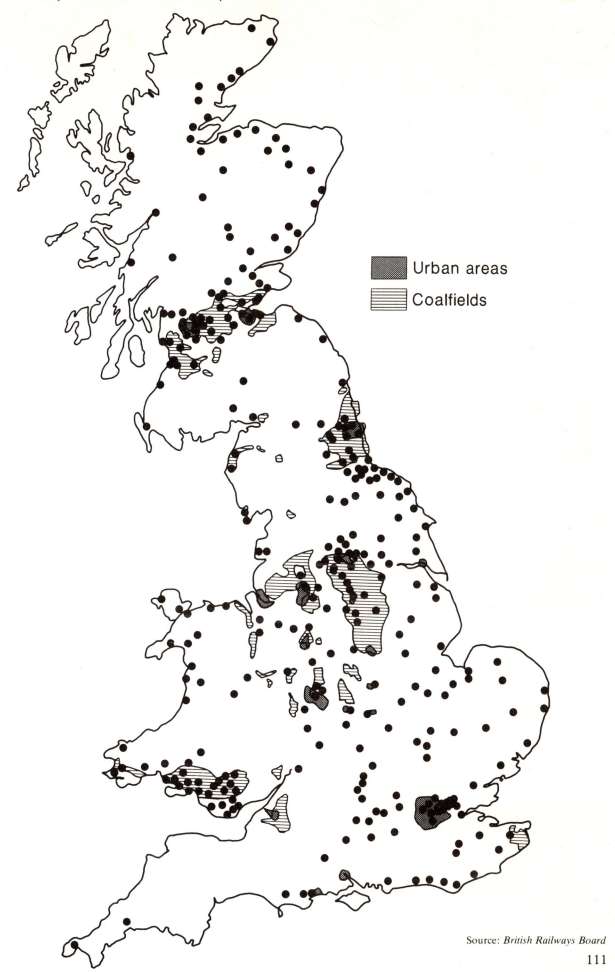

Urban areas

Coalfields

Source: *British Railways Board*

111

14.17 Containerised transport can reduce the amount of handling and so reduce degradation of the coal. This is particularly applicable to anthracite where degradation is very costly. Intermodal transfers between road, rail and ship are facilitated and cost less than conventional intermodal transfers. This method of transport is most likely to be cost effective on long distance journeys (over 150 miles), where a coal customer requires small amounts of coal (15–40 tonnes). The one or two containers required can form part of a British Rail container-train load and be dropped off at the customer's or a N C B siding. This may be particularly applicable to replacement of British Rail's 16 tonne wagons which are being phased out over the next few years. They are currently used by small industrialists and domestic coal merchants. In addition, high sided containers, in contrast to unevenly loaded tipper lorries, are unlikely to spill their load and they can be stacked enabling cleaner storage of coal at depots and in small industrialists' yards, thus reducing the need for unsightly piles of coal or storage bunkers.

14.18 There are, however, some disadvantages. Containerised transport is much more expensive than tipper lorry transport because of the higher cost of the vehicles, and the cost of the loading and unloading equipment. Stocking coal in containers ties up large amounts of capital in the containers and for small industrialists the cost of equipment to move the containers might be too great. Also, not all containers are high sided and so the spillage problem may still arise. About 85% of the tasks carried out by N C B and N C B-hired tipper lorries can be categorised as short-haul within a 25 mile radius of base/operations. This often necessitates a tipper being employed on a variety of tasks during the course of a working week, for example, movement of fuel to washeries and to stocking sites; delivery to industrial consumers; and dirt disposal. The same flexibility could not be achieved with a container type of vehicle. Thus, container vehicles would not always be fully in use.

14.19 A potentially attractive aspect of this kind of transport is "two way containerisation" by which the coal container makes return journeys carrying ash. The idea of any return load is appealing because it would significantly lower the cost of transport. However, opportunities for return loads for coal and ash traffic are limited because of the nature of the material and hence the type of wagon used, the routes taken by the coal and need for wagon cleaning. The same problems occur with road transport although this type of transport is more flexible and return loads with materials other than ash are fairly common. There is also an imbalance in the two traffics – the ash content of industrial and domestic coal is normally between around 5% and 10%. Thus between 10 and 20 times as many wagons would be needed for the coal than for the ash.

THE ENVIRONMENTALLY PREFERABLE MODE

14.20 We have seen that transport of coal is dominated by rail and road. Although other modes of transport (coastal shipping, waterways and conveyor) can make an important contribution locally to alleviation of environmental impact, viewed on the national scale the critical issues centre upon the relative contribution of rail and road. We have concluded for reasons which we discuss later that in most circumstances, where both rail and road are available, rail has environmental advantage over road for the transport of coal. We therefore recommend that, in general, rail should be the chosen method of transport and that, in particular, power station siting (with the exception of stations with coastal or waterways access) must be compatible with the most effective use of rail transport for the provision of power station coal.

14.21 Road transport should not, however, always be considered as invariably environmentally worse than rail transport. We have, therefore qualified our recommendation by saying that rail is preferable in most circumstances. Where private sidings are available, rail has the environmental advantage because it can carry the traffic direct from origin to destination. Motorways, however, like main-line rail are largely away from centres of population and to this extent environmental effects from lorries on motorways need be no worse than those from main-line trains, though this depends to some extent on our motorways being kept in a state of good repair and on the resulting cost. In addition, motorways act as a method of dispersing lorries around an area because lorries use different exit and entry points. The involvement of rail transport for part of the journey, which involves the use of coal depots, has the effect of concentrating lorry traffic. This intensifies environmental impact in the vicinity of the depot. Where customers are situated on the periphery of an urban area, the use of a central coal depot might lead to more urban traffic than would result from direct delivery by lorry. Thus, where there is choice between direct delivery by road and a journey part by rail and part by road, there could be environmental advantages and economic reasons such as the avoidance of costly inter-modal transfers in favour of the former.

14.22 However, with this qualification we regard the maximum use of rail to be essential to minimise environmental impact from coal transport. We therefore fully endorse the view expressed by the Association of Metropolitan Authorities (A M A), in its report on road and rail freight (1980) that "it should be the objective of all local and central government authorities to increase this type [coal and other bulk movement] of rail traffic by encouraging all users to extend their railhead facilities". We recognise that the rail infrastructure will not be extended significantly in the foreseeable future but, as the A M A itself pointed out, "a greater use of the existing network can be achieved if strategically located sites for railhead development are identified. Planning authorities together with British Rail, should examine the potential for development of such sites, whether they be greenfield with the possibility of easy access to the existing rail network, former British Rail depots, goods yards, or sidings". We would underline the importance of the role of local authorities if rail transport is to be maximised.

14.23 Our recommendation is in line with the Report of the Armitage Committee (Department of Transport 1980) which concluded that "there can be little doubt that the overall impact of rail freight on the environment is substantially less than that of road freight even when account is taken of the much greater amount of freight which road carries". In arriving at our recommendation, we have compared the impact of rail and road taking account of energy efficiency, safety and health, noise and vibration, dust and spillage. We have considered the importance of adequate investment in the rail freight system and we have taken into account the contribution and impact of alternative modes.

Rail and road: comparative impact

(a) Energy efficiency

14.24 In general, rail is more efficient in the use of energy for M G R and similar trainload traffic. For other traffic, it is more open to question whether rail has a large energy advantage. It is, of course, possible to compare the amount of energy used by road and rail per tonne-mile. On this basis, average energy use by lorries is about three times greater than by rail. But this is not to compare like with like since lorries and freight trains perform very different functions; and the functions which the lorry alone can perform, especially local collection and delivery, need more fuel than other kinds of road haulage. A study by the Advisory Council on Energy Conservation (A C E C) (Department of Energy 1977d), attempts to circumvent this difficulty by comparing energy use for similar traffic flows. It was concluded that: "The conveyance of freight traffic in merry-go-round trains and by other trainload quantities is very efficient in energy terms as compared with road conveyance". This is illustrated in Table 14.1 below.

Table 14.1   Rail and road transport: energy consumption

MJ per tonne kilometre

Mode	Probable average	Likely range
Rail		
M G R	0·60	1·1–0·4
Other trainload	0·65	1·2–0·4
Wagonload	0·96	1·7–0·6
Road		
Coal lorry (14 ton, six wheel rigid truck)	—	2·5–1·4

Source: Department of Energy 1977d

Although it will be apparent that particular traffic can vary considerably from the probable average, the overall picture is one of greater energy efficiency for the carriage of freight by rail. This results from lower rolling resistance and more disciplined movement of rail vehicles. Rail's greatest potential energy advantage over road, however, lies in the greater ease with which it would be possible through electrification to use the most economical energy sources.

(b) Safety and health

14.25 There are inherent difficulties in comparing the safety of road and rail. Precise evaluation is extremely difficult on account of such factors as differences in accident reporting procedures and the difficulty in allocating road accidents to different types of vehicles. We have considered the 1976 topic report on safety produced by British Rail, together with representatives from the University of Loughborough and the Cranfield Centre for Transport Studies, as one of their Environmental and Social Impact Studies (British Railways Board 1976). The conclusions of the study were, however, based on 1973 data, the latest available at the time. The Armitage Committee have produced more recent estimates although on a less detailed basis. Paragraph 179 of the Armitage Report examines a transfer of freight from road to rail which would lead to an increase of 50% in rail freight (around 6,000 million ton miles which would of course be much larger than could be achieved by coal freight alone). If this reduction took place entirely on motorways, reducing lorry traffic there by about a fifth, about 15 lives a year might be saved. If the reduction were spread equally over all roads, about 40–50 lives per year might be saved. Thus the figure would probably be within the range 15–50 lives per year. Table 18 of the Armitage Report shows that lorry involvements in fatal accidents per 100 million vehicle miles fell by 43% between 1970 and 1979.

14.26 The conclusion we draw from the evidence made available to us is that rail transport is relatively safe because it is confined to its own exclusive tracks (except when they cross public roads) and because of the sophisticated guidance and control systems that have been developed. In adjusting the Armitage approach to coal movements, an important consideration is the type of lorry that would be used. With present weight limits, the maximum pay load would be little more than 20 tonnes. An expansion in the use of coal by industry of 30 million tonnes per year, if conveyed over an average distance of 25 miles, would thus give rise to about 75 million additional heavy lorry miles per year (including the return journey) with current maximum weights. The Department of Transport told us that the total fleet of such vehicles (that is with a maximum payload of 20 tonnes) presently does about 3,000 million miles and is involved in about 200 fatal accidents per year. Assuming the risk of accident involvement of these vehicles is similar to the existing fleet, about 5 additional fatal road accidents could be expected. On the other hand, if all this coal were transported by rail, the Department of Transport calculated that the level of hazard would be reduced to about 1 fatality per year.

14.27 Many steps have been taken since the mid-1960s on lorry safety and these have contributed to an improving accident record. They include safety features in lorries themselves; the mandatory annual testing of lorries; the testing of drivers to a higher standard; restrictions on drivers' hours; the quality licensing system; and the construction of more motorways and other high quality roads. Modern heavy lorries are generally subject to stringent safety requirements, in particular in relation to braking, and the Department of Transport told us that further tightening of standards can be expected in the future.

14.28 The transport of coal by road and (non-electric) rail inevitably involves some degree of atmospheric pollution by the combustion products of the fuel, though this can be reduced greatly by regulation and proper maintenance. Visible diesel smoke is more a nuisance than a health hazard, but minute quantities of polycyclic hydrocarbons are released to the air and these, it is currently believed, must involve a correspondingly minute risk of lung cancer. The total contribution of all road traffic is, however, unlikely to amount to as much as 1% of the total risk of the disease. Other combustion products of diesel fuel do not, so far as is now known, significantly affect health.

(c) *Noise and vibration*
14.29 Both rail and road transport give rise to noise and vibration; they are probably two of the most significant environmental impacts attributable to rail transport; and, in the case of road transport, they can reach serious proportions particularly in the vicinity of urban roads. A 1977 Study (Harland D J and Abbott P G 1977) concluded that road traffic noise is the most serious form of noise received in homes in the U K, and that it disurbs more people than all other forms of noise. This conclusion was fully accepted by the Noise Advisory Council in their Report in 1978 (Noise Advisory Council 1978) on noise from road and rail transport.

14.30 The most comprehensive survey undertaken to assess the response of householders to railway noise was carried out in Britain by the Institute of Sound and Vibration Research of Southampton University, funded by the Science Research Council and the British Railway Board (Walker J G and Field J M 1977). The results emerging from analysis of the survey information (the final report has yet to be issued) indicate that annoyance increases steadily as the noise level increases and that there is no threshold for the onset of annoyance. A comparison of reaction to

railway noise with reaction to noise from other transport modes shows that for given noise levels, railway noise is less annoying than from road transport sources. The relatively greater acceptability of railway noise increases as the noise level at which comparison is made increases.

14.31 A further advantage attributable to rail is that, for various historical reasons relating to patterns of urban development, rail noise affects residential areas less than road traffic noise. This is largely concerned with the physical proximity of the noise source, roads being nearer to more centres of population compared to rail. All these conclusions are borne out by the Noise Advisory Council who have concluded "In noise terms, rail transport has undoubted advantages: noise levels are frequently lower, the distribution of population is in its favour; and public response appears generally more favourable than to road traffic noise" (Noise Advisory Council 1978).

14.32 There has been much less research on the effects of vibration. Both rail and road transport generate two kinds of vibration: ground-borne vibrations originating in the track surface and air-borne vibrations caused by the sound emitted. The effects of traffic induced vibration on people depend on many factors and some people are more susceptible to vibration than others. The possible effects of such vibration on human health and comfort are unknown. The Department of Health and Social Security recognises the need for research to see if there is any substance in claims that such vibrations, especially of low frequency, can harm health and sense of wellbeing. As far as damage to buildings is concerned, it appears that air-borne vibration is less significant than ground-borne vibration, though it may be more noticeable to people. The effect of ground-borne vibration on buildings is a matter of unresolved controversy. The most authoritative evidence suggests that traffic vibration does not cause direct failure. Its most important effect is that it may be one of a number of factors that can start damage. People perceive, and are annoyed by, vibration which does no structural damage to buildings. There is particular evidence about the effect of vibration on very old buildings, and there is circumstantial evidence of damage by traffic-induced vibration.

(d) *Dust and spillage*
14.33 Dust nuisance can arise from both rail and road transport; spillage is more normally associated with lorries. Dust can be prevented by sheeting the load or by completely covering it as happens in the rail transport of ash where fully enclosed wagons, similar to those used for bulk cement transport, are used. The load can also be dampened to prevent dust blow. Dust and spillage could, of course, arise from both road and rail transport; however, it is lorry transport which is a particular source of complaints in this area. It is maintained that tarpaulin covers are not used, wheels

114

are not washed properly or not washed at all and lorries are over-filled. This has, however, to be put into perspective; in 1978, Government carried out 30,000 spot-checks on lorries for over-loading; only 2,500 (8·3%) were found to be overloaded. However, lack of attention to all these matters has led to dust and dirt being deposited on roads and roadside houses giving rise to bitter complaints about the road transport of coal.

(e) *Powers to reduce impact from road transport*
14.34 Powers exist which attempt to reduce the potential impact from road transport. One alternative is lorry routeing or banning by which specific lorry routes can be designed or lorries can be banned from particular roads or classes of road. The issues raised are complex and go far beyond our discussion of coal. We have, however, noted the conclusion of the Armitage Committee that neither national lorry routeing nor a general system of excluding lorries from particular categories of roads would be an effective or practicable means of reducing the environmental effects of lorries in any reasonable time span.

14.35 The Armitage Committee did, however, conclude in favour of local lorry routeing and local lorry bans. They regarded restrictions by local authorities on the roads which lorries may use to be "one of the most important ways in which the impact of lorries is reduced. These local restrictions are not a panacea. They cannot solve the intractable problems which exist where there is no reasonable alternative route. Local regulation however avoids the difficulties of applying universal solutions. Its great strength is that it can be tailor-made to fit local circumstances by those who have detailed knowledge. There is general agreement about the need for local authorities to have power to control where lorries go."

14.36 Of particular relevance to coal (and ash and spoil) carrying lorries is the question of the enforcement of measures to prevent dust and spillage. When looked at in the context of the totality of the coal production and use cycle, the episodes which cause problems in this area are relatively minor. We outlined some of these in paragraph 14.33 above. However, according to our evidence the annoyance which can be caused by these incidents is considerable. It is therefore important that control measures are adequate and properly enforced.

14.37 The current legal requirements are that every load carried on a motor vehicle or trailer must be secured, that is, loose loads must be sheeted and all loads must be physically restrained, other than by their own weight, if necessary, to ensure that no danger or nuisance is caused to persons or property by the load shifting or any part of it falling or being blown from the vehicle or trailer. The regulations also provide that no motor vehicle or trailer should be used for any purpose

for which it is so unsuitable as to cause or be likely to cause danger or nuisance. (Regulation 97 of The Motor Vehicles (Construction and Use) Regulations 1978, SI No 1017). These Regulations do not specifically stipulate that loads carried in open vehicles, such as tippers and skip vehicles, should automatically be sheeted; this is a matter for the judgement of the owner operator or user of the vehicle.

14.38 At present, there is no specific legislation requiring the wheels of vehicles to be washed as a means of avoiding the fouling of the public highway. However, planning consents for new mine authorisations and for new opencast sites, normally have a condition that all reasonable steps should be taken to ensure that all vehicles leaving the land should not emit dirt or deposit slurry, mud, coal or other material from the land on the public highway. This is a legal requirement, and in practice it means the installation and proper maintenance of wheel washing facilities.

14.39 The N C B's current practice is that sheeting should be used for Board owned vehicles, and those directly hired by the Board, in the following circumstances:
   (i) when an Area, Opencast site or plant is requested to do so either by a Local Authority or the Police following consultation between the parties concerned;
   (ii) in those cases when it can be seen that a load constitutes either a danger or nuisance within the meaning of the Regulation referred to above.

The Board have installed wheel washing facilities at a large number of collieries and opencast sites. In addition, the Board have increased their road sweeper fleet in an attempt to keep entrances and exits free of mud and spoil.

14.40 There are, however, a number of complications involved in both the practice and its enforcement. The time factor involved in sheeting a lorry can greatly reduce productivity and hence increase cost. N C B maintain that the increase in cost per trip, if sheeting were to be imposed on a vehicle currently unsheeted, would be around 33%. Moreover, the cumulative effect of sheeting with its consequential reduction in throughput would mean increasing the size of the vehicle fleet, adding to road congestion and fuel consumption. There are also enforcement problems, particularly in the case of sheeting of lorries owned by contractors at opencast sites and other coal lorries not owned by the N C B. Fewer problems have been experienced with wheel washing. However, due to the washing process taking a few minutes, there is occasionally vehicle congestion and driver impatience.

14.41 We are very much aware of the degree of annoyance created by lorry dust and dirt. However, there are no simple solutions. We are not in favour of wider and more detailed regulatory powers in an area of such relatively minor day to day activity. Nevertheless, we recommend that current best practice is put onto a more formal footing and applied in a more consistent manner. We would not advocate that sheeting and wheel washing should be obligatory in every instance. Costs can be high and it would therefore be unreasonable to demand load sheeting and wheel washing regardless of environmental need. Thus washed coal would not need to be covered to prevent dust, though proper supervision would be required to prevent over-fill; and a colliery with clean metalled roads would not require wheel washing facilities. We recommend, however, that N C B draw up a code of practice for all their own vehicles leaving their sites. Such a code would be available to all management personnel and to local authorities who could then play a role in monitoring its implementation on the public highway. Adherence to the code could also be a condition of contract for N C B outside contractors. The code of practice would include which loads should be sheeted and in what circumstances wheel washing facilities should be installed. It should lay down stringent rules of tidiness and cleanliness at and around N C B premises with provision for appropriate N C B inspection. There is the prospect in the longer term that research will produce automated load sheeting devices with consequent savings in time and effort. We understand that the N C B's Road Transport Service has some automated devices under evaluation trial. We recommend that N C B associate themselves with and support research in this area to make sure that full account is taken of their requirements.

Rail investment

14.42 A key factor in the environmental acceptability of increased coal production and use will be maximum use of rail transport. It is important therefore that investment in railway track and rolling stock should not be a constraint on the transport of coal by rail. On the basis of British Rail's estimate of their share of the future coal market (based on projections of coal demand ranging from 128–170 million tonnes) they have calculated that, for the lower end of the range, investment in wagon building would need to be doubled from the present projected levels from the mid/late 80s onwards, and that substantial further investment beyond their present plans would be required for locomotives. At the other end of the range, the wagon building programme would need to be quadrupled by the next decade.

14.43 Adequate provision for rail investment will be essential to ensure its continuing availability for coal transport as the market increases. We therefore recommend that investment in railway track and rolling stock should proceed at a pace compatible with the plans of N C B and C E G B for the mass transport of

116

coal. Steps should be taken to ensure that, where profitable opportunities for shifting coal appear, such investment by British Rail is given proper priority, and the investment programme of British Rail, as approved by the Government, should be large enough to make such proper provision possible.

Alternative modes

14.44 In addition to rail and road, other modes of transport are involved in the transport of coal. These include coastal shipping, waterways, pipelines, conveyors and aerial ropeways. Largely because of the highly developed nature of our roads and, in particular, of our rail system, the contribution of these modes is small and their use is rarely economic. They do not therefore have the scope for making any great ameliorative impact on the environmental effects of the transport of coal generally. These other modes, however, have an important role to play in reducing environmental impact in the locality in which they operate.

(a) *Coastal shipping*
14.45 Coastal shipping is largely used between collieries on the East coast and the Thames power stations. In 1978, the total inwards movement of coal, coke and briquettes (foreign imports and coastwise inwards) at the ports in Great Britain was 6·5 million tonnes; the total outwards movement (exports and coastwise outwards) was 9·1 million tonnes. This mode of transport has many environmental advantages in enabling coal to be transported with a minimum of impact. However, inter-modal transfers can make this method uneconomic and it tends therefore to be attractive largely for transport to and from coastal and estuarial installations and collieries.

14.46 Any increase in this method of transport, including increased imports and exports, would of course be constrained in its upper limit by port handling facilities for coal. We understand from the National Ports Council, who point out that their figures are only estimates which will have a range of values depending upon specific load conditions, that the current capacity for handling coal imports is around 10·5 million tonnes of which only 50% (all British Steel Corporation (B S C) facilities) can handle ships in excess of 50,000 dwt (dead weight tonnes). In addition, there is spare capacity of about one million tonnes at the B S C bulk terminal at Immingham which currently handles only imported ore; this terminal also accepts vessels in excess of 50,000 dwt. There are other bulk berths which do not currently handle coal or are not specifically designed to do so. The capacity of these berths to handle coal obviously depends on the tonnage and the nature of their existing traffic. However, on the basis of current throughputs, these ports have berths which could possibly accommodate up to 1·5 million tonnes of coal inwards, though in ships of less than

10,000 dwt. They all have a rail connection at quay. In addition Bristol (Portbury Dock), which can accept vessels up to 70,000 dwt, could handle about two million tonnes of coal given about two years for the installation of handling equipment. Thus maximum import facilities could therefore amount to some 12 million tonnes immediately, increasing to 14 million tonnes in two years.

14.47 The National Ports Council estimate that the lead time required for the construction of new berths and facilities for additional imports of up to 20 million tonnes of coal would be about 4 years, including, say, one year for planning procedures. The total costs for the construction of new facilities capable of a throughput of 10 million tonnes would be about £50 million. This covers all costs including marine works, mechanical handling equipment, stockyard, and rail connection to quay, for facilities capable of receiving ships of 150,000 dwt. Facilities able to handle 20 million tonnes of coal per annum would cost about £75 million and those able to handle 40 million tonnes per annum would cost about £100 million.

14.48 The cost of modifications to berths which currently handle coal exports would depend on the method of loading. If the loading method is by conveyor or hoist the cost would be up to about £20 million for two unloaders and modification of equipment, depending on the size of the operation. Export berths at which coal is loaded by conveyor or hoist have a total capacity of about 20 million tonnes, of which Immingham accounts for about six million tonnes.

(b) Inland waterways

14.49 Transport by inland waterways is also environmentally advantageous. However, this mode currently carries only small amounts of coal, less than two million tonnes a year, and is virtually confined to the Aire and Calder Canal in Yorkshire and the Port of Goole, where a number of collieries are located adjacent to navigable waterways on which coal-fired power stations have been built. Waterways transport is very energy efficient: British Waterways Board research has concluded that use of oil in inland waterways commercial craft is over four times more efficient than road in terms of tonne/miles per gallon and 25% better than diesel-fuelled rail carriage. Further advantages are the high safety factor and negligible amounts of noise and vibration, though there can be dust problems in loading and unloading operations. There are, however, distinct constraints in the economics involved. This would be especially so where either the despatch or reception points were not on the waterside and where inter-modal transfers would be necessary. In these circumstances, we can recommend only that new coal projects should always be carefully looked at, on a case by case basis, to examine the possibility of their supply by waterway.

(c) Pipelines

14.50 Pipeline transport of solids usually takes the form of a slurry, small size solids mixed with water, propelled hydraulically along the pipeline. Other possibilities include using a liquid other than water in which to suspend solids, or pneumatic capsule pipeline whereby solids are carried in wheeled trucks (capsules) enclosed within a duct or pipe and propelled by a forced draught of air. There are no examples of coal transport by pipeline in this country; the 1960s experiment supplying Wakefield power station from a nearby colliery was abandoned chiefly due to problems of de-watering the slurry. The only example where pipelines are used in the coal-related field is the transport of ash from power stations to disposal lagoons. Such pipelines currently carry ash up to 10 kilometres from the power station, although a projected scheme envisages a pipeline of 25 kilometres. Pipeline systems are, however, more common in the United States where the usually more economic alternative of an extensive rail network is not available. Such pipelines can be as long as 440 kilometres (in Arizona), transporting 5 million tonnes of coal per annum, or the 150 kilometre pipeline in Ohio, carrying 1·5 million tonnes of coal per annum. Research on pipelines is currently being undertaken in this country by the Departments of Environment and Transport's Transport and Road Research Laboratory (T R R L).

14.51 Pipelines are quiet and unobtrusive and have many attractions for transport through environmentally sensitive areas. Environmental impact is mainly connected with possible pollution arising from the water used in slurry pipelines. The cleansing of this water would probably require lagoons covering a substantial area. There are other important constraints. Capital cost is high and flexibility is low. Origin and destination have to be fixed; maximum tonnages and flow volumes are invariable. Where installations are involved requiring an uninterrupted supply of coal, extensive stocks are required and possibly an emergency back-up system in addition. Large quantities of water are needed, and in the case of coal, thermal drying is necessary on reception, requiring additional energy use and adding significantly to the cost of this method of supply. We therefore concluded that pipelines were generally not feasible in the U K.

(d) Conveyors and aerial ropeways

14.52 Delivery from collieries by conveyor (surface or underground) and overhead ropeway accounts for 7 million tonnes per annum (1980/81 figures). These methods, which are normally restricted to situations where colliery and consumer are adjacent, are mainly used for power stations (for example Longannet power station in Fife which is coaled by four separate mines feeding coal on to a centralised conveyor system). Other uses include the supply of coal to smokeless fuel plants and coke ovens and the disposal of ash to land close to power stations. In the latter case, where open conveyors are used, the fine dust is conditioned by mixing with water to improve handling and prevent wind-blown dust nuisance.

14.53 These methods of transport can be both economically and environmentally attractive, although there are environmental drawbacks in the form of visual intrusion and dust. There are, however, several constraints on their use. Although there have been significant technical improvements in conveyors over recent years, the method is restricted to distances of a few miles; beyond this, costs rise quickly and efficiency falls. In addition, they lack flexibility with regard to source of supply and there can be problems of obtaining the necessary right of way over land not owned by the N C B or the consumer.

FAIR COMPETITION BETWEEN THE MODES

14.54 Several written and oral submissions drew our attention to the issue of fair competition between the various modes of transport. This area goes much wider than coal and to this extent is outside the scope of our Report. However, we felt that some comment was appropriate in the light of our recommendation that, in most circumstances, rail should be preferred to road for the environmentally more acceptable transport of coal.

14.55 Environmentally more damaging modes of transport should not be at an economic advantage over more benign modes through hidden subsidies – for example, by not paying their full share of operating and maintenance costs. Heavy lorries are most criticised in this regard. It would be inequitable if lorries used free or subsidised roads while the railways paid for their track and other costs in full. We therefore fully support the Armitage Report's recommendation that lorries in general, and each class of lorry, should pay in taxation at least the road track costs which they impose. Making lorries pay their track costs does not however make the terms of competition between road and rail fair. This depends as much on the rail side of the equation. The allocation of costs within rail operations is, in practice, as difficult as within the road system. A system of avoidable costing is applied to rail freight under which it pays only the costs which would disappear if rail freight services were discontinued. If the same system were applied to lorries, their road track costs would be reduced. Moreover, rail freight has been subsidised in recent years. Between 1975 and 1978 "non-passenger" rail, of which freight is the larger part, received total grants of over £100 million. Rail freight also enjoys a continuing subsidy from the write-off of previous capital investment.

14.56 Judging fairness of competition between road and rail is therefore a complex issue. All the advantages certainly do not lie with the lorry as is sometimes

assumed. Nevertheless, it would be wrong to increase whatever distortions exist in competition between the two modes by allowing certain classes of lorry to get away with paying less than their allocated road track costs. We therefore strongly support present Government policy of ensuring that there is fair competition between the various transport modes. However, in line with our recommendation preferring rail to road, we would also fully endorse the Armitage Committee's conclusion that on environmental grounds we would be content if any financial anomaly in competition between road and rail acted in favour of rail.

Section 8 grants

14.57 Grants are available to encourage rail rather than road transport under Section 8 of the Railways Act 1974, to meet up to 50 per cent of some of the capital costs involved in the construction of new rail connections. Section 8 grants are discretionary and are given where there is evidence of worthwhile environmental benefit to localities resulting from the removal of roadborne freight; where the investment proposal would not go ahead without the grant; and where new traffic or existing traffic would otherwise be lost to rail. Thus the objective is to achieve, within the limits of commercial viability, environmental benefits which could accrue from the use of rail rather than road transport. Ninety grant-aided projects are in operation or under construction at a cost of around £28 million.

14.58 We fully acknowledge the importance of these grants in encouraging more environmentally acceptable transport of coal. We particularly welcome and endorse the Armitage Committee's recommendations that to make these grants more effective they should be increased to up to 80 per cent where the environmental benefits justify such a high rate and that the standard grants should be increased to 60 per cent. We also welcome the statement in the House of Commons by the Secretary of State for Transport on 17 June 1981 (Hansard, Column 1087) that "waterways facilities could in certain circumstances attract traffic which would otherwise go by road and that grants should therefore be available to waterway users on the same basis as Section 8 grants are for users of the railways". He said that the Government would want to introduce legislation to this effect at the first opportunity and we strongly support this.

STORAGE AND HANDLING OF COAL AND ASH

14.59 Adequate storage and convenient handling of coal and ash will be extremely important to the environmental acceptability of increased coal use. Indeed, given the high standards of convenience and reliability of oil and gas-fired boilers, it is likely that developments in improved methods for storage and handling of coal and ash will be critical in stimulating conversion to coal. Such improvements will need to encompass coal storage in limited space; fully automated coal and ash handling to allow clean operation with a wide variety of coals at reasonable costs; and the development of clean and reliable smaller scale systems for use in smaller industrial operations.

14.60 It is feasible to handle, store and feed coal by fully automatic plant, but the more difficult technology of handling bulk particulate materials is much less well developed than that for liquids and gases. The handling, storage and conveying behaviour of coal is profoundly affected by its particle size, particle size distribution and overall moisture content. Thus, handling systems which are satisfactory and reliable for one grade of coal may be completely unsatisfactory for other grades or for the same grade under slightly different conditions. Claims that handling and storage systems will work for all or many different grades of coal should be viewed against this background. Coal users will need to be aware that the type of coal to be handled determines the optimum conveying method to be adopted in particular installations.

14.61 Pneumatic conveying systems are likely to be of increasing importance in coal handling in view of their flexibility of layout, cleanliness and ease of control. Although these systems have been used for many bulk particulate materials for over 50 years, the relatively recent development of "dense phase" conveying over the last decade or so has given fresh impetus to pneumatic conveying which is taking an increasing share of the handling market. The dense phase system uses compressed air to push the coal along the delivery pipe. The coal forms a slug travelling at low velocity. The other main pneumatic system is the "lean phase" system which uses large volumes of air, at high velocity and low pressure per unit of coal transported, which aerates the coal so that all the particles are in suspension.

14.62 The dense phase system has several advantages. Due to the low velocity of the coal slug, degradation is reduced; consequently abrasion and wear on the pipe system is also at a minimum. Erosion of pipes can be a severe problem for lean phase systems conveying abrasive materials such as coal and ash. The low volume of air used virtually eliminates the problem of separating the air from the coal. In addition, as most dense phase systems use compressed air, advantage can be taken of the same source of air to operate pneumatic valves which can divert coal from one bunker to another in multi-boiler installations. When these valves are used in conjunction with level controllers, the whole system becomes fully automatic. This ability to control coal levels effectively in overhead bunkers means that much smaller bunkers can be used.

14.63 Any automatic handling system requires that the coal can be discharged from storage bunkers without manual intervention. The technology to design bunkers for reliable and controlled discharge, again for a specified grade of coal, has been known for about two decades and there is a wide variety of discharge aids on the market to assist discharge for difficult materials. However, although conventional hopper discharge gives rise to frequent, time consuming and costly problems, the more advanced systems are as yet only applied by a minority of users and plant manufacturers.

14.64 Automated ash and grit pneumatic removal systems are also available not only for the larger more elaborate boiler installations but also for smaller boiler houses with more modest requirements. Systems can be below atmospheric pressure and therefore any leaks in the system are inward, allowing adequate time for repair without contaminating the boiler house environment. Where economic considerations do not permit the installation of a completely automatic system, there are a range of lower priced alternatives where, for example, the ash and grit can be conveyed manually to the disposal pipe which then discharges it into bins or skips fitted with cyclones to prevent air pollution. Also available are mobile suction units for boiler and boiler tube cleaning which can handle both hot and cold ash. This not only simplifies boiler cleaning but also reduces the danger of fine ash being inhaled when manually raking out ash.

14.65 There is a lack of awareness of the most up to date technology even by many of those presently using coal or supplying handling equipment. Increased knowledge leading to the application of existing technology will be essential if convenient, reliable and environmentally acceptable coal handling is to be more widely adopted. We recommend that there should be a vigorous marketing drive both to persuade present coal users to adopt this technology and convince potential users that coal and ash handling can be done cheaply.

14.66 Problems of coal storage are less conducive to technological solution since storage in a limited space of irregularly shaped solids is always less efficient than gas or liquid storage. One possibility being examined by N C B is the use of coal storage in silos. These could be used to reduce the area required for coal stocking, though at higher capital cost. Advantages would include the elimination of dust losses from the stockpile especially during stacking and reclaiming. However, silos would have to be designed to give reliable, blockage free and complete discharge and to resist the sometimes very high stresses on the walls imposed by the coal. Allowable foundation pressures, especially on reclaimed land, would limit silo heights or make necessary expensive piled foundations. The possibility of spontaneous combustion of the coal would also have to be considered.

14.67 Coal stocking at collieries can also be problematic, especially during periods of low demand when there may be a sudden need for extensive, though temporary, coal stocking. Such stockpiles can be unsightly, especially if they are sited in conspicuous locations, as was suggested in some evidence we received and as we saw for ourselves in South Wales. We commented on the unplanned use of colliery sites for stocking in paragraph 7.30. As our recommendation in Chapter 7 makes clear whilst we recognise that stockpiling is a necessary and integral part of coal production, we consider that there is room for improvement in pre-planning, particularly of emergency stocking sites, so that they are properly screened to minimise visual intrusion. In addition, attention to "good housekeeping", is applicable here. N C B could give a lead by ensuring that their subsidiary National Fuel Distributors avoids using sub-standard materials, for example railway sleepers for bunker construction, and that proper attention is paid to site tidiness.

14.68 The whole area of storage and handling of coal and ash is one where there is much scope for research and development. The N C B have initiated a comprehensive programme of work at their Coal Research Establishment, aimed at the improvement of existing equipment and the development particularly of new pneumatic and hydraulic systems. Work completed thus far comprises the successful development of a low-cost lean phase pneumatic conveying system; the testing of a commerical dense phase pneumatic coal conveying system which proved generally satisfactory; and the assessment of various available mechanical conveyors which are not marketed specifically for coal handling.

14.69 Current N C B work on coal and ash handling includes the assessment and development of hydraulic conveying (sluicing) systems; the development of pneumatic ash removal from boiler houses; collaborative development (with manufacturers) of dense phase pneumatic conveying systems for the feeding of coal direct from reception to furnace, automatically and in consistent small increments; an assessment of silos for coal storage; and the assessment and development of containerisation, both for coal delivery and ash removal, possibly eliminating the need for on-site handling of coal and ash.

14.70 The environmental acceptability of future coal use, and, in particular, of a large scale re-entry of coal into the industrial market, will hinge critically upon the widespread commercial deployment of improved storage and handling technology. This will involve a collaborative effort by the N C B and equipment manufacturers. The importance that we attach to this is such that we would recommend that the N C B should ensure that research and development in this crucial area is awarded appropriate priority within its overall research and development budget.

Chapter 15   Coal conversion to substitute fuels

INTRODUCTION

15.1 Coal conversion is the process whereby an alternative fuel is manufactured from coal. Three principal fuels can be produced:

  (i) gas (including, principally, a substitute for natural gas, and also other gases according to the process adopted);

  (ii) liquid fuel (again the precise output, such as petrol, heavy fuel oil and the range of by-products, can vary according to the process);

  (iii) coke and solid smokeless fuel.

A range of chemicals can also be produced from coal, but to the extent that these processes involve similar environmental impacts to gasification, we have not identified any particular environmental issues arising.

CONVERSION FOR GASIFICATION AND LIQUEFACTION

15.2 A distinction needs to be drawn very clearly between coal conversion to S N G and conversion to liquid fuels. Processes are already available for the production of S N G from coal, and although commercial production of S N G has not yet been undertaken, significant parts of S N G technology are already used in other commercial processes. The gasification of light distillate oils to make town gas or synthesis gas for the manufacture of petro-chemical products has been established technology for 20 years; indeed, the great expansion of the gas industry in the 1950s and early 1960s (before the discovery of North Sea gas) was based upon it. It is a very efficient process with low capital costs. The conversion of this gas to methane (S N G) is technically straightforward but is not at present widely practised. Some oil-based S N G is, however, produced as a means of supplementing peak supplies. Under these conditions low capital costs are more important than the fuel costs. In the long term, when S N G will be needed on base load, the lower fuel costs of coal will be more important than the much higher capital costs of its gasification. Further research and development into coal gasification is still in hand, for example, in widening the range of coals that can be used; but if a commercial demand for S N G were to be expressed immediately, the technology is available to meet it.

15.3 The same does not apply to coal liquefaction. The one existing commercially applied technique, in South Africa, is only justified by the particular resource base of that country, and cannot be regarded as sufficiently efficient for any widespread development. A range of other processes is now being developed or tested in the U S A, U K and Europe, but progress is not so far advanced as in the case of S N G.

15.4 In the S A S O L process in South Africa, the coal is first completely gasified and a range of products (liquid and gaseous fuels) is then synthesised by means of the Fischer-Tropsch process (see paragraph 15.10). The residues and treatment methods are thus generally very similar to those of gasification. The newer direct liquefaction processes under development are generally based on reacting the coal with a special solvent and hydrogenating it to make a variety of simpler liquid products, leaving some solid residue. Many variations of process details are possible, with a range of yields of solid, liquid and gaseous products of different types. A common feature of these processes is that there is not a unique product, like methane (S N G), but a complex mix of organic compounds, with a higher proportion of aromatic and cyclic hydrocarbons and nitrogen compounds than petroleum. Some of these compounds may be carcinogenic or mutagenic, but the proportion is likely to be very small. Nevertheless, further work is required on identifying the potential hazards, if any, involved in manufacturing and using these liquefied products and in devising techniques for avoiding exposure to them. Because of the time needed to demonstrate not only the technology of manufacture but also the environmental acceptability of the processes and the products, the large scale adoption of coal liquefaction will require more development than S N G manufacture.

15.5 As far as the U K is concerned, the government has stated that if oil is to be used increasingly for premium applications, coal will play an important part in displacing oil from the non-premium markets (Hansard 22 May 1981, Col.566). At the present prices of coal and oil it would not be economic to undertake coal liquefaction. The future economics of these processes will be very dependent on coal/oil price relativities. It is thought that commercial development could become viable by some time in the 1990s the world's major coal producers (U S A and Australia) but that it is unlikely to be viable in the U K until well into the next century. It would appear, therefore, that any environmental difficulties which could arise are likely to be detected in commercial plants abroad well before they will be needed here. Meanwhile, we support the initiative of the N C B in developing two new liquefaction processes and we welcome the government's decision (Hansard, 22 May 1981) in principle to provide support for a pilot plant to be built at Point of Ayr, North Wales.

15.6 A further distinction is that commercial U K demand for S N G is expected to occur earlier than for synthetic liquid fuels. From the evidence submitted to us it appears that although the timing of need and the likely scale of demand for S N G are very uncertain, capacity will be needed in the U K to manufacture S N G and that this could be economic using indigenous coal. It also appears that within the timeframe of our Study a commercially expressed demand for liquefaction of indigenous coal is less likely. We have therefore directed our attention primarily to the problems and prospects for S N G, and these are considered in more detail in this Chapter.

PRINCIPLES OF COAL GASIFICATION

15.7 The original method of manufacturing gas from coal was by destructive distillation, or carbonisation. In this, up to half the weight of the coal is driven off as a mixture of tar and combustible gases, leaving a residue of coke. This method is now employed only where coke is the desired product, and gas is then one of the by-products.

15.8 Modern processes aim to gasify the coal completely, leaving only the ash as a residue. A variety of processes are available, but they all use the same principle of reacting the coal at high temperature with a mixture of oxygen and steam at atmospheric or higher pressures. The output from the gasifier is mostly hydrogen, carbon monoxide, carbon dioxide and methane, together with some tar and small amounts of gaseous sulphur and nitrogen compounds. If air rather than oxygen is used for gasification, then the nitrogen from the air passes through unchanged and heavily dilutes the final product. This reduces its calorific value to about one half or one third. Thus low calorific value gas produced from air-blown gasifiers, although cheap to produce, is not economically attractive for long distance transmission. Its use is likely to be limited to supplying loads close to the gasifier, for example in power stations or bulk industrial use. When cleaned, such gas is a suitable fuel for gas turbines and could therefore be used for combined cycle electricity generation.

15.9 The medium calorific value gas which is obtained from oxygen-blown gasifiers would normally be a mixture of carbon monoxide and hydrogen, with a small proportion (depending on the gasifier) of methane and carbon dioxide. It is similar to the former

town gas. The type of gasifier employed and the final stages of gas processing can be used to vary the proportions of the main gases according to the requirement.

15.10 A medium calorific value gas which comprises a mixture of equal parts of carbon monoxide and hydrogen is known as synthesis gas. It is used as a starting point for a number of important chemical processes. These include the Fischer-Tropsch process for the manufacture of synthetic liquid fuels, and the chemical synthesis of methanol (and related products) and ammonia. For these applications a low methane content in the synthesis gas is desirable. However, when S N G (that is, methane) is the objective, then the opposite is the case. A range of different types of gasifier is available, which have different characteristics in the proportion of methane in the output.

15.11 Whatever type of initial gasifier is employed, subsequent process stages are needed if large scale production of S N G is to be undertaken. These subsequent stages involve gas purification and quenching, shift conversion, acid gas removal, and methanation.

15.12 In the purification stage hot gas from the gasifier is usually quenched by bringing it into contact with a large volume of circulating water-based coolant. The tars, if any, are condensed; particulate matter is precipitated; and phenols, ammonia and other trace contaminants dissolve in the water. Gas leaving the quench cooler is further cooled in waste heat boilers which raise steam for process and power.

15.13 It is then necessary to increase the hydrogen content of the gas at the expense of carbon monoxide, in order to obtain the correct ratio for the synthesis of methane. This is the main purpose of catalytic shift conversion. In addition to the conversion of carbon monoxide, the catalyst hydrogenates most of the unsaturated hydrocarbons, higher phenols and organic sulphur compounds present in the crude gas.

15.14 The acid gas removal stage has two purposes. First, it removes the gases containing sulphur. These are undesirable in the final product, and they also need to be removed to protect the catalyst in the subsequent methanation stage. Secondly, the inert carbon dioxide must be removed so that the S N G produced is compatible with natural gas.

15.15 In the final methanation stage the purified gas is passed through a catalytic methane synthesis stage, where hydrogen and carbon oxides react. This produces methane and steam. The methanated gas is then cooled, the steam condenses, and the water is re-used.

15.16 In scale and appearance a plant to convert coal to S N G will resemble a modern oil refinery rather than the traditional gasworks. An artist's impression of what such a plant may look like is at Figures 15.1a and 15.1b, and a photograph of the B G C Westfield gasifier is at Plate 15.1.

EFFICIENCY OF GASIFICATION PROCESSES

15.17 Inevitably, synthetic products will be more expensive than the natural fuels, as they involve further refining of the raw material and as energy losses are involved in the processes; at present S N G is estimated to be of the order of 2–3 times as expensive as natural gas. Thus both cost and efficiency considerations indicate a degree of caution in assessing future possibilities.

15.18 With this in mind we have looked in more detail at the energy efficiency considerations of S N G processes. We have compared S N G with other energy sources, for the purpose of domestic heating. (This is detailed in Appendix 1.) These calculations show that an S N G-fired central heating boiler in an "average" house in winter would provide useful heat to the extent of 38–52% of the energy in the original coal. A coal-fired central heating system would be more efficient, as it would use 50–70% of the energy in the coal.

15.19 On the other hand the use of S N G is considerably more energy efficient than off-peak electricity, which only uses 19–27% of the energy in the coal in a storage heater, or 25–28% in direct electric heating. In addition, coal is less clearly an efficient fuel under summer conditions for hot water heating only (20–30%, compared with 27–28% for an instantaneous electric water heater or 38–42% for an instantaneous S N G-fired water heater); or in a simple open fire, where it uses only 20–35% of the energy value of the coal.

15.20 Although the subject of heat pumps is not directly related to our Study on coal as such, it is clear from our analysis of end use efficiencies of coal-derived electricity, S N G, and direct coal burn, that heat pumps can be of substantial importance. Thus the efficiency of winter heating from an S N G-fired central heating boiler could increase from 38–52% to 66–73% if a heat pump is used; the efficiency of direct electric heating could rise from 25–28% without a heat pump, to 75–84% with one. Although a few experimental heat pumps have been installed in Britain since 1927, it is only recently that packaged equipment has become commercially available; and electrically driven heat pumps are now operating as commercial installations in various shops, offices and public buildings.

15.21 We have drawn two conclusions from our analysis of comparative efficiencies. The first is that S N G is sufficiently energy efficient to justify its eventual use, despite the fact that direct coal burn can

be even more efficient in most space heating applications. In addition, S N G is also likely to be more environmentally benign in use than direct coal burn as we discuss below. Secondly, we would see energy advantages if domestic scale heat pumps could also be developed; current cost estimates for heat pumps compared with boilers do not yet look promising and this emphasises the need for further research. Thus we recommend that Government take all necessary steps to see that such development is progressed.

15.22 In Chapter 19 we consider the role of Combined Heat and Power and District Heating Schemes (C H P/D H). In our view S N G and C H P/D H should not be seen as mutually exclusive. The conditions under which C H P/D H schemes would be viable are more constrained than for S N G, and outside these conditions S N G will eventually be required. Possible areas of competition between the two would be limited to cities with high density heat loads and here the use of an alternative fuel in a C H P/D H area would detract from its viability.

COALPLEXES

15.23 Coal is a versatile raw material which can be used for the manufacture of a whole range of chemicals as well as for conversion to synthetic gas or liquid fuels. The early stages of several processes can be common, for a range of final products; thus synthesis gas (see paragraph 15.10 above) is a very flexible intermediate, being a possible step towards methane (S N G), light hydrocarbons, hydrogen, ammonia, methanol and other chemicals.

15.24 Up to the present most studies and pilot plant designs have been directed towards plant optimised to produce a single output. This, however, need not be the only option. It would be possible to design a coal refinery which produces simultaneously electrical power, liquid fuels, gas fuels, chemical by-products, and coke (that is, a "coalplex"). Work done by the Coal Research Establishment of the N C B has suggested that the capital cost of plants designed to produce two products could be significantly lower (up to 20%), per unit of energy output, than for plants producing only one of these products. The matching of products to markets and to fluctuations in demand would, however, be a major complicating factor.

15.25 It is likely that a coalplex would evolve over time, rather than be designed and built as a single entity. It is difficult enough to envisage the right commercial, institutional and financial circumstances for a "straightforward" S N G plant, let alone the more complex coalplex. It has been suggested (Grainger L 1980) that a centre for the production of synthesis gas could form the basis of a coalplex by subsequent expansion of its activities.

Figure 15.1a   Artists impression of S N G installation

Artist Impression of SNG Installation

Source: B G C

Figure 15.1b   Key diagram of S N G installation

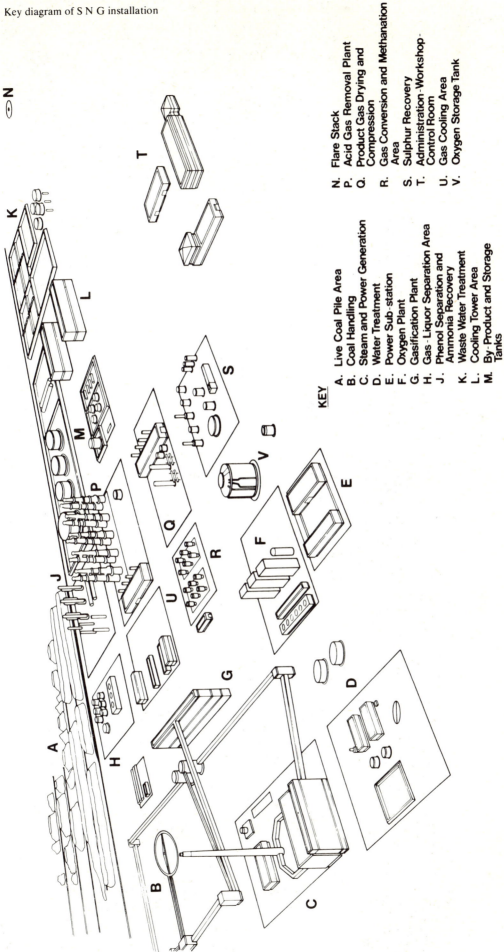

KEY

A. Live Coal Pile Area
B. Coal Handling
C. Steam and Power Generation
D. Water Treatment
E. Power Sub-station
F. Oxygen Plant
G. Gasification Plant
H. Gas - Liquor Separation Area
J. Phenol Separation and Ammonia Recovery
K. Waste Water Treatment
L. Cooling Tower Area
M. By-Product and Storage Tanks

N. Flare Stack
P. Acid Gas Removal Plant
Q. Product Gas Drying and Compression
R. Gas Conversion and Methanation Area
S. Sulphur Recovery
T. Administration -Workshop- Control Room
U. Gas Cooling Area
V. Oxygen Storage Tank

Source: *B G C*

15.26 Clearly a great deal of technical and economic analysis is needed before the coalplex concept could be translated into practical reality. The direct environmental consequences are likely to be analogous to the consequences of the individual plants, in terms of appearance and design of the equipment, and the treatment of the residues. There is, however, a broader concern: the extent to which the possibility of locating a coalplex close to another major project, for example a new coal mine, should be considered at the inquiry into that project. There is understandable apprehension that the approval of one major project could presage the later introduction into an area of other major plant.

15.27 In our view, unless the prospects for such "downstream" plants are very clear, the possibility of their subsequent location should not be considered at the stage of decision on a new mining project. The characteristics of downstream plants such as coalplexes and the likelihood of them requiring to be sited in any particular locality, would be far too uncertain for judgments to be made so far in advance. An application for a mining project should be decided on its merits. The appropriate time to decide whether downstream projects should proceed on a particular site is when a firm proposal has been made, and the downstream project has been designed sufficiently to permit its proper assessment.

CONVERSION WORK UNDERTAKEN ABROAD

15.28 There is considerable interest in coal conversion abroad; and we have particularly noted the work being undertaken or proposed in West Germany and in the United States.

15.29 Both the need, and the reason, for West German interest are clearly evident. 50% of its energy needs are met through imported oil. The implications of this for its security of energy supply are clear. At the same time West Germany is a market leader in coal conversion technology, and has had considerable experience with it. In 1928 Lurgi developed a gasification process, and in the 1930s Koppers experimented with techniques of hydrogenation, partly building on the work of Bergius undertaken as early as 1913. In 1931 Fischer successfully created oil from coal.

15.30 Techniques based on this pioneering work are in widespread use today. The current British Gas work on S N G is based on Lurgi gasifiers; German Shell have associated with Krupp-Koppers in developing the Shell-Koppers process for production of synthesis gas from coal; and the Fischer-Tropsch process of producing oil from coal is the only one in commercial application in the world at present, in South Africa.

15.31 Certainly the scale of effort in West Germany in this field exceeds that in the United Kingdom. Research and development expenditure on coal conversion in Germany is approximately four times that of the United Kingdom. It includes two operational gasification projects, two liquefaction projects under construction, and proposals for a further 10 projects under consideration. The demonstration gasification plant at Oberhausen-Holten has already had 18,000 hours of operation. The German expectation is that fully commercial technologies for S N G production will be available for market introduction in the mid-1980s and for liquefaction by 1990.

15.32 However, there are also substantial uncertainties surrounding this programme. The cost of pilot and demonstration plants is high, and budgetary difficulties must raise doubt about whether many of the proposed projects will ever go forward to construction. Cost factors are currently prompting re-assessment of the 25% West German share (jointly with the United States and Japan) in the proposed commercial scale liquefaction plant at Morgantown in West Virginia; that plant has been planned to turn 6,000 tonnes of coal per day into 20,000 barrels of oil, by 1984. There are also doubts about the cost of gas from coal; it could be up to three times as expensive as natural gas. There is therefore a reluctance for Germany to commit itself now to a large scale domestic programme. On the other hand, the protagonists of coal conversion in Germany are keenly aware that the price rises of oil and natural gas since 1973 have narrowed the price differential between the natural and synthetic products.

15.33 A further complicating factor is that, if gasification and liquefaction of coal become major energy sources in the late 1980s and 1990s, Germany will have to change its policy on coal imports. Up to now, imports from non-E E C countries have been kept low, essentially to protect domestic producers. German produced industrial hard coal costs well over twice the price of imported coal, and even that ratio is achieved only through a system of complex subsidies. The scale of coal demand for commercial operation of conversion processes could lead to significant pressure to review current policy on coal imports.

15.34 In the United States, the need to reduce oil dependence is central to energy strategy. The rapid development of a synthetic fuels ("synfuels") industry was seen as critical to the national interest by the Carter Administration. An ambitious synfuels programme was launched in June 1980 under the Omnibus Energy Security Act. This provided for the creation of a Synthetic Fuels Corporation to give financial assistance to private industry, to foster the commercial production of synthetic fuel from coal, shale and tar sands. The U S has abundant supplies of coal which account for 90% of all known indigenous fuel reserves. A national synthetic fuel production goal was set. This aimed at achieving commercial production by 1987 of at least 500,000 barrels per day of synthetic crude oil, increasing to at least 2 million barrels per day by 1991. The Act also required that a broad diversity of technologies should be pursued, and that the necessary infrastructure should be established to enable the longer term national production goal to be achieved.

15.35 The U S synfuel programme as conceived by the previous Administration appeared to us to have many of the characteristics of a crash programme. Environmental controls tended not to be seen as part of the process design, but rather for retrospective application. This, together with doubts whether adequate attention is being given to the problems of scaling up, pose major uncertainties. There are also conflicting views in the United States on the environmental implications of the programme, and on whether existing pollution control techniques proposed for the plants will be adequate to meet existing U S standards.

15.36 As we understand the current position the new Administration have adopted a radically different approach from their predecessors to the synfuels programme. The Administration will continue to support long-range, high-risk and potentially high pay-off research and development, but will withdraw support for demonstration projects and the development of commercial applications, which they believe should be left to the private sector. Responsibility for some projects may be transferred from the Department of Energy to the Synthetic Fuels Corporation in which case they would qualify for price and loan guarantee support, but the private sector would have to put a far higher proportion of their own money at risk.

15.37 In other countries where special circumstances exist, commercial conversion plants to produce synthetic gas and oil can already be justified. New Zealand, rich in natural gas but not oil, is in process of developing a plant based on Mobil technology for the conversion of methanol to oil; the methanol will be derived from their natural gas. South Africa, of course, has development plans for yet a third phase of its S A S O L plant for conversion of coal to oil. In short, even outside the market leaders of the U S and Germany, it is clear that there is no shortage of world-wide interest in the development and application of techniques to produce synthetic gaseous and liquid fuels from other hydrocarbon sources, and that in the long-term, coal is seen as perhaps the most significant of those alternative sources.

CURRENT EXPECTATIONS ON THE SCALE AND TIMING OF THE U K NEED FOR S N G

15.38 The likely scale of U K commercial demand for S N G is beset by a range of uncertainties, such as the scale of total future energy requirements; the amount of natural gas landed in future in the UK; the price and availability of imported Liquefied Natural Gas (L N G); the price and availability of UK coal; and competition from other energy sources including C H P and D H, on cost, availability and overall efficiency.

Possible magnitude of commercial demand

15.39 In order to gain an idea of the possible order of magnitude of any future S N G manufacturing requirement, it may be reckoned that about 20–25 S N G plants of the size described in paragraph 15.50 could be needed to satisfy present total U K gas demand. This gas demand is for the energy equivalent of 71 million tonnes of coal, although the scale of coal input to conversion plant to produce the equivalent quanity of S N G would need to be around 100 million tonnes to take account of conversion inefficiencies. This would be a very substantial demand for coal, additional to its established markets. The scale of the coal requirement alone is therefore likely to limit the extent to which S N G can completely substitute for the natural product.

15.40 It is, however, highly unlikely that S N G will be required to substitute for its natural equivalents over the whole range of its present uses. Much of the consumption of gas in non-premium industrial use is open to competition from direct coal burn; the attractions of direct coal burn increase as oil and gas prices rise, and its costs will necessarily be less compared with the higher value added synthetic product. Thus, the longer term market for S N G is likely to depend particularly on demand for premium uses, on competition between gas, electricity and coal in the domestic heating market and on the extent to which conservation measures can influence total energy demand.

Timescale of demand for S N G

15.41 The timescale for the transition to synthetic gas supply is also very uncertain. It is not clear when such a transition should start, nor how long it is likely to last. In the Department of Energy's "Energy Projections 1979" it is assumed that S N G from coal could be the major source of any additional gas supplies which may begin to be required around the turn of the century. Neither of the two original cases (that is 2·7% and 2% economic growth rates) project a requirement for S N G as early as 1995. However, in the higher of these two cases, up to 2,000 million therms could be required in the year 2000, and in the lower case up to 200 million therms. These figures are not particularly meaningful taken in isolation, since both cases include a subsequent rapid growth in S N G production. A quite different picture is painted by the 1% case. Under this scenario, commercial scale production is not estimated to be required by 2000, and assumptions are not made about when any subsequent need may arise.

15.42 The N C B expect that only small quantities of coal will be used in coal conversion processes before 2000, but that the subsequent growth in these markets will probably be rapid. The British Gas Corporation do not envisage a major U K need for coal based S N G before the turn of the century, although a start may be made to phase in some production capacity before then. They consider that L N G and S N G from oil could provide alternative sources in the years up to 2000, based on well established technology.

15.43 Evidence submitted from a range of other bodies also suggests that major U K need may first become apparent at about the turn of the century, with a possible rapid increase thereafter. Thus, given existing fuel supply expectations, a substantial commercial demand for synthetic fuel cannot be expected in the U K for at least 20 years. This period could be extended if future North Sea or other discoveries or access to Norwegian gas exceed present expectations.

15.44 In short, it is clear that only a limited number of S N G plants would be needed to cater for even the maximum feasible demand. In practice, 10 sites available in the years immediately following the turn of the century should be ample to meet any currently foreseen need. However, the exact scale of siting requirements for these plants will need to be monitored as clearer information of likely need emerges with the passage of time. For present purposes, we recommend that long term contingency planning should proceed on the basis of forecast need for up to 10 sites for S N G plants in the first decade of the next century. This figure should be reviewed as more accurate assessments become available.

ENVIRONMENTAL CHARACTERISTICS OF S N G

Characteristics of the fuel

15.45 A particular virtue of burning S N G derived from coal rather than burning coal itself is that the S N G is much cleaner at the point of combustion; sulphur, particulates, and smoke are not emitted and no ash is created. The problems of sulphur control and ash disposal are effectively transferred from the point of combustion to the conversion plant. These problems can be far more easily and efficiently dealt with at a centralised conversion plant than at a multiplicity of points of individual coal burn.

15.46 Secondly, there is undoubted consumer convenience in using S N G rather than coal. The combustion of S N G can be turned off and on at will, and remotely if necessary; the rate of combustion can be more accurately controlled with an instant response; and no physical handling of the fuel is required by the customer.

15.47 Thirdly, the equipment to burn S N G will be compatible with existing equipment designed to burn natural gas. Any other form of fuel would require different equipment; considerable cost and inconvenience could result to customers. Any major re-equipment programme could cause organisational difficulties if the demand for the new equipment occurred over a short period; on the other hand, any transition from natural gas to S N G could be undertaken gradually. Changes of equipment could involve not only the combustion equipment itself, but also major changes in flue or chimney requirements.

15.48 Fourthly, there are clear environmental as well as economic advantages in using an existing pipeline network for the transmission and distribution of the fuel to the final customer. This network is a major asset of the gas industry, for it avoids a whole range of problems involved in the transport, storage, and handling of a bulk fuel. The national transmission system has all been laid in the last 15 years, and 40% of the distribution system has been laid or relaid in the last 20 years. Most of the system will be fully amortized by the time S N G becomes a significant supply, and the cost will be one of maintenance and minor replacement rather than major new investment.

Characteristics of S N G production plants

(a) *Siting of plant*

15.49 In general terms there are five principal characteristics which determine site requirements:
  (i) the size of the necessary land take for each site (80–120 hectares);
  (ii) access to large quantities of coal, for example, up to 15,000 tonnes per day;
  (iii) availability of large quantities of water, possibly of over 50,000 cubic metres per day for indirectly cooled installations;
  (iv) access to disposal facilities for solid waste;
  (v) safety considerations.

15.50 A requirement for large quantities of S N G would favour the use of large plants which should give economies of scale in capital and running costs, although this has yet to be investigated in detail. For site evaluation purposes, multiples of 50 and 100 million cubic feet per day (mcfd) are being used at present for coal and oil-based plants respectively. Based on American proposals, the reference coal gasification plant size has been assumed to be 250 mcfd (that is, composed of 5 units each 50 mcfd). In practice, of course, plants may well be smaller than this, depending on load requirements. The smallest plant is likely to be about 50 mcfd size.

15.51 Of the order of 15,000 tonnes per day of coal would be needed to produce 250 mcfd of S N G. In practice, the precise amount of coal needed will depend upon the quality of the coal and the efficiency of the process. At best, using good quality coal and the most advanced processes, about 11,500 tonnes per day could be needed; at the other extreme, 25,000 tonnes per day of the lowest grade material (lignite) would be needed to produce that quantity of gas – although, of course, the U K itself does not have commercial reserves of lignite.

15.52 British Gas envisage that a 250 mcfd coal-based S N G plant would operate at a high load factor of 330 days per year. Its annual coal requirement would thus be about 5 million tonnes. It would consume the production of 10–11 (average) mines, or about half the expected output of the new Selby coalfield. This level of coal input is approximately equivalent to that of the largest power station.

15.53 Water is needed both to provide the extra hydrogen content of methane compared with coal, and for cooling purposes. Taking the 250 mcfd plant as the norm, 10,000 cubic metres of good quality water per day will be needed for feedstock and process purposes. If towers are used for cooling, there will be a make-up water requirement of 30,000 cubic metres per day; if direct cooling is used, upwards of 1·5 million cubic metres per day would be needed to pass through the plant. The quality of water for cooling is less critical than for process purposes. These figures are somewhat smaller than for the typical 2,000 M W electrical power station.

15.54 About 80–120 hectares of land would be needed for an S N G plant with an output of 250 mcfd, including an area for peripheral landscaping. This scale of land requirement is similar to that for coal-fired power stations with cooling towers.

15.55 It is possible that conversion plant could be regarded as hazardous installations. If so, this could involve a consultation zone of up to 2 kilometres around the plant, and a safeguarding zone of up to 1 kilometre in which developments such as schools and housing would be restricted, although industrial developments could be permitted. These requirements would not necessarily preclude urban siting, although the requirements could be hard to satisfy in urban areas. Individual projects will, of course, need to be judged on their merits, once details of the layout and processes are available for a particular locality.

15.56 Major uncertainties surround the scale of the likely supply of new sites. The analysis we have done does not show conclusively that there is likely to be scarcity of potential new sites, but it does indicate that a considerable possibility exists. It is clear that an initial area of search for sites for large scale development is restricted, by topography, existing development, and conservation policies, to about 25–30% of the total land surface of England and Wales and to 20% or less of the coastline. In practice, substantial parts of this

area of search would be eliminated on other grounds and the real area of search is more restricted than these figures indicate; although no doubt some of the policy constraints could be modified if need were shown to be sufficiently pressing.

15.57 In addition to the prospects for new sites, there is some possibility for the re-use of sites currently held for gas purposes. The British Gas Corporation has identified 25 of its sites, mainly urban, which they consider should be held for S N G purposes, in view of their special suitability. In almost all cases, parts of these sites are currently used for some operational activity. Some of the sites would need expansion to accommodate coal based S N G plant, and indeed capability of expansion has been a significant selection criterion. 12 of the sites are suitable for coal gasification; 9 are more suitable for oil gasification largely on account of their smaller size; and 4 sites are seen as providing for strategic storage of oil feed-stock for S N G manufacture. Only 2 of the sites are at present large enough to accommodate the 250 mcfd plant used as the standardised example of plant size. The total capacity of the 12 sites suitable for coal gasification is only about one-quarter of a possible ultimate requirement.

15.58 We are aware that there are both significant advantages and serious drawbacks if urban sites were to be used for S N G plant. Whether or not it is physically practical to use urban sites will depend on the availability of suitable sites – it can be assumed that there will be greater pressures for alternative uses in urban areas; on the size of plant that will be needed – it will be particularly hard to fit a large plant into an urban area; and requirements for safeguarding zones.

15.59 If in the event it is practical to use urban sites, then there will be major advantages in using services such as roads, rail-links and underground mains, which would have to be newly provided in any new location. Secondly, it would be easier to use the waste heat from an urban-sited plant for district heating or other purposes, than in the case of a rural location. On the other hand, major disadvantages could arise on grounds of amenity (for example, visual amenity and nuisance from fugitive emissions) and on grounds of hazard; and, if safeguarding zones were needed around plants, they could unduly limit the range of other activities that could take place over a significant area.

15.60 We have concluded that it is not possible to generalise on the suitability of urban sites for S N G plants. We consider that such matters will ultimately have to be decided with respect to individual projects. There does not seem to be any overall argument that would eliminate the use of urban sites for this purpose; but, in practice, the additional constraints and requirements posed by urban siting would severely limit the occasions when it could be adopted.

15.61 We have concluded that there is some urgency in assessing siting requirements, despite uncertainty about how in detail the build-up of demand will take place. However, this type of siting problem is by no means unique to S N G plants. Similar issues arise in the siting of a whole range of energy and other heavy industrial activities. Thus, although our analysis of this subject began with S N G plants, we quickly realised that these were only one example of a much broader problem of planning and siting for large projects. We have included specific recommendations on siting problems in Chapter 21.

(b) *Residues from the plant*
15.62 Coal gasification involves sophisticated and large scale plants. The processes involve handling very large amounts of material, employing high temperatures and in reducing conditions; both highly corrosive substances and those which may produce health threats are formed during processing; a large quantity of solid waste is produced; and liquid and gaseous waste will require controlled treatment before final discharge. Assessment of residues is only possible at present in the broadest terms. Estimates must depend on the coal feedstock available and the conversion processes selected. Few of the current processes are at a stage of development which would permit reliable quantification of all pollutant and toxic hazards.

15.63 That being said, however, there is a substantial consensus of evidence that the processes involved are properly controllable in appropriately designed plants and that the wastes generated are convertible to environmentally acceptable forms using modern treatment techniques. It will, however, be necessary and prudent to ensure operation and monitoring of pilot and demonstration plants, to clarify the types and quantity of effluents; to confirm the parameters of the treatment techniques which are required for discharge; and to determine the most efficient means to avoid or control fugitive or accidental emissions and to minimise possible odour nuisance.

15.64 The disposal of solid waste needs to be specifically addressed. Large quantities will require disposal. These represent the inorganic or non-used constituents of the coal and are a function of the feed rate of the coal rather than of the type of process. A 250 mcfd size S N G plant would produce 2,000 tonnes per day of slag, 230 tonnes of sulphur per day and 5–10 tonnes per day of sludge. The quantity of slag from one plant would be sufficient to cover 30 hectares to a depth of one metre in one year of operation (270 acres to a depth of one foot). There would also be 100 tonnes per day of ash if coal were to be used for steam raising. The disposal or subsequent re-use of the sludge, ash and sulphur are not expected to raise any unusual problems, or problems that are unique to the coal conversion industry; although the sludge from effluent treatment is

likely to contain trace elements which will require disposal to waste rather than to agricultural re-use. The slag from the B G C Lurgi slagging gasifier is inert; it is very resistant to leaching; it will not be blown by the wind; it contains very little carbon; it is formed under oxidising high temperature conditions which do not favour persistence of hydrocarbons; and tests so far have not revealed any polycyclic aromatic hydrocarbons. The potential of the slag for cultivation has not yet been examined.

15.65 The likely opportunity for the re-use or disposal of solid wastes is thus a matter of concern, given the possible scale of the operation and the amount of solid wastes likely to result. This is particularly so when added to the similar problems likely to arise with solid mine waste and power station ash disposal. We recommend that B G C and N C B should undertake the necessary research into the characteristics and means of disposal of this solid residue.

15.66 Demonstration plant the size of one stream of a full commercial plant, will thus be needed both to establish general technical and commercial viability and to confirm the adequacy of environmental safety. It should be possible to collect information necessary for the environmental analysis well within the timescale necessary to assess other technical, engineering, or commercial factors. Environmental effect investigations would not then be on a "critical path" in the development of a new process, unless the appearance of very unusual or unexpected pollutants demanded development of new abatement techniques. Existing knowledge of the pollutants does not suggest, however, that there should be any emissions beyond the capability of existing abatement technology.

15.67 If it was intended to use in the U K any process which had been operated at a demonstration scale or at a full commercial size elsewhere, then little further demonstration work in the U K need be contemplated to cover environmental engineering aspects. Adequate information should be available from plant contractors, process licensers and previous users to quantify all pollutants, to design ancillary plants for their control as necessary, to plan disposal of residues and to provide an environmental impact statement for an entire commercial plant.

15.68 Certainly there is a great deal of interest abroad in developing, refining and testing suitable processes and techniques. In many cases there has been substantial experience with pilot plants. There are a range of plans for the further development of these projects. In some cases we have noted an intention abroad to proceed direct from pilot scale (for example 250 tonnes per day coal input) to full commercial scale (for example 6,000–7,000 tonnes per day coal input) without any intermediate demonstration scale. In our view this is imprudent and we would not recommend such a course of action in the U K. Both environmental and technical factors are too much at risk in scaling up of this magnitude. Nevertheless, if such projects do go ahead abroad, they will provide the opportunity for the U K to learn from the experience gained in designing and operating them; and it could render unnecessary the need for any separate U K based demonstration plant.

15.69 In addition to the work being undertaken abroad, there is extensive development work by B G C in the U K at Westfield, Fife. The unit here is a full-scale commercial Lurgi Gasifier which B G C have modified for slagging operations (see Plate 15.1). The next step required is longer term operation, preferably at a somewhat larger size, and under increasingly commercial conditions of continuous operation. British Gas have informed us that, to this end, they are planning extended runs and are building a larger gasifier. They are also examining the possibility of building a full-size single module production unit – perhaps 50 mcfd output – at a date early enough to allow consolidation of the technology before large-scale production is needed. Such a plant would be extremely valuable in eliminating many of the uncertainties of scale-up and allowing full assessment of environmental aspects. We are also aware that private companies with interests in the U K are independently pursuing development work and refinements to S N G processes.

15.70 We have concluded that there is no certainty which of the many processes now under development would be used in any future U K programme of S N G manufacture. We have noted that B G C do have a pilot scale plant in the U K and we recommend that they should use it to examine environmental matters. Work is proceeding in other countries, meanwhile, on a bigger scale and at a faster pace than we presently need. We will be able to learn from that.

15.71 We conclude, on the basis of the evidence submitted to us, that serious environmental damage is unlikely from S N G plants, provided that due care is taken in their design and operation. However, before any process is commercialised in the U K, there must be relevant experience at a full demonstration scale (about 50 mcfd). Thus we recommend that Government should make it clear to potential developers of S N G plant in the U K that commercial development of such plants will not be permitted until satisfactory experience with demonstration plant can be established. The work should indicate the type and quantity of residues from the processes, including ash and slag, and the methods proposed to neutralise them before discharge to the environment. It should also identify the reliability expectations of the equipment under commercial conditions and the likely frequency and severity of accidental or fugitive emissions, including an assessment of likely nuisance effects.

15.72 That experience could come from abroad, but only if the foreign plant has been designed to U K health and safety and environmental standards. If not – and that seems likely – there will have to be a British demonstration plant. If we are to provide for the first commercial use by the late 1990s, the demonstration plant would have to begin only a few years from now. Unless B G C can assure the Government that foreign demonstration will be sufficient to meet U K requirements, we recommend that plans should be drawn up for an early demonstration plant in the U K.

Chapter 16   Combustion technology

16.1   We have examined thus far in Part III current and prospective patterns of coal burn, future market prospects for coal, and the transport requirements of particular markets. As a final element in this descriptive background to our assessment of the environmental implications of coal use, we now examine coal combustion technology. This Chapter therefore describes (1) the properties of coal, (2) the processes of combustion and (3) the range of industrial and domestic coal combustion equipment in current use and under development. It illustrates how the various residues are formed; these are discussed individually in more detail in Chapter 17. Their effects are considered in Chapter 18. The overall implications of coal burn in urban areas are discussed in Chapter 19.

COMPOSITION AND QUALITIES OF COAL

16.2   The major constituents of coal are carbon, hydrogen and oxygen. Around 90% by weight of clean, dry British coal is carbon, 4–5% hydrogen and 1–4% organically combined oxygen, together with about 0·5–2% each of nitrogen and sulphur. In addition, mineral matter (ash) is always associated with coal. The proportion of mineral matter and its composition can vary widely, but typically the major constituents are aluminium and silicon, amounting to some 3% of the coal. The other minor constituents of the mineral matter include chlorine, calcium, iron, magnesium, sodium, potassium, titanium and manganese. Traces of all the remaining naturally-occurring chemical elements can be found, usually in concentrations of a few parts per million or less.

16.3   Coal is a complex substance of indeterminate chemical composition, formed by various natural geological processes over millions of years from a random mixture of organic materials from primeval forests. Therefore, it is not a uniform material and variations of its characteristics occur in different parts of Britian and even in different seams within the same mine. The important characteristics which affect its use are calorific value, volatile content and coking qualities. However, the mining and preparation techniques employed at the colliery can, to some extent, affect other properties such as its size and the content of ash, moisture and sulphur. Generally, coal types and preparation techniques have to be selected to suit the intended applications.

Calorific value

16.4   The calorific value of a fuel is the number of heat units liberated when a unit of mass of the fuel is completely burned. It is usually expressed as kJ/kg, Btu/lb or therms/ton. There are gross and net calorific values for hydrogen-containing fuels depending respectively on whether the heat of condensation of the water produced is included or excluded. For low-hydrogen fuels, such as coal, the difference between gross and net values is only 3–5%, but with natural gas it amounts to 10%. By convention, fuels are usually sold on the basis of their gross calorific values and boiler efficiencies are calculated on the same basis. In practice, however, it is not possible or even desirable to cool the combustion products sufficiently to condense the water vapour so the net calorific value represents the maximum practicable heat which can be used. It follows therefore that, on a conventional basis, efficiencies of gas and oil-fired boilers would be expected to be some 5% lower than coal-fired boilers of identical design. These points should be borne in mind when comparing efficiencies in the use of various methods of domestic heating as set out in Appendix I. The calorific value of British coal varies slightly with coal type, but is chiefly affected by the amount of inert material – ash and moisture – which it contains.

Volatile matter

16.5   The volatile content of coal is that part which can be distilled by heating in the absence of air, as in a coke oven. The volatile matter is a mixture of combustible gases (mainly hydrogen, carbon monoxide and methane) and condensible aqueous and organic substances. On cooling, the latter condense into fine droplets of tar and are usually removed from the gas before it is purified and used as a fuel. When coal burns, the volatile matter is responsible for the luminous flame; if it escapes unburned, the tar is visible as smoke. The devolatised residue is coke, which is largely carbon. The major use for coke is the smelting of iron ore, but it is also used as a smokeless fuel. Coals that produce tar on heating are commonly called bituminous.

16.6   The volatile content of coal varies from 5% for Welsh anthracite to 40% for some Midlands coals. When the volatile content is less than about 20%, very little tar is produced and the coal is naturally smokeless, but only limited quantities of such coals are available. Manufactured fuels are produced from the available high volatile coals in a variety of ways by subjecting them to heat treatment which removes most of the tar-producing volatiles.

16.7 High volatile coals are easy to ignite and are free burning, but tend to cause smoke. In large sizes, they can be used domestically in traditional open fires. In the smaller size gradings, the non-coking or weakly-coking high volatile coals are suitable for industrial boilers. High or medium volatile bituminous coals which are strongly-coking are used almost exclusively for coke manufacture. Low volatile non-coking coals are good boiler fuels (the traditional Welsh dry steam coals); they can also be used as naturally-smokeless open fire fuels. Fuels with very low volatile content are difficult to ignite and for domestic use a large hot firebed is needed to sustain combustion. They can therefore only be used on suitably designed appliances with deep beds and refractory walls such as improved open fires, closed room heaters or boilers.

Coking properties

16.8 Some coals, known as coking coals, become partly fluid on heating and they swell up and agglutinate or cake. Prime coking coals make strong coke which is necessary for iron-making purposes. Prime coking coals are available in limited quantities and are reserved for the manufacture of metallurgical coke; coals with more limited coking properties may be used for the manufacture of domestic coke or used directly as fuel for pulverised fuel fired boilers. Some coking coals can be very difficult to handle on some types of automatic stokers because of their stickiness.

Size

16.9 Large coal is now used almost exclusively for domestic purposes. Graded coal, that is coal screened between a specific range of sizes (usually not more than 5 centimetres and not less than 2·5 centimetres) is used for some of the smaller types of industrial grates. Larger boilers, however, tend to use the cheaper smalls, that is coal with a specified upper size limit (usually below 5 centimetres) but no lower size limit. In modern power stations, the coal is dried and pulverised to a fine powder and then burned in a stream of air like an oil or gas burner. The initial characteristics of the coal are much less critical for this type of combustion; the most important characteristics are low cost and consistent quality.

Ash and moisture content

16.10 As produced, coal always contains moisture and mineral matter, which forms ash on burning. Generally there is between 5% and 20% of moisture and between 2% and 20% of ash. Power station coals contain about 16% ash, industrial coal around 10%, and domestic

coals 5%. It is not possible to separate non-combustible material from coal with precision and the preparation of very clean coals entails rejecting more coal with the discarded material. Coal washing also increases its moisture content and wet, fine coal is very difficult to handle. For each application there is an optimum ash and moisture content which balances the cost to the producer in terms of coal cleaning and losses of rejected coal against the savings to the user in terms of transport of inert material, disposal of ash and convenience.

Sulphur content

16.11 The sulphur content of coal is important for some applications such as the manufacture of metallurgical coke for which a low sulphur content is normally necessary. But for most industrial purposes, the sulphur content of British coals is not a constraint on their use. The sulphur content of British coal (mostly 0·5–2%) is relatively low by international standards (0·5–11%).

MARKETS BY TYPE OF COAL

16.12 The relative size and composition of the markets for coal can be seen in Table 16.1 and diagrammatically in Figure 2.2. Figures are given in Table 16.1 for 1955, 1978/79 and 2000. Tables 16.2 and 16.3 show the grades of coal supplied to power stations and to the industrial market respectively, in 1977/78. Most of the industrial coal is burned in larger plant as is indicated in Table 16.4. About a quarter of this coal is used for firing brick and cement kilns, but the majority is burned under boilers of one sort or another.

Table 16.1 Annual coal consumption in the U K

(million tonnes)

Market Sector	1955	1978/79	2000 (forecast)
Power Stations	44	83	90
Industry[1]	50	9	40
Domestic/commercial [2]	48	13	20[3]
Railways	12	0	0
Sub-total (heat & steam)	154	105	150
(% of total)	(73%)	(86%)	(88%)
Coke Ovens	27	15	20
Gas Works	28	0	0
Manufactured Fuel Plants	2	3	—[3]
Total	211	122	170

Source: N C B

[1] Includes Colliery Consumption
[2] Includes Miners Coal and others
[3] Coal for Manufactured Fuel Plants and for coal conversion is included with Domestic/Commercial for 2000

Table 16.2 Grades of coal supplied to power stations (% of Total)

Grade	1977/78
Graded and large	0·4
Washed smalls	6·2
Part-treated smalls	57·9
Untreated smalls	34·1
Slurry and others	1·3
Total	100·0

Source: N C B

Table 16.3 Grades of coal supplied to the Industrial Market

(% of Total)

Grade	1977/78 Actual	2000 Projected
Graded and large	33	20
Washed smalls	23	25
Part-treated smalls	28	55
Untreated smalls	15	
Slurry and others	1	
Total	100	100

Source: N C B

16.13 Table 16.5 shows trends in domestic solid fuel consumption. A steep decline in usage of bituminous coal has occurred, but smokeless fuels have maintained a steady market, although the mix of fuels has changed. There are difficulties associated with the present methods of manufacture of smokeless fuels, such as the environmental pollution at the plant, as well as difficulties with the availablility of suitable feedstocks and with high processing costs. The N C B have indicated to us that they do not expect overall production to increase. If coal were to recapture a larger proportion of the domestic market, it would appear necessary for bituminous coal to be used, but in a new range of appliances designed to reduce smoke emissions and to operate at high efficiency.

Table 16.4 Industrial coal usage by size of plant

(million tonnes)

Size of Plant m Btu/hr	1977/78 Actual	2000 Projected
0–5	1·6	7·8
5–50	1·7	9·2
50–300	7·5	23·0
Total	10·8	40·0

Source: N C B

Table 16.5 Supplies of solid fuel to the U K Domestic Market 1957–1979

('000 tonnes)

	1957	1967	1972	1977	1979
BITUMINOUS COAL					
Miners' coal	5230	3520	2110	1750	1610
Other house coal	30230	19200	10600	7770	7150
Sub-total	35460	22720	12710	9520	8760
NATURALLY SMOKELESS					
U K anthracite – miners coal	90	90	90	120	160
U K anthracite – others		1170	1120	990	1110
U K dry steam coal	1410	480	380	240	220
Imports – anthracite	—	—	290	270	250
Sub-total	1500	1740	1880	1620	1740
MANUFACTURED SMOKELESS					
Homefire	—	—	120	230	260
Phurnacite	380	750	680	740	600
Gas coal	2610	2150	150	—	—
Hard coke[1]	420	1450	1000	1280	1150
Other fuels	440	2110	2020	1230	1280
Sub-total	3850	6460	3970	3480	3290
Total smokeless	5350	8200	5850	5100	5030
Total solid fuel	40810	30920	18560	14620	13800

Source: Derived from Department of Energy 1980

[1] Mainly N S F Produced Sunbrite

134

COMBUSTION OF COAL

16.14 When bituminous coal is heated, it undergoes a progressive decomposition in which volatile gases and hydrocarbons are emitted and coke remains. In combustion, primary air passes through the burning fuel bed carrying away the volatile matter. At the same time an adequate supply of secondary air must be supplied to the flame to ensure complete combustion, otherwise smoke may be formed. There is a multitude of designs of grates and furnaces for burning coal but an essential consideration with them all is to control the flows of primary and secondary air so that complete combustion is achieved with as little excess air as possible. As the excess air is discharged to the chimney at a higher temperature than it enters the furnace, it causes loss of heat and efficiency.

16.15 The combustion residues consist of gases, solid ash and fine particulate material. The gases are largely carbon dioxide and water vapour, plus small amounts of sulphur dioxide, nitrogen oxides, hydrogen chloride and carbon monoxide. Other gaseous compounds are present in only trace amounts. Some of the ash is in the form of very fine particles which are carried with the flue gas stream and emitted to the atmosphere. Smoke, which is a mixture of small particles of carbon and tar, is a product of incomplete combustion. The smoke from a low temperature source, such as a domestic open fire, contains a large proportion of tar. Smoke from high temperature sources, such as large boilers, and, we understand from the new types of smoke-reducing domestic appliances, contains a much smaller proportion of tar partly because it is decomposed by strong heating into carbon, and partly because the appliances are designed to ensure that combustion of volatile matter is as complete as possible.

COMBUSTION EQUIPMENT

16.16 At present coal is burned in three main types of boiler plant of different scales:

(a) *Power stations and large industrial boilers*

These produce high pressure (up to 160 bar) super-heated steam (up to 566°C) in water tube boilers for use in turbines for power production. The size range of units available covers a coal consumption of some 3–300 tonnes an hour, in pulverised form. In a power station, maximum work is extracted from the steam and it is finally rejected at a temperature which is too low for further significant utilisation. On the other hand, in industrial plant medium and low pressure steam is often supplied from the turbines for process use. When a suitable balance between the power and heat requirements can be achieved, a highly efficient overall use of heat can be obtained. The formation of smoke due to incomplete combustion is not a problem with pulverised fuel burners. Chimneys are needed to provide adequate dispersal of flue gases and residual fine particulates, their height depending on the amount of sulphur dioxide in the flue gases.

(b) *Medium to small industrial, commercial and public sector boilers*

There is a range of types of boiler producing medium or low pressure steam or hot water which are used for a variety of processing and heating requirements for small factories, offices, and institutions. They are mostly of the shell or sectional type and burn 0·1–3 tonnes of coal an hour. They usually have grates with automatic stokers and they use washed small coal, as received. These systems produce less fine ash than pulverised fuel burners and more bottom ash. They cannot economically support elaborate systems for particulate removal from flue gases but employ simpler systems which are adequate because less fly ash is produced. These installations, when operated properly with appropriate grades of coal do not give rise to smoke except on limited occasions such as a start-up. The height of chimneys is controlled by local authorities in accordance with the Chimney Height Memorandum (Ministry of Housing and Local Government 1967) to ensure that the dispersal of combustion residues is adequate. Gaseous emissions from stoker-fired and pulverised fuel systems are very similar apart from the matter of scale. Boilers for these applications have suffered severe competition from packaged oil-fired boilers in recent years, but new designs, some of them based on fluidised bed combustion of coal, are becoming available which the N C B expect to be increasingly competitive.

(c) *Domestic fires, heaters and boilers*

Although many unimproved open fires are still in use outside smoke control zones, they are gradually being replaced by improved open fires (which are suitable for burning coke) and by open fires with high output back boilers, both of which are more efficient. Closed room heaters and boilers are the most efficient appliances, but traditionally they have only been suitable for low volatile fuels (coke, anthracite and phurnacite). A new generation of high efficiency appliances is being developed by the N C B and the manufacturers which, it is hoped, will burn bituminous coal sufficiently smokelessly to be acceptable in smoke control zones. Domestic appliances use the largest size of fuel with the lowest ash content. Because of the difficulties of ensuring adequate combustion under all conditions, they also produce the most smoke even when using so-called smokeless fuels or when burning coal in special smoke-reducing appliances. But less nitrogen oxides are emitted because of the lower combustion temperatures. Because domestic chimneys are low and numerous, their contribution to local pollution levels can be quite large.

135

16.17 In the following sections more detailed descriptions are given of the various types of combustion equipment commonly used for these applications.

Pulverised fuel boilers

16.18 The combustion system of a pulverised fuel power station boiler is shown in Figure 16.1. The fuel is injected into the boiler furnace through some 30–60 burners in a stream of primary air which transports it from the mill in which it has been ground. The suspension issues from the burner to mix in a controlled manner inside the furnace with a stream of secondary air, blown through the burners from a wind box on the furnace wall. The furnace walls are lined with evaporator tubes in which steam is raised. Banks of superheater and reheater tubes at the top of the furnace serve to increase the steam's temperature further before it passes to the turbo-generator. Peak temperatures within the flame zone may be as high as 1,750°C and heat transfer from the flame is predominantly by radiation. Evaporation of water within the furnace wall tubes keeps their temperatures tolerably low in spite of very high radiant heat fluxes. Combustion gases leave the furnace section at about 1,300°C and cool to 700–800°C as they negotiate the pendant superheater and reheater tubes at the top of the boiler. The normal maximum steam temperature of 566°C is set by the properties of the construction materials of the boiler and other plant.

16.19 The remaining stages of the boiler consist of banks of serpentine steam or water tubes to which the flue gases transfer heat mainly by convection. In the final stage (the economiser) feed water is raised nearly to boiling point before it passes on to the evaporator section. After leaving the economiser the flue gases give up further heat to warm the combustion air on its way to the furnace. At this stage, the remaining burden of ash and other solids in the flue gas is removed by electrostatic precipitators. Some deposits of solid material may also accumulate in the air heaters and in the later stages of the boiler. They are dislodged periodically, while the boiler is on load, by "soot blowing" with jets of water, steam or air.

16.20 Finally, the flue gases are discharged to the atmosphere via the stack, which is made tall enough to ensure that adequate dilution occurs before the plume can descend to ground level. The gases must discharge at temperatures of about 120°C to give the plume sufficient buoyancy to achieve proper dispersion from a stack of reasonable height.

16.21 In most U K plants two fans supplement the draught created by the warm gas in the stack. A "forced-draught" fan sucks air from the boilerhouse roof space and drives it through the air heater, wind-box and burners into the furnace. An "induced draught" fan sucks the combustion products from the furnace, through the air heater and precipitator before discharging the residual gases to the stack. By maintaining a suitable balance between the two fans, the boiler can be kept at slightly below atmospheric pressure to stop combustion products leaking out.

16.22 A 2,000 megawatt coal-fired power station consumes up to 20,000 tonnes of coal and 200,000 tonnes of air each day. At the same time, it expels some 40,000 tonnes of carbon dioxide and 6,000 tonnes of water vapour to the atmosphere. Depending on the mineral content of the coal, there can also be up to 4,000 tonnes of ash to be disposed of. Up to 20 tonnes of ash per day leave the stack as a very fine dust. This consists mainly of glass-like particles of fused mineral matter, along with traces of unburnt fuel and soot. The main coal impurity is, of course, sulphur, typically present in concentrations of around 1·5%. It burns to give sulphur dioxide and up to 1% of this may oxidise further to yield sulphur trioxide which combines with water vapour to form sulphuric acid. Most of the latter is neutralised by the alkaline oxides in the ash and flue gas, leaving sulphur dioxide as the major pollutant in the flue gases. In 24 hours, the station emits some 600 tonnes, representing about 1,200 ppm of the flue gas volume.

16.23 In high temperature flames, some nitrogen from the coal and the air will be oxidised. From these two sources, the flue gases may acquire between 400 and 700 ppm of nitrogen oxides, mainly nitric oxide. There may also be some 200–500 ppm of hydrogen chloride, formed from salt in the coal. Other minor ingredients of the flue gases are carbon monoxide (40 ppm) and very much smaller traces of partially burnt hydrocarbons.

Industrial boilers

16.24 The industrial and commercial market covers an extremely wide range of boiler types and outputs. They can be divided in a very general way into three constructional types:
(a) Cast iron sectional boilers for hot water or low pressure steam and burning up to 0·25 tonnes of coal an hour which are often fired by underfeed stokers.
(b) Fabricated steam-raising shell boilers which are usually fired by chain grate or coking stokers. Those with a single furnace tube burn 0·1–0·25 tonnes an hour and those with twin or triple furnace tubes burn up to 1·5 tonnes an hour. Similar sized boilers give 50–100% more output when oil-fired.
(c) Water tube boilers are used for high pressure steam when coal consumption is above 3·5 tonnes an hour. The larger sizes are site constructed and fired by site-built chain or coking stokers or by pulverised fuel.

Figure 16.1 Coal fired power station: schematic combustion system

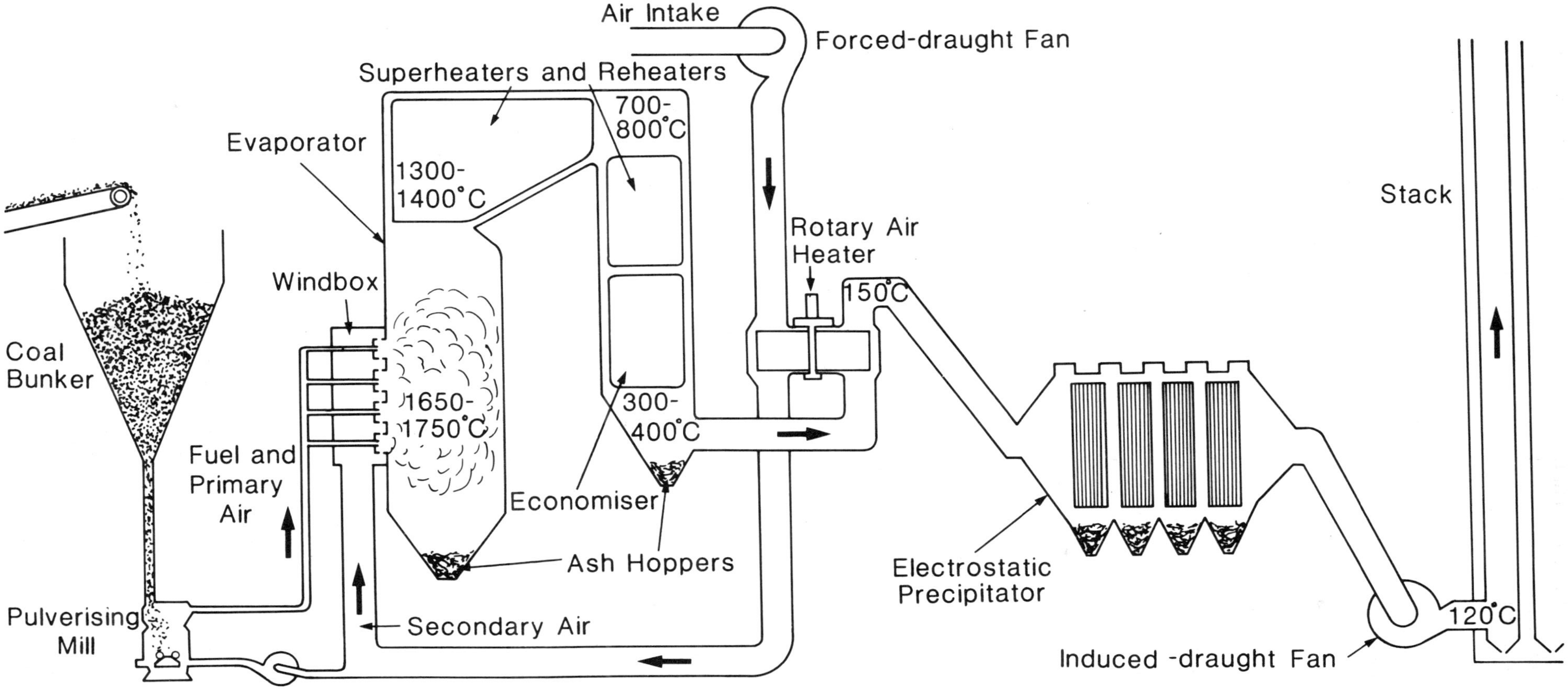

Source: C E G B 1977

137

16.25 All types have reached a high point in design and efficiency, and the main trend over the last decade has been in reducing both capital cost and the space taken up, by increasing the heat release in the combustion chamber and the heat transfer rate over the secondary surfaces. Developments of packaged forms of water tube boiler have brought this into the range formerly exclusive to the shell boiler.

Stokers and burners for industrial boilers

16.26 *Mechanical stokers* established themselves in watertube boilers before the start of this century, their use becoming more widespread and extending down to smaller sizes of plant over the next half century. They were only superseded by pulverised fuel firing after the Second World War, but were then rapidly introduced into the shell boiler field. Stokers incorporating fixed fuel beds can be reliable and easy to operate and control. They have the advantage over pulverised fuel firing of burning the coal as delivered without need for grinding and the fire can be banked when not operating. Acceptable levels of grit and dust emissions can be achieved by simple equipment such as cyclones. The coal coking properties are important and the coke that is formed – which can vary from large fused masses to light friable powders – largely determines the completeness of combustion and it also affects the discharge of the ash. The size of the coal is important for mechanical stokers. The important differentiation is between graded coal, having a limitation on both maximum and minimum particle sizes, and "smalls" which have a specified upper size only. Unwashed coal is cheaper than washed coal but its high and often variable ash content limits the types of stokers on which it can be used. The composition of the ash may be significant if the ash fusion temperature is abnormally low or if the chlorine content of the coal is abnormally high.

16.27 All modern stokers should be able to meet the requirements of present Clean Air legislation. Some smoke emission during light-up from cold occurs however with many stokers (and is allowed for in Clean Air legislation) while some stokers may give intermittent smoke in operation. All stokers emit some grit or dust particles into the gas stream, the quantity increasing with the more highly rated appliances. The particle size of this emission is such that a large proportion of it can be caught by simple, and hence reasonably inexpensive, equipment. In the following paragraphs we describe briefly the types of stoker in common use.

16.28 *Gravity-feed stokers* are used to burn anthracite or coke in small boilers where their simplicity and cheapness offset the higher cost of the fuel. The fuel flows from a hopper above the combustion region down an inclined grate. The bed is held in position by the ash which collects at the bottom of the grate. This ash is removed periodically by a ram or scraper.

16.29 *Underfeed stokers* are the simplest and cheapest class of stoker for coal consumption in the range 0·03–0·15 tonnes an hour. The operation is illustrated in Figure 16.2. Coal is forced up into the bottom of a burning bed with the result that the burning coal is moved up and eventually falls away sideways, burning away on the outer edges of the bed from where the ash or clinker is normally removed periodically and manually.

Figure 16.2 Underfeed stoker

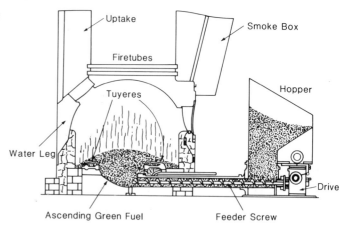

Source: *Ministry of Fuel and Power 1944*

16.30 *Spreader stokers* feed the coal on to the grate by distributing it on to the top of the bed through the furnace space, as illustrated in Figure 16.3. The coal on the bed is moved horizontally across the area of the grate so that the ash is brought to one point where it is discharged from the bed.

Figure 16.3 Spreader stoker

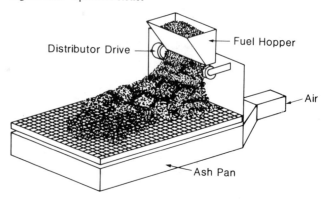

Source: *N. Berkowitz 1979*
(copyright © *1979 by Academic Press Inc*)

16.31 In the *coking stoker* the fuel is first deposited on a coking plate where the volatile matter is partially released to burn above the bed. It is then pushed by the incoming coal along and down on to grate bars where the coke burns, the ash being discharged over the end of the grate (Figure 16.4).

Figure 16.4   Coking stoker

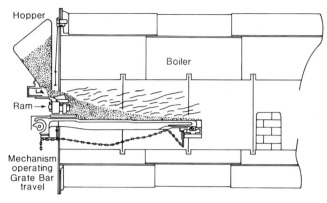

Source: *Ministry of Fuel and Power 1944*

16.32 In stokers of the *travelling-grate type* the coal is fed on to the grate under gravity from a bunker, the bed depth being controlled by a guillotine as shown in Figure 16.5. The coal is then carried through the furnace on a moving grate, the residual ash and clinker being discharged over the far end. To maintain ignition at the top of the bed, an arch of refractory material is used to radiate heat down on the bed from the coal burning further along the grate. Chain grates have been built in all sizes up to units for use in watertube boilers burning 10 tonnes of coal an hour. Probable future applications will however be mostly limited to shell boilers.

Figure 16.5   Travelling-grate stoker

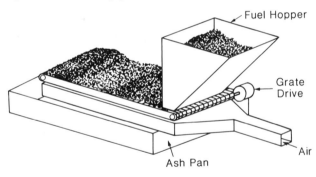

Source: *N. Berkowitz 1979*
(copyright © 1979 by Academic Press Inc)

Fluidised bed combustion

16.33 Fluidised bed combustion (F B C) of coal involves burning it in a bed of its own ash or other granular refractory material kept in a state of turbulent motion by the upward flow of combustion air through the bed. (Figure 16.6). Because of the thermal inertia provided, it is tolerant to changes in fuel characteristics and will handle a wide range of fuels, including low grade fuels. To prevent fusion of the ash, the temperature of the bed is kept below 900°C. A potential advantage of F B C is its ability to remove sulphur from the fuel by the addition of limestone or dolomite to the bed. To remove a high proportion of the sulphur, however, an amount of limestone considerably greater than that required for reaction

with the sulphur must be added and this doubles the quantity of ash. F B C concepts fall into two distinct categories according to the depth of the bed – deep beds and shallow beds.

16.34 *Deep bed fluidised combustion* usually involves burning crushed coal in a bed of its own ash in which boiler tubes are immersed to keep its temperature below the melting point of the ash. This type of system is favoured for water-tube industrial boilers which can benefit from the high rates of high transfer achievable in fluidised beds. The main industrial thrust in the U K on deep bed F B C is being made by Babcock & Wilcox (U K). For several years they have operated a converted water-tube boiler as an atmospheric pressure demonstration unit. They are licensees of the technology developed by the N C B, B P and National Research and Development Corporation (N R D C) and are able to offer industrial water-tube boilers with coal consumption up to 2·5 tonnes an hour with standard commercial guarantees.

16.35 Pressurised F B C is at an earlier stage of development. It is hoped that the hot exhaust gases can be fed to a gas turbine before being used to raise steam. Such a system would be suitable for a very large scale user such as a power station. A pressurised combustion facility burning 10 tonnes of coal an hour has been constructed (without a gas turbine) at the N C B's Grimethorpe Colliery under an I E A agreement entered into by the U K, U S A and Germany. A programme of testing and development has been commenced which will provide some of the data from which a full-scale prototype combined gas and steam turbine power plant can be designed. Much work, both of test bed type and engineering development, is needed to bring this technology to commercial reality and a prototype demonstration plant of commercial significance will be needed to prove not only the combustion process but also gas cleaning, system control and gas turbine performance.

16.36 In *shallow bed fluidised combustion* systems the coal burns in or on a shallow bed of refractory particles. Existing designs operate at atmospheric pressure and do not require the coal to be crushed to the same extent as for deep beds. They are most likely to be suitable for shell or sectional boilers burning 0·01–3 tonnes of coal an hour. The technique is also applicable to larger water-tube boilers but some form of bed cooling then becomes necessary to avoid ash fusion. Because the gases traverse shallow beds very quickly, they are less effective for sulphur removal than deep beds. The N C B have an active programme of development in this field and several purpose-built prototypes are undergoing field testing. Work is also well advanced on the technically more difficult "retrofit" burner systems designed to replace oil or gas burners without all of the severe loss of boiler output that conversion to conventional stoker firing would incur. Keen interest is

Figure 16.6 Fluidised bed boiler

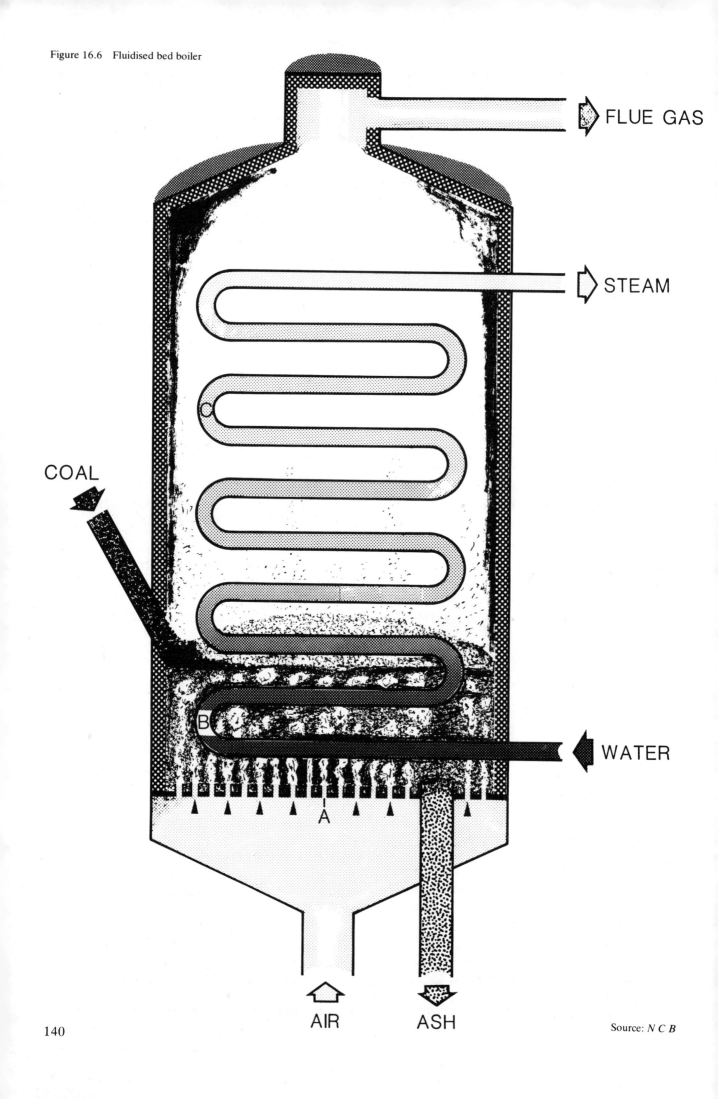

FLUE GAS

STEAM

COAL

WATER

B

A

AIR

ASH

140

Source: *N C B*

being shown by a number of industrial boiler manufacturers and several commercial models are expected to be on the market in two to three years time. As these systems are still experimental, their economics are not yet established. It is expected however, that their capital costs will be 30–50% higher than equivalent conventional oil-fired systems and that to break even coal prices would have to be 3–4p a therm less than oil.

Domestic appliances

16.37 The *simple open fire* burning coal has been the traditional British domestic heating appliance. Apart from the cheerful appearance of the fire it has few virtues. It will only burn high volatile fuels, such as wood and bituminous coal, and on average, some 3–4% of the coal is emitted as tarry smoke. It causes excessive ventilation and heat loss up the chimney. The direct radiant efficiency is around 25%, but up to a further 4% can be gained from a chimney centrally-located in the house. The addition of a back boiler can recover up to a further 10% of the heat in hot water. Little direct control of the rate of combustion is possible except by the refuelling regime and by banking it with fine coal, which causes considerable smoke emission. In smoke control zones, ordinary open fires can burn only the more expensive smokeless fuels of the reactive type (Coalite, Rexco and Homefire).

16.38 *Improved open fires* were developed 20 years ago to permit unreactive smokeless fuels, primarily coke, to be burned on open fires. A deep bed of fuel is contained in a refractory-lined grate which is effectively sealed into the fireplace and the airflow through the bed can be controlled. Their radiant efficiency is increased to 35–40%. Models equipped with high output back boilers can provide hot water for additional radiators and an overall efficiency (including chimney gains) of up to 60% is then obtainable. Figure 16.7 illustrates open fires with high output back boilers, both the conventional type (for smokeless fuel) and a new type, which can burn bituminous coal, with a similar efficiency. With the new type smoke emissions from coal have been significantly reduced but not yet sufficiently for use in smokeless zones.

16.39 *Closed room heaters and boilers* achieve the highest efficiencies because better control of combustion air and of room ventilation is possible. They are usually designed to burn coke, Phurnacite or anthracite; some designs may be suitable for bituminous coal but it is not universally recommended because the volatile matter may cause the build-up of soot in the air passages and possibly may form explosive mixtures in the flue. Overall efficiencies (including chimney gains) of 65–75% can be achieved, the higher figures applying to roomheaters with integral boilers and to independent boilers.

16.40 *Openable roomheaters* may be operated with the front door either open, so that the fire can be seen, or closed. Efficiencies are somewhat lower than for closed appliances. To produce an attractive fire, coal or reactive smokeless fuels are usually necessary, but when closed they will generally burn coke.

16.41 *Smoke-reducing appliances* are relatively novel, the first generation being marketed about ten years ago. These appliances are designed to burn graded bituminous coal so that the overall smoke emission is no greater than with a conventional open fire burning smokeless fuel. A down-draught combustion system is employed in which air enters at the top and the combustion products leave at the bottom of the firebed. Volatile matter is thus heated as it passes through the bed and is burned in a separate hot, insulated refractory chamber behind the bed. Figure 16.8 illustrates a combined roomheater and boiler which, unlike some earlier appliances, can operate without a fan. The N C B attach great importance to the development and marketing of a suitable variety of smoke-reducing appliances to provide room heating and/or hot water for a range of outputs and they are working closely with the manufacturers on a number of designs.

16.42 We received evidence from some local authorities about the unsatisfactory performance of the earlier designs, but on our visit to the N C B's Coal Research Establishment we were impressed with the progress which has been made to simplify them and to ensure that they will be able to operate smokelessly under all likely operating conditions. But, in order to work smokelessly, we consider that these appliances will need to be carefully installed and properly operated and maintained.

16.43 Compared with open fires, where smoke emissions are estimated to amount to 3·5% of the weight of coal burned, the figures which we were given for smoke-reducing appliances were in the range 0·3–0·6%, which correspond to the smoke emissions from smokeless fuels. The N C B are still determining the major components of the smoke from various types of domestic appliances and fuels but preliminary results were said to show that the smoke from smoke-reducing appliances contains very little of the poly-aromatic hydrocarbons which are found in conventional tarry smoke, but no figures were available.

MAGNETOHYDRODYNAMIC GENERATION (M H D)

16.44 M H D is a technique by which the heat and velocity of ionised combustion gases might be harnessed to generate electricity directly by passing them at high velocity (1,000 kilometres per hour) through a powerful magnetic field. The exhaust gases may then be passed to a conventional steam plant

Figure 16.7   High output back boilers

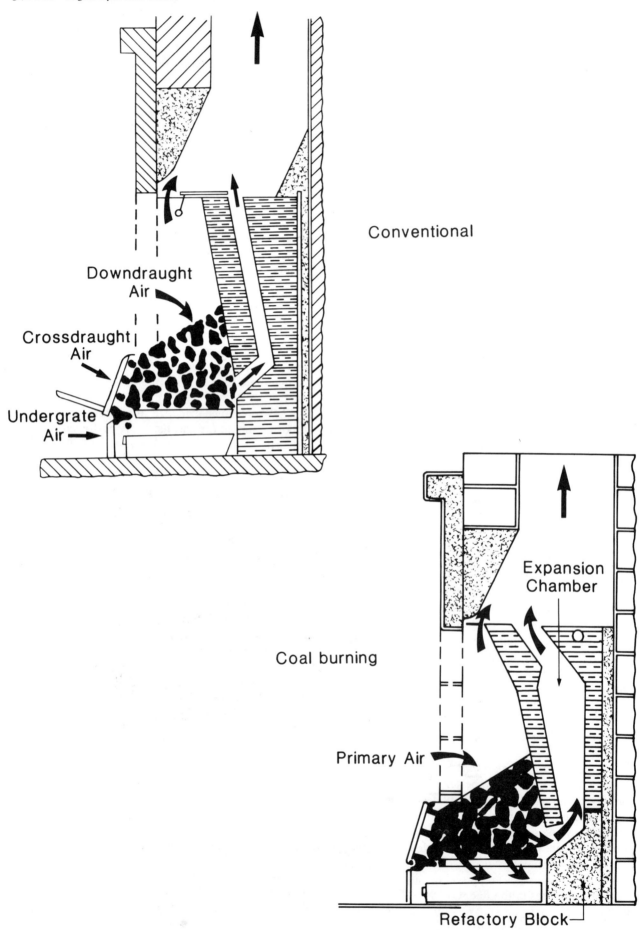

Conventional

Downdraught
Air

Crossdraught
Air

Undergrate
Air

Coal burning

Expansion
Chamber

Primary Air

Refactory Block

Source: *N C B*

Figure 16.8   Smoke reducing room heater with boiler

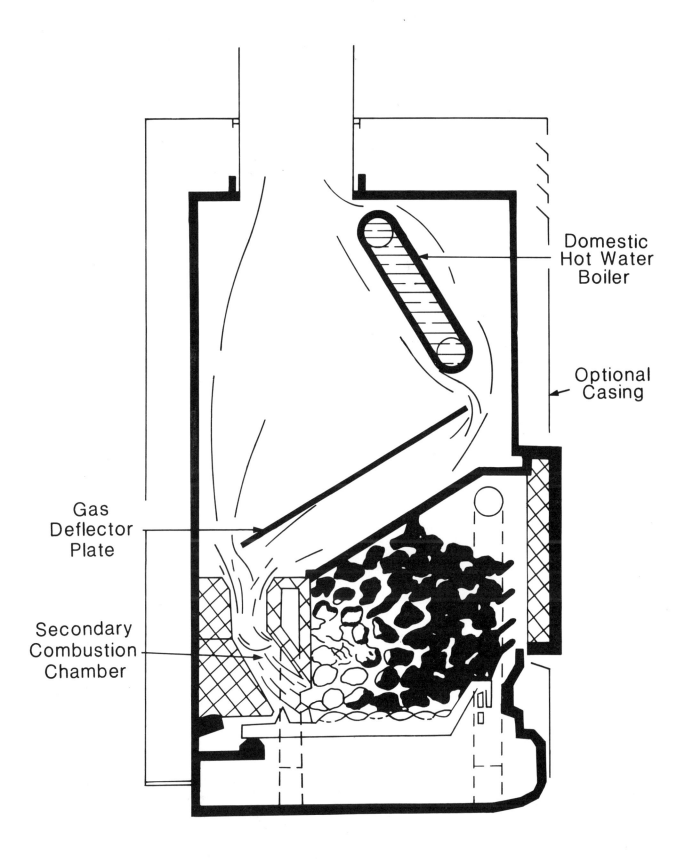

Domestic
Hot Water
Boiler

Optional
Casing

Gas
Deflector
Plate

Secondary
Combustion
Chamber

Source: *N C B*

coupled to turbo-alternators and overall the efficiency of generation may be raised to around 50%. Very high temperatures (2000–3000°C) are needed to ionise the gases and make them sufficiently conducting and this will require the air to be pre-heated considerably or to use oxygen for combustion, or both. To enhance their conductivity the gases are usually "seeded" with alkali metal salts which for economic reasons need to be extracted and recycled. To achieve the maximum output of electricity very high intensity electro-magnets are needed, for which superconducting coils cooled by liquid helium will be required.

16.45 M H D is still only at the experimental stage and the plants usually burn clean gaseous or liquid fuels. For large-scale economic application in Britain the system must use coal as the basic fuel. This does not necessarily imply that coal itself should be burned in the combustion chamber since a devolatilisation or gasification process could be employed. For direct coal burning the ash presents a significant problem as it may erode or corrode not only the combustor but also the downstream components such as the M H D duct, air preheater and the steam generator.

16.46 The U K carried out a major research programme on M H D in the 1960s but it was terminated because of the formidable technical problems involved. The C E G B do not consider that M H D systems could make a significant contribution to our electricity output during this century. A collaborative U S–U S S R programme has been pursued for several years but the U S A are now reducing their effort. Where natural gas is used in power stations, as in the U S S R, there will be fewer problems in developing M H D on a commercial scale.

COAL – OIL MIXTURES

16.47 Coal–oil mixtures are a technique for supplementing fuel oil by incorporating about 50% finely ground coal. Various techniques can be used to prevent the coal separating out, but the technology of production is not yet fully proven. Some problems with utilisation can be expected on water tube boilers designed for oil-firing, resulting from the deposition of coal ash. The technology is seen as having a "window of application" for a period when fuel oil prices are expected to rise considerably but existing oil-fired boilers have a lot of useful life left in them. In the long run, coal-oil mixtures should in general give way to direct firing of coal. British Petroleum are constructing a £5 million plant at West Thurrock to manufacture 100,000 tonnes a year of coal-oil mixture to test its system based on fine grinding, which it claims is the simplest for the long term stability of the suspension. The N C B, Shell and the Department of Energy have agreed to participate in a review by the I E A of the available world technology.

Chapter 17 Combustion residues of coal: Sources, quantities and control

INTRODUCTION

17.1 When coal is completely burned, the residues consist of gases and solid ash, some of which may be fine particulate material. The gases are largely carbon dioxide and water vapour, plus small amounts of sulphur dioxide, nitrogen oxides, hydrogen chloride and carbon monoxide. Other gaseous compounds are present in trace amounts. Some of the fine particles of ash may be carried in the flue gas stream and emitted to the atmosphere. Smoke, which is a mixture of soot, tarry matter and inorganic substances, is produced when bituminous coal is incompletely burned.

17.2 In considering pollutants it is useful to look at solids and gases separately. Solids need to be considered over a range of sizes in terms of whether they are small enough to be airborne; and if so, whether the airborne particles are small enough to be respirable. Solids of the largest size – ash and slag (fused ash) – are not considered as air pollutants but only as land pollutants at disposal sites, with possible implications for water pollution. Dust and grit are considered together as "particulates": they are essentially the smaller sizes of ash – possibly mixed with partially burnt coal – and can be transported by flue gases and thereby emitted to the atmosphere. Grit is not respirable; and only a small proportion of the finest dust particles is respirable. Smoke, which is largely carbonaceous material and may contain small quantities of carcinogenic substances, is largely respirable and in consequence it is more noxious than grit or dust. Trace metals and radioactive species, which are present in all sizes of particulate material, are considered separately. The gaseous pollutants are then discussed under the headings sulphur, nitrogen oxides, carbon dioxide, carbon monoxide, hydrogen chloride and other gases.

17.3 There is some confusion about the terminology used for smoke, particulates and ash. In this Report, these terms are used as follows:

Ash	The mineral residue left after the combustion of coal. It may include *pulverised fuel ash, bottom ash* and *fly ash*. The term is also used to mean the mineral content of coal (which will become ash after combustion).
Pulverised Fuel Ash	*Fly ash* extracted from the flue gas stream of pulverised fuel boilers.
Fly Ash	The fine particles of ash carried away in the flue gas stream.
Bottom Ash	Semi-fused or coarse ash which remains in the combustion chamber after combustion.
Particulates Particulate matter	Particles of solid material (including *fly ash, grit* and *smoke*) which are carried away in the flue gases after combustion. Particulates may be considered in separate size fractions as grit, dust or smoke.
Grit	Large particulate matter of size over 75 μm diameter. This largely consists of coked, incompletely burnt coal particles.
Dust	Small particulate matter in the size range 1 μm to 75 μm diameter which is mostly ash.
Smoke	Small gas-borne liquid and/or solid particles of size less than 1 μm consisting of carbon and complex carbonaceous compounds formed by the incomplete combustion of fuel.
Soot	Aggregates of carbonaceous particles impregnated with tar formed by the incomplete combustion of carbonaceous material, particularly bituminous coal.

Generally, particles of less than 15 μm remain in suspension for a long time and are capable of being inhaled; larger particles do not.

17.4 Control of pollution in the U K attempts to ensure that concentrations that would be harmful to human health are everywhere avoided or, as far as practicable, minimised. Works which are supervised by the Inspectorates (see paragraph 17.45) (which include power stations and gas works) are required "to use the best practicable means for preventing the emission into the atmosphere from the premises of noxious or offensive gases and for rendering harmless and inoffensive such substances as may be emitted". (Health and Safety at Work etc Act 1974).

17.5 Information on a wide variety of pollutants is published by the Department of the Environment in "Digest of Pollution Statistics" (1980). We have found these invaluable in identifying the impact of coal combustion with respect to other sources and in analysing the trends. We have reproduced several tables and illustrations from the 1979 and 1980 publications (see Tables 17.1–17.3, 17.6, 17.7 and 17.12, and Figures 17.1–17.4 and 17.6). It should be noted that the statistics are produced on several different bases – emissions, depositions, and urban concentrations averaged over different periods. Whilst total emissions are not without interest, it is local concentrations which are the cause of undesirable effects and which determine the areas where remedial action may be needed.

SMOKE

Source and quantity

17.6 The formation of smoke due to incomplete combustion is not a problem with pulverised fuel boilers installed in power stations and the largest industrial boilers. Similarly, medium and small sized industrial and commercial units, when properly operated with appropriate grades of coal, do not give rise to smoke except on limited occasions such as start-up. On the other hand, domestic appliances, because of the difficulties of providing adequate combustion control, nearly always produce some smoke. When bituminous coal is burned on an open fire, some 3–4% of the weight of the coal is emitted as smoke. The substitution of smokeless fuels reduces emissions to about one-fifth with only traces of tar. As we mentioned in the previous Chapter, new appliances have been developed for burning bituminous coal relatively smokelessly. The N C B have stated that these appliances produce very little tar in the residual smoke.

17.7 Background levels of smoke are monitored by the National Air Pollution Survey carried out by The Department of Industry's Warren Spring Laboratory. Figure 17.1 and Table 17.1 show that since 1960, smoke emissions from coal combustion have declined to one-sixth and that urban concentrations have declined correspondingly. The smoke has also changed in type and generally contains less tar than in former years, especially in smoke control areas. Of the remaining smoke 90% now comes from domestic sources. Table 17.2 illustrates the continuing fall in winter smoke concentrations at major urban sites in the U K. The average for U K urban areas in 1978/79 was 35 μgm^{-3} and at only 3% of the sites did average winter concentrations exceed 80 μgm^{-3}.

Control

17.8 Separate emission standards for smoke and particulates, where appropriate, are set for each process registered under the Alkali Act. Chimney emissions from industrial processes not within the scope of this Act are subject to the control of local authorities under the Clean Air Acts. Dark smoke may

not be emitted from the chimney of any commercial or industrial premises except for certain limited periods such as when lighting up. All new furnaces, except domestic boilers smaller than 55,000 Btu per hour capacity (sufficient for heating a typical three-bedroomed semi-detached house), must be capable, as far as is practicable, of operation without emission of smoke.

17.9 In areas subject to smoke control orders under the Clean Air Acts, smoke emissions from domestic chimneys are also forbidden. Householders continuing to use solid fuel must either install appliances capable of burning coal smokelessly or change to an approved smokeless solid fuel, such as coke, anthracite or various manufactured fuels. Over 8 million premises in over 5,600 control areas are currently covered by the Clean Air Act. Over 100 new orders covering about 200,000 additional premises are approved each year. In large urban areas subject to smoke control where little solid fuel is burned, for example, Central London, the principle source of smoke is now often road vehicles.

Table 17.1 Smoke: Trends in emissions from coal combustion: by source U K   million tonnes

	Domestic	Industry etc.[1]	Railways	All sources
1958	1·28	0·51	0·22	2·01
1959	1·18	0·41	0·20	1·79
1960	1·21	0·35	0·19	1·75
1961	1·13	0·27	0·16	1·56
1962	1·15	0·23	0·13	1·51
1963	1·13	0·20	0·11	1·44
1964	0·97	0·19	0·08	1·24
1965	0·95	0·14	0·06	1·15
1966	0·88	0·14	0·04	1·06
1967	0·80	0·12	0·02	0·94
1968	0·76	0·12	0·01	0·89
1969	0·70	0·08	0·01	0·79
1970	0·64	0·08	—	0·72
1971	0·55	0·06	—	0·61
1972	0·45	0·05	—	0·50
1973	0·44	0·05	—	0·49
1974	0·42	0·04	—	0·46
1975	0·35	0·04	—	0·39
1976	0·33	0·04	—	0·37
1977	0·34	0·04	—	0·38
1978	0·30	0·03	—	0·34

Source: *Department of the Environment 1980*

[1] Final energy users (ie energy other than fuel conversion industries); includes collieries, public services, agriculture and miscellaneous.

Notes: The emission factors used in calculating these figures are given below.
Domestic: Smoke is taken as 3·5% of the weight of coal burnt in domestic open fires.
Industry etc: For 1958, smoke is taken as 0·9% of the weight of coal burnt; for 1962 it is taken as 0·5%, and 1971 and subsequent years, 0·3%.
For the intervening years, intermediate proportionate values are used.
Railways: An estimate of 2% of the weight of coal burnt is used.

146

Figure 17.1 Smoke: trends in emissions and average urban concentrations

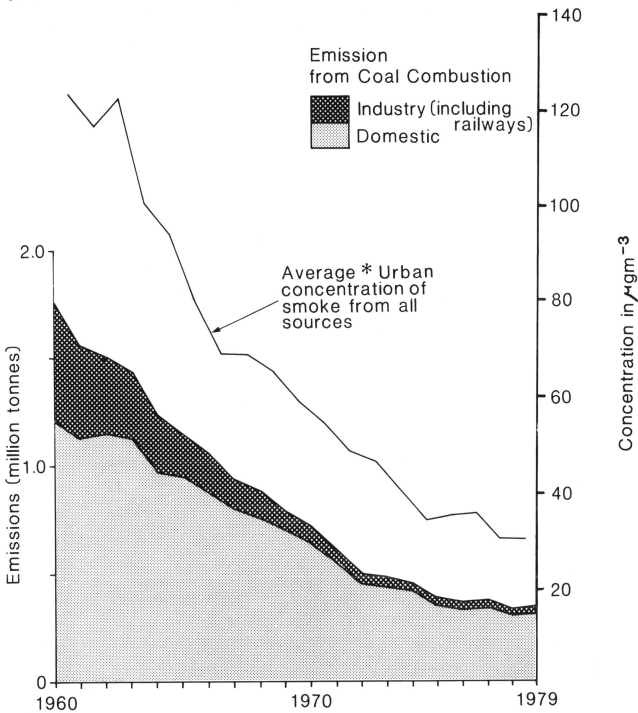

* Some actual urban smoke concentrations are given
in Table 17.2 , but see also the footnote to that table.

Source: *Department of the Environment 1980*

Table 17.2   Smoke: Daily concentrations at individual urban sites for the winter period[1]: by region and country U K

	Average concentration μgm^{-3}			Median concentration[2] μgm^{-3}			Percentage of sites with winter mean >45μgm^{-3}			Percentage of sites with winter mean >60μgm^{-3}			Percentage of sites with winter mean >80μgm^{-3}		
	1974/75	1976/77	1978/79	1974/75	1976/77	1978/79	1974/75	1976/77	1978/79	1974/75	1976/77	1978/79	1974/75	1976/77	1978/79
North	45	45	34	37	39	31	44	42	20	31	27	8	11	14	2
Yorkshire & Humberside	58	65	50	45	51	42	50	65	42	32	35	27	18	26	13
East Midlands	45	52	42	40	49	36	37	55	33	24	37	13	5	11	6
East Anglia	34	32	29	31	31	27	18	9	—	—	—	—	—	—	—
Greater London	35	37	30	32	34	29	18	20	8	8	3	3	—	1	—
South East (excluding Greater London)	24	25	23	22	25	21	2	4	2	<1	1	—	—	—	—
South West	24	25	21	26	23	23	—	3	—	—	—	—	—	—	—
West Midlands	45	47	29	37	39	35	36	40	30	16	23	9	11	9	4
North West	51	52	38	47	49	36	53	60	27	29	28	8	11	10	<1
England	41	44	36	34	38	31	31	38	22	18	20	9	7	9	4
Wales	27	32	25	25	30	25	8	13	5	3	8	3	—	3	—
Scotland	39	48	34	39	45	34	33	50	19	11	26	5	—	9	2
Northern Ireland	44	99	50	39	72	48	33	73	50	11	60	39	—	47	11
United Kingdom	40	45	35	34	38	31	30	38	21	16	20	9	6	9	3

Source: Department of the Environment 1980

[1] October to March

[2] The median is that value above and below which half the observations lie.

Notes: The figures refer to concentrations in urban areas. They are obtained by averaging over all sites in urban areas for which valid winter averages were recorded. Since the regional figures are not based on precisely the same sites from winter to winter, the figures are not strictly comparable between winters.

The ratio of the annual mean concentrations to the winter mean concentrations of all those sites in the U K recording both valid annual and winter means in 1978/79 (ie about 800 sites) was 0·74.

Table 17.3   Respirable particulates[1]: emissions from fuel combustion: by type of consumer and fuel U K

(a) By type of consumer

	1977		1978		1979	
	Thousand tonnes	Percentage contribution to total emissions	Thousand tonnes	Percentage contribution to total emissions	Thousand tonnes	Percentage contribution to total emissions
Domestic	366	67	330	64	337	64
Commercial/public service[2]	13	2	12	2	12	2
Power stations	25	5	26	5	28	5
Refineries	5	1	5	1	5	1
Other industry[3]	69	13	67	13	68	13
Rail transport	10	2	10	2	10	2
Road transport	60	11	63	12	64	12
All consumers	549	100	512	100	525	100

(b) By type of fuel

	1977		1978		1979	
Coal	398	72	361	70	372	71
Solid smokeless fuel	44	8	43	8	43	8
Petroleum						
Motor spirit	26	5	28	5	28	5
Derv	34	6	35	7	36	7
Gas oil	19	3	19	4	18	3
Fuel oil	23	4	23	4	23	4
Refinery fuel	5	1	5	1	5	1
All fuels	549	100	512	100	525	100

Source: Department of the Environment 1980

[1] Refers to particulates <15μm in size. Larger particles are not regarded by some authorities to have a significant effect on human health.

[2] Includes miscellaneous consumers.

[3] ie other than power stations and refineries. This category includes agriculture.

Notes: The soiling capacities of particulates are not the same for every fuel. For example, particulates from combustion of diesel fuel have about 3 times the soiling capacity of particulates from coal combustion.

PARTICULATES

Source and quantity

17.10 As explained earlier, particulates may encompass a range of particle sizes and compositions, their common feature being that they are carried out of the combustion chamber by the flue gases. Some particulates are derived from the ash; some are incompletely burned coal and some (smoke) may be created from the incomplete combustion of volatile matter distilled from the coal. The ash which remains in the furnace is considered in the next section.

17.11 When graded coal is burned on industrial grates, the ash particles are comparatively large and most of it accumulates within the furnace, from which it is periodically removed and dumped. Some of the smaller particles together with a small proportion of unburned fuel are carried off with the flue gases as grit and dust. Grit is larger in size and is readily separated from the gases by simple equipment. Dust is more difficult and costly to remove and a small proportion of the finer dust is emitted with the flue gases. Particulates greater than $15 \mu m$ in size are too large to be respirable. Although they can be a nuisance, they are not hazardous to health. In the pulverised fuel systems of power stations, the ash is very fine and up to 90% of it may be carried out of the furnace as fly ash, only 10–30% remaining as bottom ash. Elaborate gas cleaning equipment is fitted which can collect up to 99·5% of the ash. Because the gases are emitted from tall chimneys, the remaining fine particulates are well dispersed. The particulates emitted from domestic fires are mostly smoke, but they also contain a small fraction of fine ash.

17.12 Table 17.3 lists total emissions of respirable particulates in the U K from all sources and all fuels combined. In 1979 the major source for all fuels was domestic, which accounted for 64%, industry and power stations accounting for only 13% and 5% respectively. The major sources by fuel type were coal and solid smokeless fuel (both mostly from domestic sources), which accounted for 71% and 8% respectively. It is noteworthy that the order of particulate discharge is the inverse of the order of ash production given in Table 17.4, (power stations 12·5, industry 1·2 and domestic 0·7 million tonnes a year).

Control and disposal

17.13 There are four main techniques for dust removal. In increasing order of efficiency and cost they are: cyclones, wet scrubbers, electrostatic precipitators and bag filters. Even small industrial plant will have cyclones, but only very large industrial plant and power stations can afford electrostatic precipitators or bag filters.

17.14 Cyclones swirl the gas stream so that a combination of centrifugal and gravitational forces separate out the particulates. Cyclones are cheap, simple, relatively maintenance-free, reliable, low in energy losses and can handle high particulate loadings. They are usually used as a first treatment to remove the coarser particles before the gases enter more elaborate particulate removal systems. However, they are effective only down to particle sizes of 5–10 μm and their overall collection efficiency is around 80–90%.

17.15 Wet scrubbers wash out the particulates with water sprays and remove around 99% of material larger than 2 μm. They are relatively high in energy consumption due to gas and water pressure losses. Water clarification plant is needed to recover the particulates. Careful supervision and maintenance is required to minimise plant corrosion, water effluent problems and entrainment of mist into the flue gases.

17.16 In electrostatic precipitators a high voltage is applied between slender wires or rods and metal plates or tubes in narrow vertical passages. A charge is imparted to the particulates which causes them to be attracted to the metal plates where they adhere loosely. The accumulated layer of particulates is periodically shaken loose by vibrating the plates so that the particulates fall into hoppers for disposal. Energy consumption is minimal and pressure losses are slight. They are usually used as the final stage of dust collection and efficiencies of over 99·5% can be achieved.

17.17 Bag filters provide the most efficient method of dust collection. A typical bag house for a large installation contains thousands of bags each about 30 centimetres in diameter and 10 metres long. Dust is removed periodically by a combination of reversed air flow and shaking. Although the captial costs are similar, the running costs of bag filters are about twice those of electrostatic precipitators. Both types can remove dust down to 0·1–0·3 μm in size, with bag filters being marginally more effective for the smallest particles.

17.18 Electrostatic precipitators are used in all modern British coal-fired power stations. The most recent C E G B specification calls for an efficiency of 99·5%. This satisfies the current requirement by the Inspectorates that the particulate content of flue gases emitted from new power stations shall not exceed 0·115 gm^{-3} (measured at 15°C, 1 bar and at 12% carbon dioxide content). A few plants, designed to the standards of a decade or more in the past, have fallen short of expectations under some conditions and are currently being modified to improve performance. The C E G B consider that such shortcomings can be avoided in future designs and that electrostatic precipitation will continue to prove a satisfactory form of dust control for future coal-fired power stations. There is little use of bag houses in this country because of their greater running costs. According to C E G B, typical stack emission of particulates from a modern power station is about 0·6% of total particulates produced, or 0·1% of the weight of coal burnt. For a 2,000 MW power station, this means up to 20 tonnes of particulates per day. The majority of this is less than 20 μm in size and is dispersed as if it were a gas. Less than 5% of the larger particulates are deposited relatively near the power station and the maximum deposition rates usually occur at distances of 4–5 kilometres.

17.19 Particulates are deposited down wind from the chimney over distances of many kilometres. Monitoring of particulate deposition around power stations is undertaken by the C E G B and the South of Scotland Electricity Board who regard as a theoretical basis for the design of new plants that, if they operate to the required emission standards, then they contribute, at worst, about 10% of total particulate deposition measured at open country sites. The majority of the depositions in rural areas is wind-blown dust and soil.

ASH

Source and quantity

17.20 The mineral matter in coal is chemically altered by combustion and becomes ash. Ash content and ash production by market sector in 1975/76 are given in Table 17.4. The great majority of ash comes from power stations. The handling and disposal of pulverised fuel ash from large boilers does present some problems. This ash is very fine and is produced on a very large scale. A 2,000 MW power station produces over 2,000 tonnes of ash per day on average, over 90% of which is very fine fly ash.

Table 17.4   Ash production in the U K in 1975/6

million tonnes

	Coal consumption	Ash content	Ash production
Power stations	73·1	17·1%	12·5
Carbonisation (coke etc)	17·5	6·0%	1·1
Industrial etc	12·1	9·6%	1·2
Domestic and manufactured fuels	15·0	4·7%	0·7
Others including export	1·5	6·3%	0·1
Total	119·2	Weighted average 13·2%	15·6

Source: N C B

Control and disposal

17.21 The Inspectorates require that particular care be taken to ensure that removal of pulverised fuel ash from arrestment plant is achieved without significant emissions. Dry ash must be transported in sealed containers, with the displaced air during the filling operation being vented to the arrestment plant. Conditioned (dampened) ash must be transported in properly covered containers to prevent drying out of the surface. Where ash is to be permanently stored on site, this must be carried out in a controlled manner to prevent dust blow. Control of the siting of new ash dumps is effected through the Town and Country Planning Acts.

17.22 About 40% of pulverised fuel ash is used commercially. Proportions in excess of this have been achieved, but levels are dependent upon rates of external demand arising from the civil engineering and building industries. Table 17.5 provides a breakdown of typical commercial applications. The remaining ash is dumped in land fill sites. Where it is possible and economic, the ash can be used to fill up worked out excavations or to recover water-logged land. For example, ash from several Midlands power stations is transported to the Peterborough clay pits. More often the ash is dumped in the vicinity of the power station.

17.23 Two disposal methods are used: lagoons and dry ash dumps. Ash mixed with water in a ratio of 2:3 is piped as slurry to lagoons created by bund walls and/or existing topographic features. When the lagoons are filled, they are allowed to drain and another lagoon can be created on top, if required. For dry ash dumps, the ash is conditioned with water to reduce dust nuisance and transported to the dumping site. The choice of disposal method is dependent on local circumstances and the relative economics.

17.24 Most dry ash dumps are covered by top soil and returned to agricultural or recreational use. Lagoons are drained and treated similarly when dumping has finished. Dust emission is controlled during the dumping process by careful management aimed at minimising the exposure of the unprotected ash surface to drying and weathering. The final surface treatment (for example, top soiling) is planned to follow closely behind the tipping face. For surfaces liable to long exposure, techniques for binding surface layers by spraying a thin polymer coating have been developed. Where top soil is too expensive, special planting and cropping procedures have been developed which can convert the dump to limited agricultural productivity over a number of years. Organic wastes such as sewage sludge can be applied to speed up the process. Techniques for planting mature trees on ash dumps for landscaping purposes are also available. For lagoon systems, dust control is achieved by the growth of bulrushes and other hardy marginal plants in the floating ash layer. The main potential hazards of ash arise from dust blow and from contamination of surface and underground waters either by leaching of certain compounds from the ash by rainfall or by physical erosion of the ash surface during run-off.

Table 17.5 Typical commercial applications for pulverised fuel ash from power stations

Use	Percentage of total ash make
Specialised road fill	15
Building block manufacture	13
Fill for building sites	4
Lightweight aggregate for concrete	2·5
Cement substitute for concrete	1·5
Other commercial uses	2·3
Total	38·3

Source: *C E G B*

17.25 Ash contains a wide range of elements but only about 2% of the ash is soluble in water and around 1% of the soluble material consists of elements which are potentially harmful. The high alkalinity of the liquid medium inhibits the solution of many of the heavy metals. The volumes of water required to transport ash to lagoons carry away much of the material that is immediately soluble. This also applies to ash stored in lagoons before transportation to the final disposal site. Experiments by the C E G B have demonstrated that when rain falls on such saturated ash, percolate appears quickly and the initial leachate contains much lower concentrations of dissolved solids compared to percolate from dry ash dumps.

17.26 In the case of dry ash dumps, the volume of water required to saturate the dump before any leachate appears can be equivalent to several years of normal rainfall. The highest pollution levels occur in the initial leachate and they decline considerably thereafter. The World Health Organisation (W H O) recommended standards for drinking water are initially exceeded slightly for only a few of the potentially toxic elements, but further dilution in the aquifer reduces the concentrations to acceptable levels. A further factor which reduces the concentration of soluble material is the presence of gravel. Absorption by gravel reduces the concentration of most elements to about 15% of the initial concentration.

17.27 Hydrological data is necessary to select appropriate management strategy and dumping rates. In addition, according to the C E G B evidence, wells or boreholes should be avoided within the immediate vicinity of ash dumps; but this precaution is unlikely to be needed more than a few hundred metres from the site. Minor local problems from disturbance by noise and dust around ash dumps can occur, but are essentially no different from those normally associated with any large industrial enterprise. Little concern about particulates and ash was expressed to us. More attention was devoted to the availability of disposal sites.

17.28 There are minor problems with ash disposal from small boilers. However, the ash produced is mainly bottom ash. Although this is generally rather heterogeneous, the quantities produced by individual units are not large enough to cause serious problems. Some care must be taken in ash handling and transport to prevent emissions of dust and grit.

17.29 Because fluidised bed combustion operates at a lower combustion temperature than stoker-fired boilers, it is expected that the ash produced will be more consistent in quality and of smaller particle size than that from stoker-fired boilers. If limestone is added to the bed to capture sulphur then the solid residues will include both unreacted and sulphated lime. There are established commercial markets for pulverised fuel ash and, although it is possible that fluidised bed combustion ash will have similar potential, further testing and research work is needed to assess these possibilities. This will not be possible until the ash is available in larger amounts than at present. However, disposal problems for ash from fluidised bed combustion may arise, particularly if large amounts are produced in urban areas. But as for pulverised ash and municipal rubbish in general this is likely to be largely a question of finding suitable sites for disposal.

TRACE ELEMENTS AND RADIOACTIVITY

Trace elements

17.30 The concentration of a great many trace elements in coal is close to their average concentration in the earth's crust. Almost all known elements have been identified in coal, most of them in only minute amounts. Since most elements remain in the ash after combustion, their concentration becomes enhanced due to the removal in combustion of the carbon and moisture. Enrichment of some elements in the fine particles of pulverised fuel ash occurs as a result of the condensation of elements which are volatile at combustion temperatures. Small amounts of trace elements are emitted to the atmosphere either as vapour or are contained in solid particles which escape collection in the gas cleaners.

17.31 The potentially toxic effects of some trace elements resulting from coal combustion have only recently been investigated in detail and only a limited amount of work has been carried out. Both the United Kingdom Atomic Energy Authority and the C E G B have carried out an exercise in the Selby area to establish concentrations of trace elements in different size ranges of airborne particulates at ground level. This work to date has revealed that when standard non-directional collectors were used, it was not possible to identify any component of trace elements in the samples that could be related specifically to power station operations. The relative proportions of elements contained in the samples were similar to those in other parts of the country where there are no power stations. When directionally sensitive collectors were employed near Drax, some small evidence of enrichment was found. This was not regarded as significant in relation to public health. The C E G B work is continuing, but they consider the results to date show no cause for concern. We are in agreement with this conclusion but also agree that investigations should continue. As was discussed earlier (paragraphs 17.25–17.27) investigation has also shown that there is no problem from leaching of any toxic trace elements from ash disposal sites.

Radioactivity

17.32 Radioactivity is a particular aspect of trace elements. As already noted coal contains most trace elements, including radioactive species, in much the same proportion as other rocks. Therefore coal is about as radioactive as any average rock or soil, that is to say, hardly at all. On combustion, most of the radioactive elements are concentrated in the ash, which is therefore more radioactive than coal itself, but the ash is still no more radioactive than granite, or several other common rocks including many shales, basalts and clays.

17.33 The exposure level attributable to the radioactive elements in coal has been studied by the C E G B and the National Radiological Protection Board (N R P B) in Britain (Corbett J O 1980; Camplin W C 1980) and by others in the U S A. The C E G B compared the exposure of power station workers and members of the public living close to power stations with the average exposure of the total U K population to natural background radiation, taking account of the appropriate internationally accepted safety standards. Their figures were based on estimated ground level concentrations, assuming a 1,000 M W power station operating continuously and emitting particulates at the maximum allowable rate. They concluded that, on these assumptions, it would take about 100 years to increase by 1% the radioactivity of typical surface soil, even making the most pessimistic assumptions about absorption and retention.

17.34 Power station workers have the highest potential exposure from inhalation of dust and from ash handling operations. The C E G B 's data suggest that the maximum potential exposure for such workers is unlikely to exceed 150–200 microsieverts per annum, that is about 11% of the average natural dose or 1% of the limit recommended by the International Commission for Radiological Protection (I C R P) for occupational workers not subject to personal radiological monitoring on a routine basis. It is less than the variation in natural exposure resulting from living in different parts of the country.

17.35 Exposure of the general public is discussed in the section on health effects in the next Chapter. It is concluded there that exposure resulting from emissions

of radioactive material from coal-fired power stations is extremely small and not a cause for concern. Similar results have been obtained from studies in the U S A (for example, Myers G K, 1978). More research work needs to be carried out in this area but none of the figures quoted to us suggested significant hazards.

17.36 In terms of radiation dose to the public, the emission of radioactive material from coal-fired power stations is in fact comparable with the very small amounts arising from the normal operation of present nuclear power stations including emissions from the nuclear fuels reprocessing cycle. In both cases the average dose equivalent received by members of the public is far less than the natural variations in the annual dose received from all natural sources. These conclusions apply to normal operating conditions but so far as coal is concerned the radiation cannot increase appreciably as a result of malfunctioning of the plant.

WATER POLLUTION

17.37 Water pollution does not generally result from coal combustion. Thermal and other pollution from cooling water is a problem of energy conversion. Mine water pollution relates to coal extraction. There are only two points where coal combustion could cause water pollution. The first is drainage from pulverised fuel ash disposal sites. But as noted above this is not a serious problem because of the insolubility of most potentially polluting components, coupled with absorption in the sub-soil and dilution by ground water. The second is pollution by contaminated water from scrubbers used for particulate or sulphur control, which are only used on a small scale so the quantities involved are relatively unimportant. If it became necessary in the future to carry out flue gas desulphurisation (F G D – see paragraphs 17.64–17.68) on a large scale, as is currently the case in the U S A and Japan, water pollution could become a problem. It is a factor which should be included in any consideration of the advisability of flue gas treatment.

SULPHUR

Source and quantity

17.38 Sulphur is present in coal partly as organic and partly as inorganic compounds. The organic compounds are diverse and complex; the principal inorganic compound is iron pyrites, an iron sulphide. On combustion, nearly all of the sulphur is converted into gaseous sulphur dioxide, apart from a proportion which remains in the ash as a sulphate. On average, 10% is retained in ash from industrial boilers and 20% in the ash from domestic fires. About 1% of the sulphur dioxide is oxidised further to sulphur trioxide before emission. Some sulphur compounds are also emitted by natural biological and geological sources and a further contribution is imported by winds from elsewhere.

Table 17.6   Sulphur dioxide: Trends in emissions from fuel combustion: by type of consumer and fuel U K

(a) By type of consumer   Million tonnes

	Domestic	Commercial/ public service[1]	Power stations	Refineries	Other industry[2]	Rail transport	Road transport	All consumers
1970	0·52	0·39	2·77	0·24	2·12	0·03	0·04	6·12
1971	0·46	0·34	2·80	0·25	1·94	0·02	0·05	5·86
1972	0·37	0·31	2·87	0·26	1·76	0·02	0·06	5·65
1973	0·36	0·29	3·02	0·29	1·78	0·02	0·05	5·81
1974	0·35	0·26	2·78	0·30	1·59	0·02	0·05	5·36
1975	0·30	0·24	2·82	0·26	1·48	0·02	0·05	5·17
1976	0·28	0·24	2·69	0·28	1·44	0·02	0·05	5·00
1977	0·29	0·24	2·74	0·27	1·39	0·01	0·05	5·00
1978	0·26	0·23	2·81	0·29	1·37	0·02	0·05	5·02
1979[P]	0·26	0·23	3·01	0·29	1·39	0·01	0·06	5·26

(b) By type of fuel   Million tonnes

	Coal	Solid smokeless fuel	Petroleum						All fuels
			Motor spirit	Derv	Gas oil	Fuel oil	Refinery fuel	Total	
1970	3·34	0·25	0·01	0·03	0·16	2·07	0·24	2·52	6·12
1971	3·04	0·20	0·02	0·03	0·16	2·16	0·25	2·62	5·86
1972	2·65	0·18	0·02	0·04	0·20	2·30	0·26	2·83	5·65
1973	2·89	0·17	0·02	0·03	0·20	2·20	0·29	2·75	5·81
1974	2·56	0·16	0·01	0·04	0·17	2·11	0·30	2·64	5·36
1975	2·68	0·15	0·01	0·04	0·17	1·85	0·26	2·34	5·17
1976	2·74	0·13	0·02	0·04	0·16	1·63	0·28	2·13	5·00
1977	2·79	0·13	0·02	0·04	0·14	1·61	0·27	2·08	5·00
1978	2·75	0·13	0·02	0·04	0·14	1·66	0·29	2·14	5·02
1979	2·99	0·13	0·02	0·04	0·14	1·64	0·29	2·13	5·26

Source: *Department of the Environment, 1980*

[1] Includes miscellaneous consumers.
[2] ie other than power stations and refineries. This category includes agriculture.
P Provisional

17.39   The total quantity of sulphur dioxide released in the U K each year from all combustion sources is around 5 million tonnes, of which 3 million tonnes comes from coal. Estimates from Warren Spring Laboratory are given in Table 17.6 which shows emissions by type of fuel and by source. The N C B estimate future sulphur dioxide emissions to amount to 6 million tonnes per year by 2000, of which 4·25 million tonnes (70%) will come from coal. Warren Spring Laboratory forecast total sulphur dioxide emissions of 4·5–5·2 million tonnes per year by 2000 of which 2·9–3·4 million tonnes (65%) will come from coal. The differences are due to different assumptions about the growth of G D P and hence of fuel consumption. The increase in emissions is not as great as the forecast increase in coal burn because much of the coal will replace heavy residual fuel oil, which on a thermal basis has a higher sulphur content.

17.40   Figure 17.2 shows trends in emissions since 1960 by chimney height. Peak emissions of 6.12 million tonnes were reached in 1970. There has been a fairly continuous decline since then. Year to year fluctuations reflect variations in total fuel consumption which is affected both by the weather and national economic activity. Whilst emissions from low and medium level sources have fallen by nearly half since 1960, those from high level sources have increased by more than half. The upper line in Figure 17.2 shows that average urban concentrations reached a maximum in 1963 and have declined steadily since to less than half of these maximum figures. Table 17.7 gives some figures for daily concentrations at urban sites for the winter period. The effect is illustrated in Figures 17.3 and 17.4 which show large reductions in the areas of the U K which lie within the contours of 50–100 μgm^{-3} winter mean concentrations between 1972/3–1975/6.

Figure 17.2   Sulphur dioxide: trends in emissions and average urban concentrations

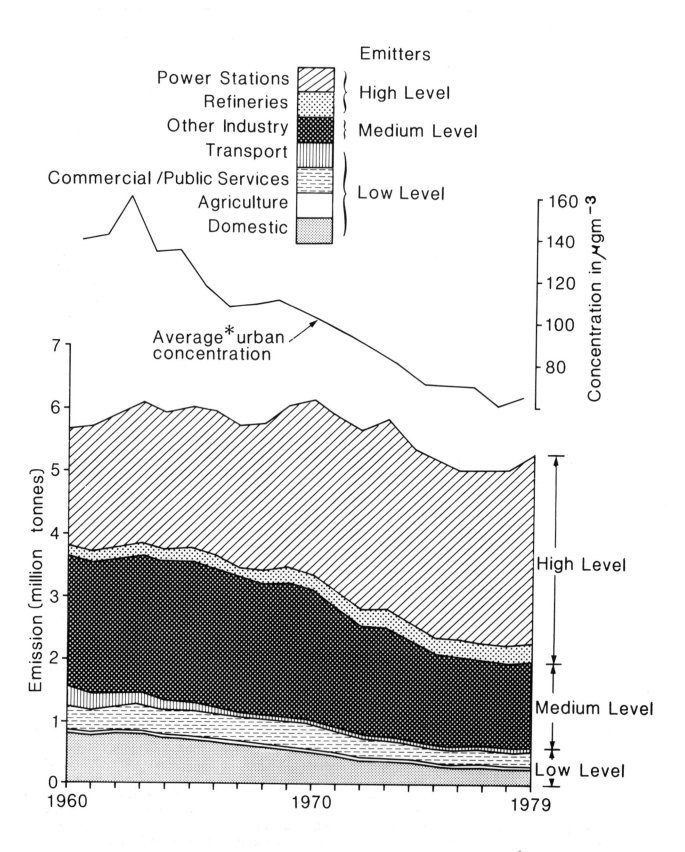

*Some actual urban sulphur dioxide concentrations are given
 in Table 17.7, but see also the footnote to that table.

Source: *Department of the Environment 1980*

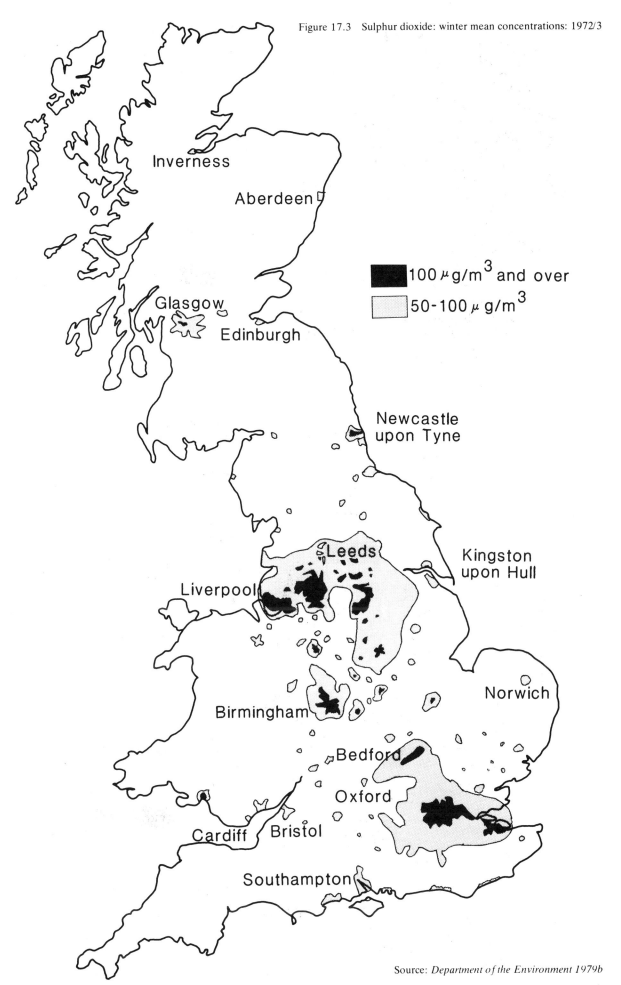

Figure 17.3   Sulphur dioxide: winter mean concentrations: 1972/3

100 μg/m^3 and over

50-100 μg/m^3

Source: *Department of the Environment 1979b*

155

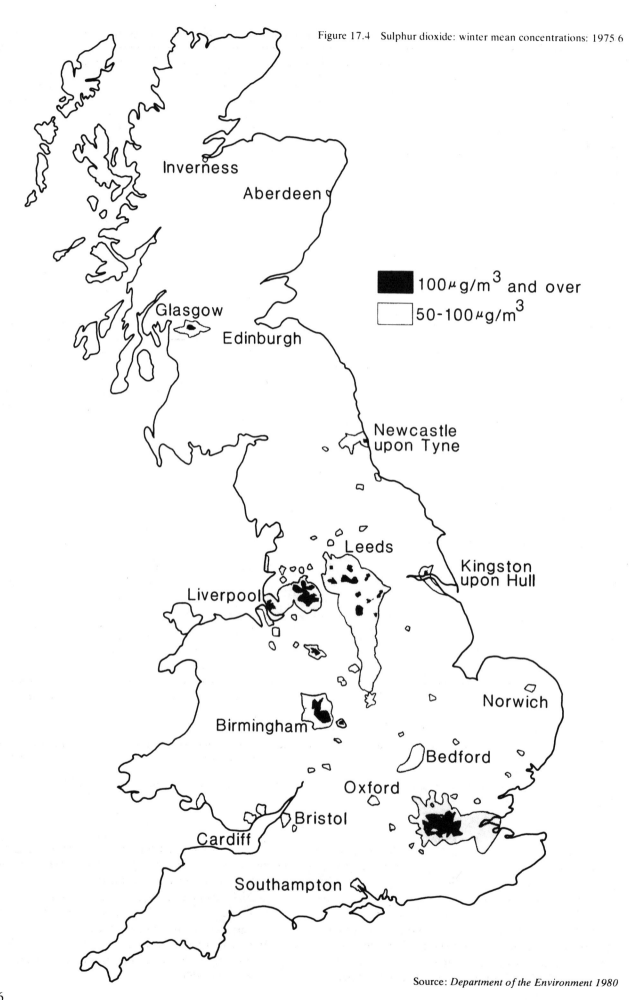

Figure 17.4   Sulphur dioxide: winter mean concentrations: 1975 6

100 μg/m³ and over

50-100 μg/m³

Source: *Department of the Environment 1980*

Table 17.7 Sulphur dioxide: daily concentrations at individual urban sites for the winter period[1]: by region and country

	Average concentration μgm^{-3}			Median[2] concentration μgm^{-3}		
	1974/75	1976/77	1978/79	1974/75	1976/77	1978/79
North	68	68	57	60	59	53
Yorkshire and Humberside	100	104	94	96	105	95
East Midlands	73	85	84	72	80	80
East Anglia	65	70	62	62	74	64
Greater London	117	110	100	107	100	94
South East (excluding Greater London)	62	59	59	64	56	56
South West	47	44	49	48	38	44
West Midlands	87	89	86	83	93	89
North West	96	109	91	95	108	91
England	86	87	80	82	83	76
Wales	56	50	54	56	51	57
Scotland	63	68	58	51	64	55
Northern Ireland	60	76	41	47	78	41
United Kingdom	82	84	76	78	78	72

	Percentage of sites with winter mean $>50 \mu gm^{-3}$			Percentage of sites with winter mean $>75 \mu gm^{-3}$			Percentage of sites with winter mean $>100 \mu gm^{-3}$			Percentage of sites with winter mean $>150 \mu gm^{-3}$		
	1974/75	1976/77	1978/79	1974/75	1976/77	1978/79	1974/75	1976/77	1978/79	1974/75	1976/77	1978/79
North	8	69	58	23	30	20	14	20	2	2	3	—
Yorkshire and Humberside	95	91	92	82	80	70	45	59	46	6	10	4
East Midlands	84	91	92	45	56	58	12	32	27	2	1	3
East Anglia	91	100	91	18	45	—	—	—	—	—	—	—
Greater London	99	98	97	90	84	84	57	53	42	22	18	9
South East (excluding Greater London)	68	63	72	26	24	20	4	4	2	—	—	—
South West	36	37	27	11	5	6	—	3	3	—	—	—
West Midlands	84	88	92	60	65	67	36	39	30	4	1	—
North West	95	95	96	79	78	70	43	60	38	5	14	3
England	85	82	84	58	57	52	30	36	26	6	7	3
Wales	63	50	59	24	13	18	3	—	—	—	—	—
Scotland	57	65	62	27	38	26	19	21	6	1	2	—
Northern Ireland	44	73	35	22	53	—	11	27	—	—	7	—
United Kingdom	80	79	79	53	53	47	27	33	22	5	6	2

Source: *Department of the Environment 1980*

[1] October to March.

[2] The median is that value above and below which half the observations lie.

Notes: The figures refer to concentrations in urban areas. They are obtained by averaging over all sites in urban areas for which valid winter averages were recorded. Since the regional figures are not based on precisely the same sites from winter to winter, the figures are not strictly comparable between winters.

The ratio of the annual mean concentrations to the winter mean concentrations of all those sites in the U K recording both valid annual and winter means in 1978/79 was 0·82.

17.41 Because chimney height has such a great effect on dispersal rates it is difficult to relate actual ground level concentrations to emissions of the pollutant from particular sources within a given area. The Department of the Environment have however made some estimates of the contributions from different sources for the most heavily polluted large urban areas – Central London, Central Manchester, and Central Liverpool. Table 17.8 reproduces that breakdown for 1975. If we assume that the relative contributions of the particular sources are typical of all urban areas we can draw four general conclusions:

(i) high level emitters (power stations and refineries) contribute less than a quarter of the sulphur dioxide at ground levels;

(ii) at least half the sulphur dioxide is derived from oil, and for Central London the proportion exceeds 80%;

157

(iii) in the business centres of London and Manchester, the contributions from commercial establishments exceed those from industry;

(iv) domestic solid fuel contributes 10–23% of urban concentrations, whereas in terms of total national emissions of sulphur dioxide, they amount to only 5%.

This underlines the relative importance of the contribution of low level sources of urban pollution. Projections of ground level concentrations in these three urban centres have been made by Warren Spring Laboratory for the year 2000 on the basis stated in paragraph 17.39. These indicate that reductions of 16–29% can be expected in Central London, 20–33% in Central Manchester and 23–45% in Central Liverpool.

Table 17.8   Contributions to ground level concentrations of sulphur dioxide, 1975

μgm^{-3}, Annual Averages

	Central London	Central Liverpool	Central Manchester
Domestic			
Coal	13	13	10
Smokeless Fuel	—	6	8
Burning Oil	1	—	—
Commercial			
Solid Fuel	2	4	5
Gas Oil	16	3	6
Fuel Oil	50	7	42
Industry			
Solid Fuel	3	4	8
Gas Oil	3	1	2
Fuel Oil	17	24	35
Road Transport	4	1	3
Power stations			
Coal	6	13	18
Fuel Oil	8	1	2
Refineries	2	4	2
External sources	12	—	—
Total	137	81	141
Proportion from:			
Coal	19%	50%	35%
(of which domestic)	10%	23%	13%
Oil	81%	50%	65%
High level emitters (Power Stations and Refineries)	12%	22%	14%

Source: *Department of the Environment*

Control

17.42 Sulphur dioxide is included in the list of "noxious and offensive gases" for which legislation requires that the best practicable means shall be used to prevent their escape into the atmosphere and for rendering such gases where discharged harmless and inoffensive. Although major combustion sources are large sources of sulphur dioxide, the gas is relatively dilute (0·1–0·2%) in the flue gases and consequently, until recently, there have been many technical problems in removing it. The costs of F G D remain high and the gases are usually discharged from tall chimneys so that the ambient concentrations of sulphur dioxide are reduced to generally harmless levels.

17.43 In some countries however, notably the U S A and Japan, environmental policies have been pursued which require some plants (usually new power stations) to use low sulphur fuels or to install F G D equipment. The effects of sulphur dioxide and other pollutants on the environment are discussed in detail in Chapter 18. It is clearly important that any sulphur control policies which might be considered should take into account not only the possible magnitude of any detrimental environmental effects but also the costs of possible control measures.

17.44 In the rest of this section on sulphur, we outline the current U K policy for dispersing emissions and discuss the connections it may have with acid rain. This is followed by a review of the general methods available for achieving some reduction of sulphur dioxide emissions from the major sources and of the scope for and costs of their application. The conclusions on control of emissions should of course be read in conjunction with the conclusions of the subsequent Chapter on their effects.

17.45 H M Alkali and Clean Air Inspectorate is responsible in England and Wales for administration of the control of air pollution from works registrable under the Alkali Etc Works Regulation Act 1906, as amended by the Alkali Etc Works Orders 1966 and 1971 and the Control of Pollution Act 1974. In Scotland, H M Industrial Pollution Inspectorate are responsible for the administration of the Alkali Etc Works (Scotland) Order 1973. Works which are so registrable and which use appreciable quantities of coal include electricity, gas and coke works, ceramic works, cement and lime works.

17.46 For the purpose of the Alkali Act, air pollution is identified in the list of "noxious and offensive gases" which include smoke, grit and dust, hydrochloric acid gas, acid-forming oxides of nitrogen, fumes containing a wide range of common and trace elements and their compounds, carbon monoxide and sulphur oxides. As explained above, Bpm for dealing with combustion gases is normally by dispersal from tall chimneys. Since there is much more sulphur dioxide present than other noxious gases, the chimney height is determined with respect to sulphur dioxide. For registered works the calculation basis is that the mean ground level concentration measured over a 3 minute period at the point where the plume has maximum effect at ground level, under neutral atmospheric stability and in a 6 metre per second wind speed at ground level is in the range 400–450 μgm^{-3}. The point of maximum ground level concentration from a plume will shift continuously with wind strength and direction, so that the maximum concentration averaged over a day at a fixed point would be less than 100 μgm^{-3} and averaged over a year less than 5 μgm^{-3}.

17.47 A different criterion, which is called the Threshold Limit Value (T L V), is used by the Health and Safety Executive (H S E) for the protection of workers; this is currently set at 13,000 μgm^{-3}. The T L V is the time-weighted average concentration for a normal 8 hour work day or 40 hour work week to which nearly all workers may be repeatedly exposed without adverse affects. We would draw attention to the high value for the T L V in comparison with the E E C Air Quality Directive's Limit Values for the protection of human health discussed in Chapter 19. The E E C Limit Values are intended to apply when smoke is also present. These figures suggest that in the mix of pollutants in urban atmospheres smoke is of greater significance than sulphur dioxide.

17.48 The Inspectorates' mandate extends only to Scheduled Works. For the majority of new industrial boilers and furnaces which are not controlled by the Alkali Inspectorate, the minimum chimney height is regulated by the local authorities under the Clean Air Act in accordance with the Memorandum on Chimney Heights (issued by the Ministry of Housing and Local Government in 1967). This Memorandum includes detailed graphical methods for calculating chimney height using essentially the same criteria as are applied to registered works. It takes account of different situations by, for instance, requiring greater chimney heights in heavily built up areas with background pollution from existing works than in undeveloped areas with no other emitters in the locality, and by taking account of the dimensions of adjacent buildings.

17.49 Emissions from chimneys are carried downwind, expanding and diluting by atmospheric turbulence, and the highest ground level concentrations occur shortly after the lower edge of the diffusing plume reaches the ground. The maximum effect occurs only a few metres from a domestic chimney whereas it is a few kilometres for chimneys some 200 metres high such as are used in power stations. Emissions from a large number of sources of varying heights, as in an industrial area, diffuse together so that 20 kilometres downwind it is normally no longer possible to distinguish the sources. Occasionally, however, meteorological inversions occur, particularly at night-time, and in winter they may persist all day. In an inversion the normal temperature gradient in the atmosphere (decreasing temperature with height) is inverted or reduced at some height, which marks the base of an inversion layer, and it forms an effective barrier to the natural vertical diffusion of gases. Inversions are associated with still air

and so conditions are conducive to the build up of pollution from low-level sources. On the other hand, the tall chimneys provided for large individual emitters usually eject the gases with sufficient velocity and thermal buoyancy to penetrate inversion layers and they are then effectively trapped above them. For this reason, as well as the fact that such sources are not generally situated in large conurbations, they make only minor contributions to urban pollution levels.

Acid rain

17.50 Sulphur dioxide is highly soluble in water which it renders acidic. It is thus rapidly washed out of the atmosphere by rain and is absorbed by water droplets in clouds. In the atmosphere it is slowly oxidised and reacts with water to form sulphuric acid. In turn, this may react with atmospheric ammonia from biological and industrial sources to form ammonium sulphate. This is often present in the atmosphere as an aerosol and is frequently a component of haze. Sulphur dioxide is also absorbed dry by vegetation, soils and other materials. Because of the variety of methods by which it is assimilated into the environment it has an average residence time in the atmosphere of only a few days and atmospheric concentrations and ground depositions decline rapidly away from source.

17.51 In some areas, such as Southern Scandinavia, the evidence we received suggested that the natural environment was being harmed by acid rainfall. This is discussed further in Chapter 18. At this stage we merely note that most sulphur dioxide is deposited in or near the regions where it is emitted and that depositions in Scandinavia are many times smaller than those in the industrial regions of Europe. This is illustrated in Figure 17.5 which shows the contours of deposition rates for Western Europe, based on 1974 data, as determined by a co-operative international project sponsored by the O E C D (1977).

17.52 There appears to be some misunderstanding about the contribution made by tall stacks to long range air pollution, namely that large emitters using tall stacks greatly enhance the transport of sulphur dioxide for long distances. This is not so: the long distance transport of sulphur dioxide is almost independent of chimney height. The main effect of tall stacks is to delay the time and distance before the plume returns to the ground, and thus reduce the ground path for sulphur dioxide absorption by some 10 kilometres. In exceptional weather, however, it appears possible that plumes can be isolated above an inversion layer and carried relatively rapidly for some hundreds of kilometres. The C E G B are now carrying out a research programme on the movement and dispersal of major plumes over the North Sea which should clarify the physical and chemical mechanisms involved in the long range transport of pollutants.

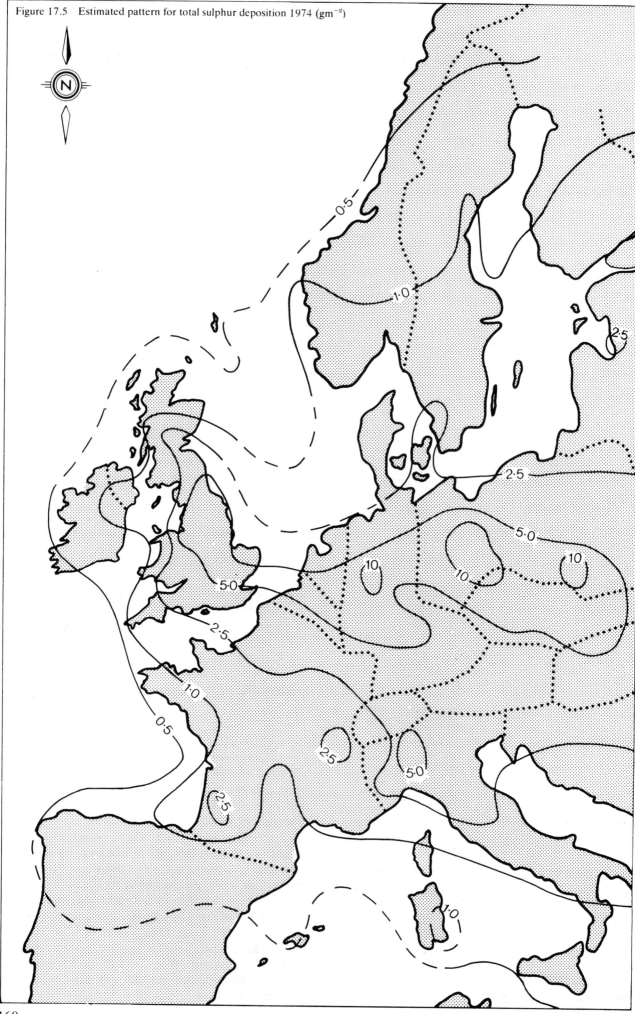

Figure 17.5 Estimated pattern for total sulphur deposition 1974 (gm^{-2})

Source: *Derived from O E C D 1977)*

17.53 The problem of sulphur emissions and depositions is of international dimensions and is being extensively investigated. Estimates from the O E C D study (1977) indicate that in 1974, when emissions of sulphur dioxide in Great Britain and Ireland were 5·36 million tonnes, 28% was deposited here, and 35% was deposited on the continent of Europe, most of the remainder being deposited in the sea. Norway was estimated to receive 2–3% of our emissions (representing 25% of total depositions there) and Sweden less than 2% of these emissions (representing 8% of total depositions). Of the sulphur deposited in Sweden, 20% originated there and similarly 12% of Norwegian deposition came from Norwegian sources. Most of the remainder came from the rest of Western and Eastern Europe and a small proportion from unknown sources probably outside Europe. U K emissions have declined since 1974 (Table 17.6) and are provisionally estimated to fall below 5 million tonnes for 1980. About 55% of U K emissions from combustion sources are due to coal (Figure 17.3).

17.54 Nitrogen oxides also react with air and water to form acids and in Europe they usually contribute about one third of the total acidity in rain. Emissions of nitrogen oxides (measured as nitrogen dioxide) are 1·9 million tonnes a year (Table 17.11 and Figure 17.6). The significance of nitrogen oxides, as is discussed in a later section, is that they are formed during combustion of all fuels. Hydrogen chloride, produced from chlorides in the coal, is another minor source of acid.

17.55 The mechanisms leading to the formation and deposition of acid rain are not yet well-enough understood to attribute its occurrence to particular source areas or to predict the effect which, for example, a policy of emission control in the U K or elsewhere might have on the environment of Scandinavia. Even if emissions from the British Isles were reduced to zero, depositions in Norway would be reduced by less than a quarter. Unfortunately, the evidence is insufficient for us to make any positive recommendations beyond the need to pursue the necessary research. We discuss this further in Chapter 18.

Options for reducing emissions

17.56 We had no initial preference on environmental grounds for the maximum reduction of emissions as opposed to more selective controls aimed at achieving minimum standards of air quality; the costs involved would inevitably fall on the public, either as energy consumers or as taxpayers. In our opinion a judgement has to be made as to whether the detriment involved in pursuing a policy of dispersal of sulphur dioxide is greater than the cost incurred in limiting its emission. The Department of the Environment and the O E C D have gone some way towards drawing up balance sheets, but there are immense difficulties in the way of producing a full and credible cost-benefit analysis.

These difficulties chiefly concern the identification and quantification of the detriment, which can have a variety of causes other than sulphur dioxide, as well as the problem of attaching monetary values to effects on health and amenity. We discuss this further in Chapter 18.

17.57 If it were decided to change U K policy by introducing emission limitation, there are several options aimed at both coal and oil. At present, around 40% of U K sulphur dioxide emissions arises from the combustion of oil, principally residual fuel oil. Apart from reducing fuel consumption, the main techniques are:
  (a) fuel substitution, including preferential use of low-sulphur fuels in those areas subject to high sulphur dioxide concentrations at ground level;
  (b) desulphurisation of fuels;
  (c) removal of sulphur during combustion;
  (d) F G D.

Techniques (a) and (b) are more easily applied to oil fuels and aimed at reducing local ground level concentrations in areas of high pollution. Techniques (c) and (d) are more suited to large combustion plant, which, because of their location and high chimneys, have little impact on polluted urban areas. They would be a means of reducing long range transport of sulphur.

(a) *Substitution*
17.58 The principal application of fuel substitution in the U K, namely of coal by oil, and of residual fuel by distillate fuel, has been to reduce smoke concentrations rather than sulphur dioxide. This has been done by the use of smokeless fuels in Clean Air Zones established under the terms of the Clean Air Acts (1956 and 1968). In addition, Section 76 of the Control of Pollution Act 1974 contains powers to enforce the use of low-sulphur oil in specified areas. There are no corresponding legislative powers available to control the sulphur content of coal. Such fuel substitution had been given earlier impetus by a general preference for cleaner and more convenient fuel.

(b) *Coal cleaning*
17.59 Coal cleaning is aimed primarily at reducing ash content but it does remove some sulphur. It also results in the rejection of some coal with the discarded material. At most, 40% of total sulphur could be removed from British coals by physical cleaning, but because of the costs and losses involved, 25–30% would be a more realistic aim. About two-thirds of power station coal undergo some treatment, but only 7% is fully washed. The N C B have studied the implications of cleaning all the coal supplied to power stations. The additional annual cost would be at least £120 million (£2 per tonne of coal) to cover the extra coal cleaning capacity, the extra waste to be tipped and the loss of combustible material in the waste. The total reduction in sulphur is difficult to estimate without

extensive testing, but an examination of coal from five different seams showed an average reduction of 3–4% with a significant loss in the extra coal discarded.

17.60 An alternative method of treatment would be to produce two streams of coal, one with a reduced sulphur content and the other ("middlings") with enhanced sulphur. The middlings could be either crushed and rewashed or alternatively burned in power stations equipped with F G D. The additional capital cost nationally would be £525–635 million over costs for total coal washing and the overall cost of coal washing would be increased to around £5–6 a tonne. Overall reduction of sulphur content of around 20% might be achievable with the middlings re-treated and this would increase to about 25% if they were burned in power stations equipped with F G D. Because of the high cost of coal washing and the comparatively modest sulphur reductions achieveable, the overall cost per tonne of sulphur dioxide abated by this technique is very high (around £1000–2000 per tonne) and we do not consider it to be a viable approach.

17.61 A major problem with the application of extensive coal cleaning methods is the disposal of the high sulphur residues. Because of their coal content, tips of the material would be liable to spontaneous combustion and special treatment would be needed to prevent uncontrolled emissions of smoke and sulphur dioxide.

17.62 The removal of sulphur by chemical methods is already practised to some extent in the manufacture of distillate oils (petrol, kerosene and gas oil). The technology can be extended to some, but not all types of residual oils, but it is not done on a significant scale at present. The demand for low-sulphur residual fuel oil is largely met by processing low-sulphur crude oil which at present commands a price premium of about 10%. If demand were to outstrip supply it would be necessary to desulphurise ordinary residual fuel oil and the additional cost has been estimated by the Department of the Environment to be over £13 a tonne. There are no economically feasible chemical methods of desulphurising coal.

(c) *Removal during combustion*
17.63 The main potential option for sulphur removal during combustion is in conjunction with F B C. If limestone or dolomite is added to the bed, sulphur present in the fuel will combine with the additive to form calcium sulphate or sulphite and be discharged with the ash. Techniques have yet to be proved for the use of F B C systems for sulphur retention on a commercial scale. They could also give rise to secondary environmental effects due to the production of limestone and to the disposal of the additional waste produced. Estimates of the amount of limestone to be added to achieve 90% sulphur retention range from 6–10 times the weight of sulphur in the coal. For a

typical 2000 MW power station about 650,000 tonnes of limestone a year would be needed and the quantity of residue for disposal, a mixture of ash and spent limestone, would be doubled. The limestone would need to be quarried and transported, and land required for disposal sites would have to be at least doubled, or even quadrupled if it were not possible to find commercial uses for the F B C ash mixture as is presently the case for pulverised fuel ash. We recommend that the C E G B and the N C B should collaborate in research on commercial use of F B C ash mixture.

(d) *Flue gas desulphurisation*
17.64 F G D is the only method which is presently employed in the world on any scale for the specific purpose of reducing sulphur emissions from coal or oil after combustion. The first major F G D plant was built at Battersea power station in 1932, but because of various operating problems it has not been in use for many years now. At present, Japan and the U S A have the largest installed F G D capacity, with lime and limestone scrubbing systems predominating. In the U S A there were 101 installations in operation or under construction in mid-1980 for use mainly in power stations, but also in some industrial plant. Investment by other countries in F G D has been far smaller, although West Germany now requires sulphur dioxide limitation on new coal-fired power stations and F G D systems are being installed.

17.65 There are now at least 16 proprietary F G D systems either being commercially offered or under development, though only a small proportion of these has been demonstrated on coal-fired boilers of any size. They can achieve over 90% reduction of the sulphur dioxide in the flue gases under optimum operating conditions, but they are expensive and have some environmental disadvantages. F G D processes can most easily be categorised by the manner in which the sulphur compounds removed from the flue gases are eventually produced for disposal: throw-away processes and regenerative processes.

17.66 In throw-away processes, the product is disposed of entirely as waste. Processes involve wet or dry scrubbing of the flue gases with limestone or lime. The product is a slurry of calcium sulphate and sulphite mixed with unreacted lime or limestone. In some cases, the calcium sulphate (gypsum) is purified and marketed, for example, in Japan where there is little natural gypsum. But in Europe and the U S A better quality natural gypsum is widely available and does not need expensive purification. The majority of present F G D installations are throw-away processes using lime or limestone.

17.67 Regenerative processes are specifically designed to recover sulphur dioxide in a concentrated stream so that, with further processing, it can be converted to marketable grades of liquefied sulphur dioxide, sulphuric acid, or elemental sulphur. The advantage of regenerative processes is that the product can be sold to offset the extra costs and to avoid waste disposal problems. Of the possible products, elemental sulphur is the most useful. It can be readily stored and transported. Sulphuric acid and sulphur dioxide raise problems of storage and can normally only be marketed within a limited radius. But the principal use for these products is for the manufacture of sulphuric acid, some of which is already made from by-products of other industrial processes such as smelting and oil refining. The widespread adoption of regenerable F G D systems could produce much more sulphur or sulphuric acid than the market could absorb.

17.68 Adverse environmental and other effects which may be associated with some F G D processes include the following:

(i) The land area required for disposal of the waste from throw-away systems is considerable. The waste produced each year by a 2000 MW power station would occupy an area of about 33 hectares (86 acres) filled to a depth of 1 metre. Chemical treatment of the slurry before disposal is essential, adding to the overall cost.

(ii) Wet-scrubbing systems can produce a cool wet plume which descends rapidly to ground level near the plant. Maximum ground level concentrations arising from the small percentage of sulphur dioxide remaining in the plume can be greater than if the plume had been emitted unwashed with its full sulphur dioxide content. To avoid these adverse effects, it is necessary to re-heat the flue gas which is expensive both in energy and financial terms.

(iii) There could be problems of fitting F G D plants into existing power station sites, though planning consent conditions for post-war power stations have included the reservation of space for the future installation of sulphur removal plant if this should be required. Each 660 MW unit of a 2000 MW power station would need 4 F G D streams, more if standby streams were necessary to maintain station availability.

(iv) A 2000 MW power plant would require about 250,000 tonnes per year of lime or limestone for most F G D systems. This quantity would need to be quarried, crushed and transported to the site and would therefore have considerable environmental impact.

Costs of desulphurisation

17.69 Estimates for the costs of partial desulphurisation of oil and raw coal have been given in paragraphs 17.59–17.62. When a greater degree of desulphurisation is required for coal, only F G D processes are available as established technology, but F B C systems are of potential interest. In both cases we have considered costs in terms of applications to power stations because it is considered practicable to fit such systems only to large units. Shallow-bed F B C systems for industrial boilers may be developed to achieve modest reductions in sulphur emissions, but it seems likely that 90% removal efficiencies will be economically achievable only in the deep-bed designs more suited to power stations.

17.70 The C E G B considered that environmental considerations would militate strongly against the use in Britain of the somewhat cheaper F G D systems which produce sludge. We agree. Initially they estimated the capital cost of installing a Wellman-Lord regenerative plant at £190 million (1978 prices) for a new 2000 MW coal-fired power station. The net effect of these capital costs together with the reduction of the station output, the loss of thermal efficiency and additional operating costs was estimated to increase the generating costs of the station by 25–30%. As a result of a recent detailed design study by consultants, more accurate estimates have become available. These indicate that, for a 2000 MW station commissioned in 1995, the capital cost would be between £106–£122 million at March 1981 prices, depending on national economic activity and price escalation in the interim. The net effect of these costs and other operating factors would be to increase the annual system costs by £33–45 million which would increase the generating costs of the station by around 10%. Apart from the firmer financial basis of the estimate, it has also allowed for recent technical improvements in the design and operation of the Wellman-Lord process.

17.71 We received varying estimates from several sources, but they were in agreement on the central point that F G D is a costly process. The N C B estimated that the total capital cost for applying F G D to all U K coal-fired power stations would be nearly £5,000 million with total additional generating costs of £1,300 million per year. Figures from the U S A Environmental Protection Agency indicated that F G D would add 10–20% to generating costs. A detailed comparison of the costs of 9 different processes is contained in a recent study (N A T O C C M S 1980) which took as its central case a new 500 MW plant in the Mid-West of the U S A designed for removal of 90% of the sulphur dioxide in the flue gas. Fuels with different sulphur content were evaluated and varying plant sizes tested. On this basis, and scaled up to a 2000 MW plant, the capital cost of a Wellman-Lord regenerative process would be around £100 million (1979 prices) which is broadly in line with the most recent C E G B estimate.

17.72 The N C B gave estimates for the additional capital and operating costs of F B C plant in a power station. These are given in Table 17.9 scaled up to a standard 2000 MW station at mid-1979 prices. The C E G B estimated that the capital costs of a F B C system, equipped for sulphur removal, and with handling facilities for the additional waste, would be about 10% more expensive than a conventional system without F G D. In comparison with their initial estimates of the costs of F G D, the overall system based on F B C appeared to be some 30% cheaper; but their later estimates of F G D costs have been reduced. The capital and operating costs of a conventional plant equipped with F G D are now expected to be similar to those of the F B C system. However, we consider that figures at this stage of development of F B C must be somewhat speculative.

Table 17.9 *Estimated additional costs of limestone addition for sulphur absorption in F B C plant for a 2000 MW power station*

	Atmospheric Pressure	Pressurised
Capital invested (£ million)	40	117
Increase in running costs in p/kWh, including capital costs at 10% and limestone supplies	0·074	0·168
1979/80 average generating costs including capital charges for a C E G B coal-fired power station in p/kWh.		1·56

Source: *N C B*

17.73 Having examined the cost of individual sulphur control technologies, we looked at the overall cost of several control options used simultaneously and their potential for reducing total sulphur emissions. To be realistic, this involves including the options for oil-fired installations also, since the presumed objective would be to limit emissions from all sources in aggregate to alleviate the problem of long range transport of pollution. Indicative figures are shown in Table 17.10 which is based on data supplied by N C B, C E G B, the Department of the Environment and the Department of Energy. Four policies were considered for simultaneous application by 2000 to different categories of users of fuel oil and coal:

(a) for power stations burning residual fuel oil, low-sulphur (1%) oil would be substituted for the normal grade (3% sulphur);

(b) similarly, low-sulphur residual fuel oil would be substituted for the normal grade for commercial and industrial users;

(c) for coal-fired power stations, 4000 MW of capacity would be equipped with F G D;

(d) for industrial users of coal, F B C systems with limestone addition would be adopted for all new installations after 1990.

17.74 We also examined options which included increased coal washing, but as noted in paragraphs 17.59 and 17.60 this is a much more costly approach so it has not been included in this comparison.

17.75 In Table 17.10 the second column states the basic unit costs of installing and operating the various control technologies. The fourth column indicates the

Table 17.10 *Summary of potential reductions in sulphur dioxide emissions and their costs*

Control technology	Unit cost (1)	Cost per tonne of sulphur dioxide abated £	2000 potential	
			Reduction in sulphur dioxide tonnes ('000) (% of U K Emissions)	Annual control cost £ million
(a) Continuous use of 1% sulphur fuel oil in power stations (2)	£5–16 per tonne	135–435	230–280 (5%)	30–100
(b) Continuous use of 1% sulphur fuel in commerce and industry (2)	£5–13 per tonne	150–400	300–330 (6%)	40–105
(c) Use of F G D on 4000 MW capacity of new power stations (3)	£220 million capital cost £50 million per annum operating cost	230	220 (4%)	50
(d) F B C in all new industrial uses of coal after 1900: limestone added (4)	£3 per tonne (notional)	250–190 (notional)	300–400 (7%)	75 (notional)
Total			1050–1230 (23%)	195–330

Sources: *N C B, C E G B, Department of the Environment and Department of Energy.*

Notes: (1) F G D costs at 1981 prices; other costs at 1978 prices.
(2) Premium on 1% sulphur oil: £5/tonne applies if no Europe-wide requirement for low sulphur fuel oil. £13–16/tonne assumes universal use.
(3) F G D designed for 90% removal: operating costs of F G D plant only – other system costs may be incurred.
(4) F B C: costs very speculative – no large scale plants built or operational at present, control costs limited to cost of limestone absorbent and its disposal: removal of 60–80% of sulphur assumed on 25 million tonnes of cleaned coal.

maximum reasonable reduction of sulphur dioxide emissions which they could achieve by 2000; the two figures correspond to the lower and higher forecasts of energy use in the Department of Energy's Energy Projections 1979. In the fifth column are the corresponding total annual costs. Figures in the third column are derived from those in the fourth and fifth columns and are a useful means of comparing the cost-effectiveness of the technologies.

17.76 We present these figures as rough estimations only, since both the costs (particularly for undeveloped technology such as F B C) and the potential for application by 2000 are uncertain; even the lower estimates of the extent of application are now considered to be rather high. In spite of these caveats, the figures are of interest, because they show not only the high costs of sulphur control, but that even with the maximum practicable application of all the techniques, only modest reductions in emissions (about a quarter) are feasible. Fitting of F G D to existing power stations could extend sulphur emission reductions, but the capital costs would be higher and the feasibility of doing this to any large extent must be questioned.

Conclusions on desulphurisation

17.77 We draw the following conclusions from the evidence presented:

(i) No single technique can make more than a minor reduction in emissions over the next 20 years. To make a significant reduction several control techniques will be necessary, the majority of which can only be practicable on large emitters. There is likely to be little impact either on urban concentrations in the U K or on acid rain depositions elsewhere.

(ii) At present the cheapest way of reducing emissions is to use low-sulphur fuel oil. But natural low-sulphur oil is in short supply worldwide and any major increase in demand would necessitate the use of fuel oil desulphurisation techniques at greatly increased cost. As discussed in Chapter 19, the use of low-sulphur oil is, however, a valuable technique for reducing local pollution by sulphur dioxide in some urban areas, but such use would have little effect on total national emissions.

(iii) The maximum reductions by the use of low-sulphur fuel oil are not large and may well be less than the Table suggests if oil is increasingly displaced by coal. However, the displacement of residual fuel oil by coal could itself lead to some reduction in sulphur dioxide emissions because on combustion most U K coals emit only 60–90% as much sulphur dioxide as the thermally equivalent amount of heavy residual fuel oil.

(iv) Addition of limestone to F B C systems might be the cheapest way of reducing emissions from coal-firing, but much further development work

is needed before the costs can be established either for power generation or for industrial coal burn and there will be environmental impacts associated with limestone.

NITROGEN OXIDES

Source and quantity

17.78 Nitrogen oxides are formed during the combustion of all fuels from the oxidation of nitrogen in the combustion air, particularly at high temperatures such as those found in internal combustion engines and power station boilers. The nitrogen in the coal itself also adds significantly to their formation. The nitrogen oxides in power station flue gas consist of about 95% nitric oxide and 5% nitrogen dioxide. Their concentration is 400–700 ppm, that is, about one-third of that of sulphur dioxide, and their rate of emission is about 8 kilogrammes per tonne of coal burnt. Nitric oxide is oxidised in the atmosphere to form nitrogen dioxide which is subsequently converted to nitric acid. The average residence time of nitrogen oxides in the atmosphere is about a week. The majority is removed by wet deposition as nitric acid or nitrates, though dry deposition may be more important in dry periods.

17.79 Concentrations of nitrogen oxides are not extensively monitored and there are no general historical measurements from which to establish trends. It is only comparatively recently that reliable analytical techniques have been determined for nitrogen oxides and so earlier figures could be inaccurate. Warren Spring Laboratory estimate that the total U K emissions in 1978, expressed as nitrogen dioxide, were 1·9 million tonnes of which about 47% was due to coal burning. Details are given in Tables 17.11 and 17.12 and Figure 17.6.

Table 17.11 Nitrogen oxide emissions by source (1978)

Domestic	2·5%
Commercial and industrial	26·4%
Power stations	42·7%
Incineration etc	0·4%
Road vehicles	25·6%
Railways	2·4%
	100%
Coal burning	47%
Oil burning	24%
Transport	28%
Other	1%
	100%
Total emissions as nitrogen dioxide	1·869 million tonnes

Source: *Department of the Environment 1979b*

17.80 Ambient nitrogen oxide levels show daily average concentrations in urban areas in the range 0·02–0·06 ppm. These levels are largely due to road

Figure 17.6 Nitrogen oxides, carbon monoxide, and hydrocarbons: contribution of various sources to total emissions 1979

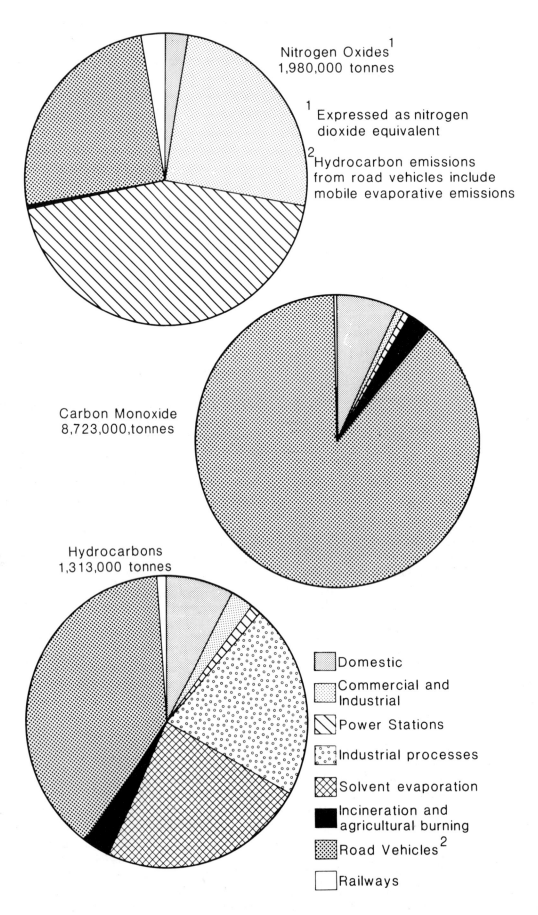

Nitrogen Oxides[1]
1,980,000 tonnes

[1] Expressed as nitrogen dioxide equivalent

[2] Hydrocarbon emissions from road vehicles include mobile evaporative emissions

Carbon Monoxide
8,723,000,tonnes

Hydrocarbons
1,313,000 tonnes

Domestic

Commercial and Industrial

Power Stations

Industrial processes

Solvent evaporation

Incineration and agricultural burning

Road Vehicles[2]

Railways

Source: *Department of the Environment 1980*

Table 17.12   Nitrogen oxides[1], carbon monoxide trends in emissions: by source U K

Nitrogen oxides										Thousand tonnes
	1970	1971	1972	1973	1974	1975	1976	1977	1978	1979
Domestic	41·2	48·6	48·4	50·7	50·5	49·0	48·6	46·6	47·4	50·9
Commercial and industrial	767·0	690·0	652·8	677·2	624·4	565·7	572·0	501·6	492·8	517·3
Power stations	778·6	755·1	720·9	806·2	728·9	766·4	775·7	786·1	797·8	868·6
Incineration and agricultural burning	6·0	6·0	6·5	6·5	6·5	7·3	7·7	7·7	7·7	7·5
Road vehicles	385·9	402·5	419·9	449·1	437·5	428·3	446·2	457·2	479·2	490·2
Railways	54·1	53·8	51·0	52·3	49·3	46·5	43·9	43·2	44·1	45·6
All emissions	2032·8	1956·0	1899·5	2042·0	1897·1	1863·2	1899·1	1842·6	1869·0	1980·1

Carbon monoxide										
Domestic	1163·4	1016·2	867·8	877·1	840·0	722·6	673·4	657·3	599·4	612·3
Commercial and industrial	104·6	94·4	88·2	92·8	89·0	83·3	85·0	74·6	73·0	68·4
Power stations	46·8	46·1	46·1	49·6	45·4	46·2	44·7	46·6	47·5	51·3
Incineration and agricultural burning	214·0	214·0	214·0	214·0	214·0	217·6	222·9	222·9	222·9	224·0
Road vehicles	5922·1	6217·7	6597·6	7027·8	6841·4	6692·4	7004·8	7194·4	7606·8	7750·7
Railways	19·6	19·4	18·4	18·9	17·7	16·8	15·8	15·6	15·9	16·6
All emissions	7470·5	7607·8	7832·1	8280·2	8047·5	7778·9	8046·6	8211·4	8565·5	8723·3

Source: *Department of the Environment 1980*

[1] Expressed as nitrogen dioxide equivalent.
[2] Including mobile evaporative emissions, which are evaporative losses from the petrol tank and carburettor.

traffic so that readings vary greatly with the proximity of vehicles. Short term peak concentrations are several times greater. Rural concentrations are somewhat smaller, generally 0·005–0·02 ppm. The general background from natural sources is smaller still. The much less reactive nitrous oxide produced by biological processes is present in much larger amounts, 0·25–0·4 ppm. The mechanisms for the removal of this latter gas from the atmosphere are not yet established quantitatively. Some is decomposed in the stratosphere by ultra-violet light and some is absorbed by the soil, but a small proportion may be converted to nitric acid. There appears to be very little evidence of any effects on human health or on vegetation at these concentrations. Total emissions of nitrogen oxides from fuel burning are about a third of those of sulphur dioxide. It seems likely that acid depositions at long distances from these two pollutants are in about the same ratios.

Control

17.81 No active control measures are adopted, or specially called for, since the dispersal techniques required for sulphur dioxide are effective also for nitrogen oxides. Should emission control ever be considered necessary, there is a number of techniques that could be adopted for partial reduction of emissions from coal and oil burning sources. Most of them involve reducing combustion temperatures by relatively simple methods such as controlled mixing of fuel and air or re-circulation of flue gases. Reduction in emissions of around 60% appears to be possible on large pulverised fuel boilers.

17.82 Nitric oxide is chemically rather unreactive and is difficult to scrub from flue gases. A variety of processes have been devised but they are relatively expensive and need careful control to avoid emitting unreacted chemicals as new pollutants. The modification of combustion conditions would appear to be the better technique for reducing emissions.

CARBON DIOXIDE

17.83 The combustion of all carbonaceous fuels including coal releases carbon dioxide, which is also a naturally occurring constituent of the atmosphere. It is a global effect. Atmospheric levels of carbon dioxide have increased from about 290 ppm in the middle of the last century to 335 ppm today and the trend continues upwards. Most estimates suggest a doubling of "pre-industrial" concentrations within the next 100 years largely as a result of the combustion of fossil fuels. Other reasons for the increase of carbon dioxide include the massive deforestation which has been taking place over the past century for agricultural and urban development. Carbon dioxide is evolved when the trees are burnt and organic matter in the soil is exposed to weathering and so decays. Carbon dioxide is removed from the atmosphere by the biocycle and by dissolution in the oceans, but these processes are relatively slow and about half the man made addition has remained in the atmosphere. This amount is equivalent to about 1% of the total amount involved in the interchanges between the atmosphere and other reservoirs. We examine the possible effects of an increase in the carbon dioxide content of the atmosphere (the "greenhouse" effect) in the next Chapter.

17.84 No practicable method of controlling carbon dioxide emission from combustion is likely to be possible on a very large scale, because the essence of the combustion of fossil fuel is the conversion of carbon compounds to carbon dioxide. Methods of dissolving it in the ocean have been considered but none seems to be very practicable. If it were deemed necessary to combat the rise in atmospheric levels, the most feasible action would entail burning less fossil fuels, increasing afforestation or increasing the use of nuclear or renewable sources of power; an alternative approach would be to moderate the adverse effects of any consequent climatic changes. Any such action should clearly be international in scope and would need to be based on the outcome of international research in which the U K is participating.

CARBON MONOXIDE

17.85 Carbon monoxide is formed by the partial combustion of carbonaceous fuel in a limited supply of air. Small amounts are always formed on the combustion of fossil fuel. In the atmosphere it is eventually converted to carbon dioxide. Of the carbon monoxide emissions in the U K, 90% arises from internal combustion engines as can be seen in Table 17.12 and Figure 17.6. Around 7% comes from domestic coal fires. Only about 1% comes from coal burning by power stations and industry. We do not consider carbon monoxide emissions from coal burning to be a problem.

HYDROGEN CHLORIDE

17.86 U K coals contain on average about 0·25% of chlorine largely as sodium chloride, common salt. On combustion over 95% of the chlorine is converted to the strongly acid gas hydrogen chloride; the remainder ends up as chlorides in the ash. Because both hydrogen chloride and chlorides have a strongly corrosive action on boilers, the acceptable level of chlorine in coal for major users is 0·3%. Where necessary this level is achieved by blending high and low chlorine coals.

17.87 As with other gaseous pollutants, control is achieved by the use of tall chimneys. There have been no problems (or even suspicions of problems) arising from hydrogen chloride emissions from coal combustion at short and medium ranges. Hydrogen chloride adds to the total acidity of flue gases and thus may contribute to the formation of acid rain (see sulphur emissions, paragraphs 17.50–17.55). Studies by the C E G B are continuing on the extent to which hydrogen chloride from power station emissions may affect the acidity of rainfalls in areas of the country where the chlorine content of coal is relatively high.

17.88 Should further controls ever be considered necessary this could probably most easily be achieved by leaching chlorides from raw coals by washing with water, though there could be problems involved in the disposal of the resulting saline water. Most F G D processes will also remove hydrogen chloride.

OTHER GASES

17.89 There is a number of other gases that occur in trace amounts in flue gas emissions. These include hydrogen cyanide, carbonyl sulphide, carbon disulphide, methane, hydrogen fluoride, hydrogen sulphide and complex organic compounds including various polycyclic aromatic compounds. The high combustion efficiencies and temperatures of large boilers ensure that most of these are produced in only trace amounts, especially the more complex compounds. Rather larger quantities of complex tarry compounds are produced by the lower temperatures and uneven combustion of domestic coal fires. We discussed these earlier in the section on smoke.

17.90 Historic measurements show that polycyclic aromatic hydrocarbon levels in Central London declined by a factor of 10 between 1950 and 1975 as a result of the Clean Air Acts. Over this period motor vehicles became a more important source of emissions than coal combustion. Polycyclic aromatic hydrocarbons are discussed further in the section on health in the next Chapter.

17.91 Apart from tars in smoke from domestic sources, there was no evidence to suggest that these trace gases were a problem. We have noted that some research studies by C E G B to check that this was indeed the case were continuing.

Chapter 18　Combustion residues of coal: effects

INTRODUCTION

18.1　In the previous Chapter, we examined sources, quantities, and the control of residues formed by the combustion of coal. In this Chapter, we examine the adverse effects on people and the environment which can be caused by these residues. Consideration of air pollution effects can be extremely complex and cause-effect relationships are frequently uncertain. In most cases, the magnitude of an effect is related to the ambient concentration of the pollutant or pollutants, but the dose-effect relationship varies widely depending on different targets and pollutants. We begin by looking at effects on people; this is followed by effects on crops and ecosystems; and finally we consider effects on the physical world.

EFFECTS ON PEOPLE

Non-malignant respiratory disease

18.2　Complaints about smoke and sulphurous fumes from the combustion of coal have been recorded since at least the 13th century. However, until the disastrous London fog of December 1952, there was little serious public concern. The famous London "pea-souper" fogs were generally regarded as major nuisances rather than serious health hazards, though some serious investigations of the distribution of atmospheric pollutants, especially smoke and sulphur dioxide, had been carried out sporadically since early this century. During the December 1952 incident, unusually adverse meteorological conditions prevented the dispersion of fumes from chimneys for several days, so that their concentration in the air built up to many times normal levels. Many people experienced respiratory problems and later it became clear that, during the episode, about 4,000 more people had died in Greater London than would have been expected under normal conditions (Ministry of Health 1954). The great majority of these excess deaths were among elderly or infirm people, many of whom had a history of respiratory illness and a short expectation of life.

18.3　The only measurements of pollution available at that time were of "smoke" (particulate matter assessed in terms of its blackness) and of sulphur dioxide. At its worst, concentrations of sulphur dioxide were around $3,800\ \mu\mathrm{gm}^{-3}$ and of smoke were around $4,500\ \mu\mathrm{gm}^{-3}$. Since that date, clear associations have been

established between high levels of smoke and sulphur dioxide combined, and increases in deaths and illness. (Gore A T and Shaddick C W 1958; Martin A E and Bradley W 1960; Martin A E 1964; Waller R E, et al 1969; and Waller R E 1979.)

18.4　We must distinguish between acute and chronic effects. Acute effects arise immediately after exposure to high levels of pollution; chronic effects develop over many years and may be the result of continued exposure to elevated levels. There is some interaction between the two types of effects with acute illnesses possibly contributing to the development of chronic disease and chronic ailments making people more liable to develop acute symptoms on exposure to elevated pollution levels.

18.5　Particulate and gaseous pollutants commonly occur together in urban air. Therefore epidemiological techniques seldom permit the attribution of an observed effect to a specific pollutant. Moreover, there is the possibility of synergistic interactions whereby the combined effect of two pollutants is greater than the sum of the individual effects. The results of animal studies suggest that this may be the case with sulphur dioxide and particulates (World Health Organisation (W H O) 1979). Sulphur dioxide on its own is less cause for concern as is instanced by the relatively high concentration of sulphur dioxide that is accepted by the H S E as the T L V for industrial workers. This suggests that sulphur dioxide and associated smoke and particulates should be considered as indices of pollution and not the specific agents that cause the effects. The true culprits may be pollutants associated with or derived from them – such as acid sulphates, chemicals contained in smoke, or gaseous pollutants absorbed on to the surface of smoke particles. The change in concentration of pollutants over the years has been of some help in unravelling the effects of smoke and sulphur dioxide pollution.

18.6　Mainly as a result of the working of the Clean Air Acts, the concentration of smoke has fallen greatly, far more than that of sulphur dioxide. It would appear that smoke is an important factor in producing acute effects. For example, in London in December 1962 there was a similar pollution episode to the one in December 1952; but on this occasion, although sulphur dioxide levels were about the same as those in 1952, smoke

Figure 18.1   Acute effects of exposure to smoke and sulphur dioxide pollution

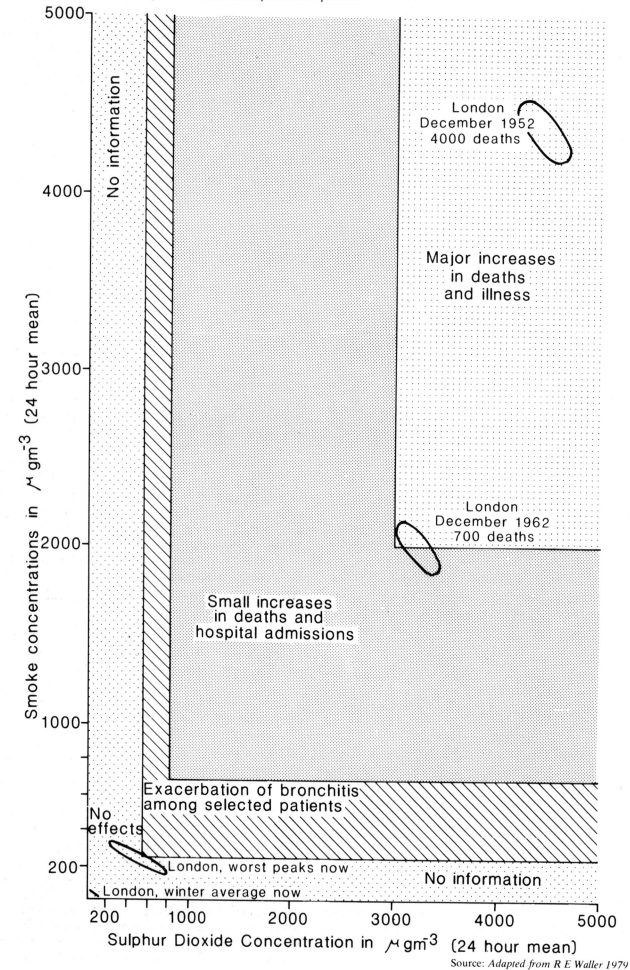

Source: *Adapted from R E Waller 1979*

concentrations were more than halved. 700 excess deaths were recorded in this episode which was less than a fifth of the 1952 figure.

18.7 Since then atmospheric pollution by smoke and sulphur dioxide has decreased still further. Episodes of high pollution have become rare, and there is no longer any evidence of acute effects attributable to occasional peaks of pollution in terms of sudden increases in mortality or morbidity. Epidemiological studies have shown that the minimum concentration of pollutants leading to any significant response among bronchitic patients was about 500 μgm^{-3} of sulphur dioxide together with about 250 μgm^{-3} of smoke. The historical development and present situation concerning acute effects are presented in Figure 18.1 (adapted from Waller 1979). Atmospheric pollution from smoke and sulphur dioxide, even under the most adverse meteorological conditions, is now below the levels which cause serious adverse effects, even among sensitive people.

18.8 The situation regarding chronic effects is more complicated. There are many difficulties involved in the establishment of dose-effect relationships and allowances must be made for many other interacting relationships such as smoking, socio-economic factors, and climate.

18.9 The first suggestion that exposure to pollution might have chronic effects on health came from epidemiological studies of differences in death rates between urban and rural areas. In general, it was found that death rates from respiratory diseases, particularly bronchitis, tended to be higher in the more polluted areas. Many studies have been carried out which concentrated on the symptons of respiratory disease among adults and children, measuring effects such as various pulmonary function indicators, phlegm production, exacerbation of colds and shortness of breath. Although cigarette smoking was clearly the major factor, a relationship between the prevalence of respiratory symptoms and high ambient levels of sulphur oxides and particulates was established.

18.10 Long term studies of groups of children have also generated useful information. Some of these studies are intended to continue throughout the lives of the groups, so by their very nature many aspects remain to be completed. They have, however, indicated an increased occurrence of respiratory ailments, such as infections of the lower respiratory tract, in areas of high pollution. It was concluded in a Report by the W H O (1972) that discernible increases in the prevalence of respiratory effects were seen when the annual mean concentrations of both smoke and sulphur dioxide exceeded 100 μgm^{-3}. In the U K, such annual figures would be expected to produce values in excess of 400 μgm^{-3} for smoke and 300 μgm^{-3} for sulphur dioxide on only 7 days a year.

18.11 Studies of chronic and acute effects continuing since the 1972 W H O report, including a revision of it in 1979 (W H O 1979), have generally tended to confirm its conclusions. A few of these studies, in particular the American "C H E S S" series, produced more disturbing views based on relationships between suspended sulphates – derived from sulphur dioxide – and various health indices in urban areas. But this series of studies has been criticised for faults in design and interpretation so that in fact few useful conclusions can be drawn from them (U S Congressional Committee on Science and Technology 1976).

18.12 We showed earlier the general decline of pollution from smoke and sulphur dioxide in Figures 17.1 and 17.3, (see also Plate 18.1). In urban areas, smoke levels are now a sixth of the values of 20 years ago and sulphur dioxide levels have been halved. Despite complicating factors, there is clear evidence that the reduction in pollution by smoke and sulphur dioxide has been accompanied by the virtual disappearance of the onset of the acute symptoms of respiratory stress that used to be related to high pollution episodes caused by unfavourable weather. It is too soon yet to be absolutely certain that there have been improvements in the long term situation regarding chronic effects, but the indications that this is so are strong. We endorse the conclusions of the W H O Report (1979) on the expected health effects of sulphur dioxide as summarised in Table 18.1 taken from that Report, with the reservation that it is doubtful whether acute effects are discernible at as low a level as 250 μgm^{-3} of sulphur dioxide together with 250 μgm^{-3} of smoke. On the basis of work in the U K (for example, Ellison J McK and Waller R E 1978), figures of 500 μgm^{-3} of sulphur dioxide together with 250 μgm^{-3} of smoke are more acceptable.

Table 18.1   Expected health effects of smoke and sulphur dioxide

Pollutant	Excess mortality and hospital admissions	Worsening of patients with pulmonary disease	Respiratory symptoms
SO$_2$	500 μgm^{-3}	250 μgm^{-3}	100 μgm^{-3}
Smoke	500 μgm^{-3} (daily averages)	250 μgm^{-3} (daily averages)	100 μgm^{-3} (annual means)

Source: W H O 1979

18.13 In other countries there has been some evidence that nitrogen oxides are involved in causing or at least exacerbating chronic respiratory disease (Ferris G F 1978). However, in the U K, general ambient concentrations of nitrogen oxides, even in large cities, are below levels at which detectable effects have been demonstrated. Local high levels are clearly associated with motor vehicles, for example at roadside sites. However, the highest ambient concentrations to which most people are exposed are in fact indoors and are due

Figure 18.2   Chronic bronchitis and emphysema: trends in mortality 1951–1979

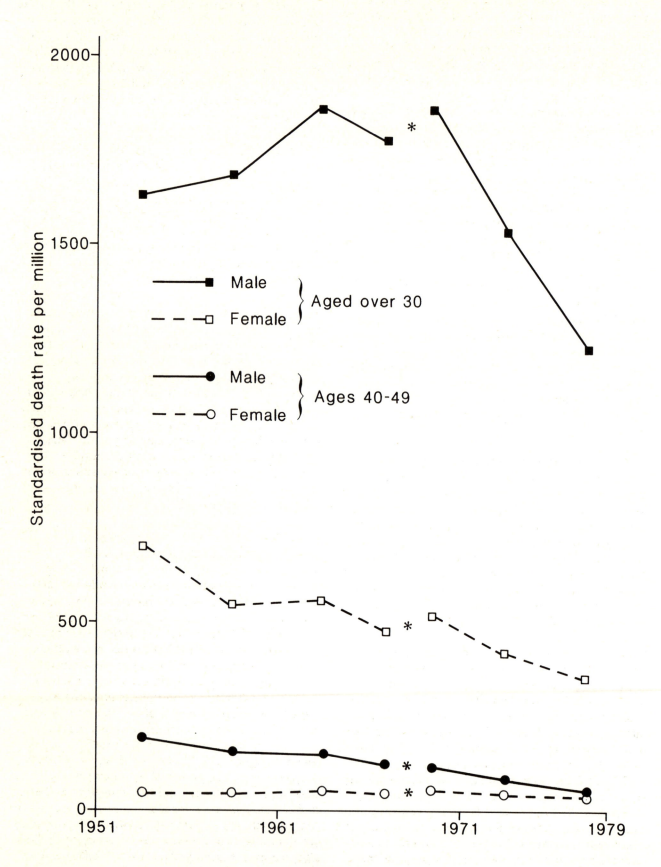

*   reclassification of health statistics

Source: *Derived from statistics published by the Office of Population Censuses and Surveys
England and Wales series D H (deaths)*

to the use of gas cookers and flueless oil, gas and propane heaters. There is little evidence that indoor concentrations of nitrogen oxide from these sources, which routinely reach levels several times greater than those outdoors, cause any effects that can be distinguished from confounding effects such as smoking, humidity, temperature, and socio-economic factors.

18.14 It is considered by some researchers that there are synergistic interactions between nitrogen oxides and other pollutants such as sulphur dioxide and smoke. If this is so, it is possible that the ambient levels reached in some locations in some urban areas could give some slight cause for concern. But the great proportion of nitrogen oxides at such sites comes from motor vehicles, with coal combustion making only a very minor contribution to ambient concentrations of nitrogen oxides in such locations.

18.15 Trends in the general incidence of mortality due to chronic bronchitis and emphysema since 1953 are shown in Figure 18.2. They are generally encouraging, but they also illustrate the complexity of the problem of establishing relations between pollution and health. For both men and women, the trend is downwards both for all ages and for individual age groups such as those aged 40–49 years. But mortality rates and the rate of decline in recent years are higher for men than for women.

18.16 The probable explanation is that for men, cigarette smoking, which is also linked with bronchitis, peaked at a higher level in the mid-1950s, some years before doing so among women. Since chronic effects reflect the results of exposure over a period of years, the graphs for women show the combined effects of a rising trend for smoking against a falling trend for pollution. For men, although the start of the graphs shows the effect of a still rising smoking trend, the end of the lines shows the effects of a decrease in both smoking and pollution.

18.17 As far as acute and chronic effects are concerned then, the indications are that current levels of air pollution in the U K are generally satisfactory with the possible exception of those in a few localities. These are the areas, identified in a recent national survey by Warren Spring Laboratory (see Table 19.3) where concentrations of smoke and/or sulphur dioxide exceed the air quality limit and guide values of the E E C Air Quality Directive. In these areas while the mean annual smoke and sulphur dioxide levels are satisfactorily below the W H O figures, daily averages above the minima associated with acute effects may still occur. As discussed in Chapter 19, measures to deal with the air pollution problem in such areas by the application of smoke control orders or restrictions on the use of ordinary fuel oil are already in hand or planned.

Cancer

18.18 Conclusions regarding the effects of pollution on the incidence of cancer are less easy to reach. For while there may be a level below which pollution does not cause acute effects or chronic bronchitis, no such level can be postulated for the production of cancer. There are, indeed, strong theoretical grounds for believing that dose-effect relationships established at high dose levels extend down to zero dose and it is normally assumed that they do.

18.19 There are three main types of compound emitted from coal on combustion which have been shown to induce cancer at high concentrations. These are certain chemicals contained in smoke, especially polycyclic aromatic hydrocarbons (P A H s), trace elements such as arsenic, and radioactive elements. There is, therefore, reason to believe that atmospheric pollution created by the combustion of coal may occasionally cause cancer.

18.20 Evidence from many sources has shown that, in the past at any rate, mortality from lung cancer was almost invariably greater in large towns than in the countryside. At first sight it seems reasonable to attribute at least some of this excess mortality to air pollution. But it was then found that the lung cancer mortality rate was extremely low among non-smokers irrespective of where they lived, so it was evident that atmospheric pollution alone had very little effect in producing the disease. It was possible, however, that some pollutants in urban air interacted with tobacco smoke to increase the incidence of cancer above that expected from tobacco alone.

18.21 The existence of such an interaction is commonly assumed but the evidence for it is inconclusive. The difficulty is that the risk of lung cancer among cigarette smokers in middle and later years of life depends not only on their current smoking habits but also on the amount they used to smoke many years before. We have therefore to take account of the differences in smoking habits between men and women living in different parts of the country in the distant past, as well as the present, before we can assess the contribution that urban atmospheric pollution may make to the excess incidence of lung cancer in larger cities. The problem is illustrated by the trends in mortality from lung cancer over the past 30 years shown in Figure 18.3. For men of all ages combined, smoking habits, as numbers of cigarettes smoked, peaked about 1960 (and probably somewhat earlier in terms of the amount of tobacco tar inhaled, because of changes in the type of cigarettes sold), but the mortality curve for lung cancer is only just beginning to decline, after reaching its peak in the early 1970s. For women, the numbers of cigarettes smoked did not peak until the 1970s so that the mortality curve for women is still rising. Distant habits become progressively less

Figure 18.3   Lung cancer: trends in mortality 1951–1979

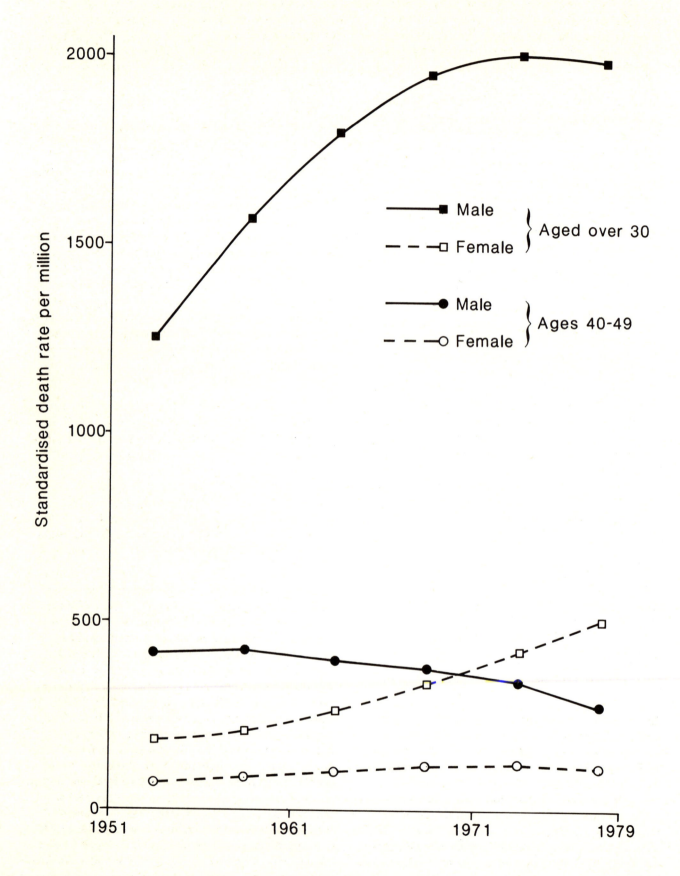

Source: *Derived from statistics published by the Office of Population Censuses and Surveys
England and Wales series D H (deaths)*

important as age decreases, and mortality at ages 40–49 has followed general smoking habits more closely, so that the curve for women aged 40–49 has started to decline now. The switch to filter-tipped and low tar cigarettes helped since it further reduced the amount of tar delivered, effectively shifting the peak years for cigarette use back somewhat earlier than the peak for actual numbers of cigarettes smoked.

18.22 The prevalence of smoking was, and still is, to some extent greater in cities than in rural areas. Smokers in towns have tended to smoke more cigarettes and to start smoking earlier in life. These differences account for much of the variation between lung cancer incidence rates in town and country; but the available information is insufficiently detailed to decide whether they can account for all of it or not. The direct epidemiological evidence is therefore of very little help in deciding what the carcinogenic effect of atmospheric pollution may be, except in so far as it sets upper limits to it.

18.23 An alternative approach is to extrapolate from occupational studies in which mortality from lung cancer has been found to be abnormally high as a result of exposure to high levels of particular pollutants. Measurements have therefore been made of the concentrations of certain carcinogens in the air, especially P A H s and arsenic. P A H s are a constituent of smoke and are usually characterised in terms of the level of benzo-a-pyrene (B A P). It is assumed that the important carcinogenic agents in air polluted by smoke are similar to those in the smoke in the retort houses of gas works and around coke ovens where occupational exposure to high levels in former years has been associated with increased rates of lung cancer.

18.24 In former years, urban B A P concentrations were around 30–40 nanogrammes per cubic metre. (One nanogramme per cubic metre is equivalent to one gramme per cubic kilometre.) The conclusions of such studies were that in the past the combustion products of fossil fuels in ambient air acting together with cigarette smoke *could* have been responsible for about 5–10% of lung cancer in big cities.

18.25 Thirty years ago smoke in London generally contained over 20% of benzene soluble material, sometimes as much as 50%. It now contains less than 10% and B A P levels in urban air now are generally only a tenth of former values. Moreover, only a proportion, often less than one half, is due to coal combustion now, other major contributions coming from motor vehicle exhausts. In this respect the N C B have reported that smoke from domestic open fires is different in type from that produced by smokeless fuels or by smoke-reducing appliances and industrial boilers. The smoke from open coal fires contains much more

tarry matter which is a source of P A H s. On this analysis, the present contribution of atmospheric B A P and associated materials from coal combustion to the production of lung cancer in the future is unlikely to account for more than one half per cent of cases of lung cancer among cigarette smokers, that is for not more than 150 cases a year. (Cederlöf R et al 1978; Hammond E C and Garfinkel L 1980; Lawther P J and Waller R E 1978; Rogot E and Murray J 1980).

18.26 A multi-element survey by the Warren Spring Laboratory in several cities in the U K has shown that the ambient levels of arsenic and all other trace elements surveyed are less than 1% of the appropriate T L V for occupational exposure and that most, including arsenic, are less than 0·1% (McInnes G 1980). The T L V s are, strictly speaking, irrelevant to the production of cancer, as they were derived for other purposes and ignore the presumed linear relationship from zero dose between dose and cancer induction. But extrapolation from industrial conditions, under which occupational cancers have been produced, suggests that the minute amounts of arsenic present in urban air are unlikely to be responsible for even one case of lung cancer a year in any large city, if they are responsible for any at all.

18.27 Radioactive elements are a special case of trace elements. We have discussed some aspects of this problem earlier (paragraphs 17.32–17.36). In fact coal contains no more radioactive elements than any ordinary rock, and so is hardly radioactive at all. Nevertheless, concern has been expressed about the amounts of radioactivity emitted by the naturally radioactive gas radon, and other radioactive elements in the particulate matter emitted by major coal users such as power stations.

18.28 In terms of activity measured as curies or becquerels the radioactivity of the emissions from a power station chimney is about the same as that emitted by the ground on which it stands. But the majority of the activity of emissions from the ground is from the naturally radioactive gas thoron, a short lived isotope of radon which is derived from traces of thorium present in all rocks and soil. Thoron has a high activity because it has such a short half life, but the quantities involved are minute and it decays to harmless materials too rapidly to result in any significant exposure. The majority of the activity from power station emissions is also derived from thoron and radon, but there is also very low activity from other longer lived naturally radioactive elements such as uranium and thorium. Larger doses can be received from much less active elements if they can maintain contact with people for long periods, and it is the dose received by people which is important, not the degree of activity of the material.

18.29 Both the C E G B (Corbett J O 1980) and N R P B (Camplin W C 1980) have estimated the dose received by maximally exposed individuals of extreme habits, so-called critical groups. The N R P B calculated that the maximum exposure for a member of the critical group could be about 230 microsieverts per year from all pathways, the most important of which is the consumption of liver from livestock grazing in the vicinity of a power station. Such a dose would be about 13% of the normal exposure in the U K from natural sources of about 1860 microsieverts. It is less than the difference between the doses received in different parts of the country from differences in local geology. It is also less than 5% of the annual dose limit of 5,000 microsieverts recommended by the I C R P for critical groups. The total dose-equivalent to the entire population of the U K from 30 years operation of a 2000 MW coal-fired power station, is estimated as 340 mansieverts, which would be expected eventually to give rise to about six radiation-induced health effects such as fatal cancers, or genetic abnormalities. Effects calculated by the C E G B are about one-third of those calculated by the N R P B, largely because less extreme, but more realistic, assumptions are made, especially as regards the diet of the critical group. Similar figures have been obtained by other investigators.

18.30 Fly ash is used for the manufacture of lightweight building blocks. Their radioactivity has not been studied extensively, but preliminary measurements by the C E G B have shown that, although they are more radioactive than an equal weight of bricks, on a volume basis they are of about the same radioactivity or less than bricks. Therefore, when used as a building material, they should contribute a dose of radiation no larger than conventional brickwork. Initial measurements indicate that this conclusion is correct.

18.31 Using the N R P B figures, the radiation dose-equivalent for an average person resulting from the operation of all the coal-fired power stations in the U K amounts to about 0·2% of the average dose of about 1860 microsieverts per year received from natural sources. This is less than the year to year variation due to changes in cosmic ray activity and other natural causes. We do not consider these figures to be of any concern.

18.32 We have concluded that, for the general public, current pollution from coal combustion residues may cause, in conjunction with cigarette smoking, a maximum of one or two hundred cases of lung cancer a year. Present knowledge, however, is incomplete and in this situation we recommend that the level of smoke pollution is kept at the lowest practicable level.

Photochemical smog

18.33 In some countries there have been problems arising from the inter-action of sunlight with nitrogen oxides and hydrocarbons in air pollution. A mixture of compounds is formed which causes irritation of the eyes and respiratory difficulties. This type of pollution is known as photochemical smog. The most notorious incidence occurs in Los Angeles where a combination of local geography, climate and high emissions of nitrogen oxides and hydrocarbons from motor vehicles, are conducive to the formation of smog in unfavourable dispersion conditions, so that high pollution levels build up in the atmosphere. The action of bright sunlight leads to the formation of compounds such as ozone and organic peroxides which have an irritant effect on respiratory systems and the eyes. In the U K, however, such photochemical smogs do not occur on anything like the same scale for several reasons: this country is further from the equator than Los Angeles so that sunlight is less bright and therefore less capable of inducing relevant reactions; also the airstream over the U K is generally more turbulent and the specific meteorological conditions required do not occur often. During the 1970s, episodes of photochemical pollution were reported in the U K, but they have not exhibited all the features of Los Angeles smog. The main sources of the hydrocarbons involved appear to have been oil refineries, and in some cases, smog appeared to have built up from unusual air trajectories passing over undetermined sources on the continent.

Dust and grit

18.34 Dust and grit are the larger sizes of particulate material and consist chiefly of particles of ash. The majority of this material is too coarse to be inhaled or to penetrate far into the respiratory tract. It comes largely from industrial rather than domestic sources, and it is mainly deposited close to the source. The main impact of such dust and grit is nuisance in the form of dirt. There are local problems in some cases, but these are generally due to industrial processes, such as cement manufacture, not from the use of coal itself. The remedy for such cases must lie in the careful siting of the plant to minimise the impact of the emissions. Provided the standards laid down by the Clean Air Acts and other relevant Acts are enforced, dust and grit emissions from coal burning are not a source of any serious general nuisance.

EFFECTS ON CROPS AND ECO-SYSTEMS

Local

18.35 At short ranges, where the concentrations of smoke, dust and grit, sulphur dioxide and nitrogen oxides are highest, there can be adverse effects on the growth of plants and on some animal species. Other pollutants from fuel combustion, such as hydrogen chloride and carbon monoxide, are too dilute. Within a few hundred metres of major emitters, grit and dust have caused problems in the past, smothering the leaves of plants so that only the hardiest varieties survive. Such serious effects no longer occur as a result of emissions from coal burning.

18.36 In former years, the effects on plants of high ambient levels of smoke and sulphur dioxide in cities were all too evident. Tree trunks were blackened and sensitive plants, including mosses, ferns, lichens, algae, and fungi were completely eliminated. But in the past 30 years, smoke has been reduced by 90% and sulphur dioxide by 50% in most cities. In many cases now the major source of smoke, especially in the growing seasons, is not coal combustion but the internal combustion engine. As a consequence of these lower pollution levels, the hardier varieties of even the most sensitive species such a lichens and algae are now returning to urban areas (Rose C I and Hawksworth D L 1981), although many delicate plants are still unable to grow within cities. But plants in cities are mainly grown for amenity purposes, and nearly all the species suitable for amenity use, as well as food plants grown in gardens, are able to thrive in most parts of most cities. The exceptional areas, where some amenity plants do not thrive so well, are largely those few central urban areas which are not yet subject to Smoke Control Orders under the Clean Air Acts. Although nitrogen oxides may be considered a problem in other parts of the world, in the U K they have never reached concentrations which give cause for concern regarding their effects on animals and plants, even in inner city areas where emissions from motor vehicles are greater.

18.37 We consider the situation regarding effects of pollution from coal combustion on plants in cities to be generally satisfactory. Measures taken to protect human health have also protected other animal life and vegetation. Accordingly, appropriate measures should be expedited to extend pollution control in those few exceptional areas where current pollution from smoke and sulphur dioxide still exceed satisfactory levels for protection of human health.

Medium range

18.38 Outside city centres, away from widespread local sources of fuel combustion, the principal cause for concern is sulphur dioxide which can affect vegetation by itself if ambient air concentrations are high enough. At high concentrations, over 500 μgm^{-3} as an annual average, development is inhibited and some species die. Such levels are not encountered anywhere in this country. Indeed, as Figure 17.34 shows, it is only in a few urban areas that the annual average concentration exceeds 100 μgm^{-3}.

18.39 The evidence on the effects of sulphur dioxide reflected contrasting views and in some cases was unable to separate sulphur dioxide effects from those of other pollutants. Experimental evidence was generally varied as was also found by the Seventh Report of the R C E P (1979). At annual averages above 200 μgm^{-3}, most studies have demonstrated significant reductions in the yields of crops. In the range 100–200 μgm^{-3}, results were varied with some showing small yield losses, some none at all. In the range 50–100 μgm^{-3}, results were highly variable and in some cases showed increases in yield, especially for plants grown in sulphur deficient soils. Peak exposures are probably more important than annual average concentrations, but the occurrence of such peaks is generally reflected in the annual average value. Even at 50 μgm^{-3} some plants can suffer microscopically detectable changes which in turn could affect photosynthetic and other growth processes. But such effects must be of small, if any, significance since they must be less than the barely detectable effects on yield from levels of 100 μgm^{-3}.

18.40 Less than 10% of crops and grazing land in the U K is exposed to annual average sulphur dioxide concentrations greater than 50 μgm^{-3}. Less than 1% is exposed to concentrations above 100 μgm^{-3}. On this basis the Department of the Environment's Systems Analysis Research Unit (S A R U) estimated possible crop losses from sulphur dioxide pollution in the U K to be about £23 million per year for worst case conditions. These calculations make no allowances for possible beneficial effects of sulphur dioxide such as fungus growth repression and the supply of sulphur to plants growing in sulphur deficient soils. Calculations were also made on the effects of future U K emissions on crops. Using the higher G D P growth cases of 2·7% per year discussed in Chapter 3, it was concluded that effects on crops would remain about the same. On the basis of the lower G D P growth case of 2% per year, there would be a small improvement.

18.41 Nitrogen dioxide concentrations in rural areas are very low, for example, less than 20 μgm^{-3} at Harwell in Oxfordshire in spite of the Didcot power station nearby. Even in the centres of large cities, the median concentrations are not large, for example 60 μgm^{-3} in Central London. Although research has indicated the possibility of synergistic action between nitrogen dioxide and sulphur dioxide on crop yields, this only occurred at much higher average levels of the two gases than are observed anywhere in the U K, for example, 140 μgm^{-3} of nitrogen dioxide together with 195 μgm^{-3} of sulphur dioxide (Ashenden T W and Mansfield T A 1978).

18.42 The effects of air pollution on agriculture were examined in great detail in the Seventh Report of the R C E P in 1979. They found the evidence of effects on crops at air pollution concentration encountered in the U K to be sparse and conflicting, with some evidence for beneficial effects resulting from exposure to low levels of sulphur dioxide for plants grown in sulphur deficient soils. The results of two years further research into various aspects of the problem have not yielded any new cause for concern and have emphasised that any effects on crops in the U K from atmospheric pollution from coal combustion must be very small. There remain problems associated with pollution arising from emissions from some processes associated with fuel combustion such as fluorine compounds from brick making. However, these types of problem do not arise from coal as such but from the materials used in the process.

Long range effects

18.43 A considerable amount of evidence was presented on the effects at long range of sulphur dioxide and other acid emissions. The principle adverse effect relates to the acidification of rain. Although such discussion is usually related to sulphur dioxide, nitrogen oxides also make a significant contribution and a minor contribution is made by hydrogen chloride. In areas of siliceous geology such as parts of Norway and Sweden, and parts of Scotland, acid rain is not neutralised by calcium compounds in the soil as it is elsewhere, and so may add to the natural acidity of soils, rivers and lakes. Excessive acidity leads to adverse ecological effects including the loss of fish, especially sensitive species such as salmon and trout. Whilst it is clear that there have been adverse effects in some areas which could be attributable to the acidification of surface waters, it is by no means clear that emissions of sulphur oxides some hundreds of kilometres away are the sole cause, as has been alleged.

18.44 We have received evidence on both the cost of damage caused by acid rain and on the costs of reducing sulphur dioxide emissions. As was discussed in paragraphs 17.69–17.76, the costs of limiting sulphur emissions cannot be determined with precision. We were given a range of £195–330 million per year as the cost of reducing by one quarter total U K emissions of sulphur dioxide from coal and oil. Similarly we have received a range of estimates for the cost of damage caused by acid deposition. A recent O E C D study (1981) estimated that a 37% reduction of sulphur dioxide emissions throughout Europe would avoid damage to fish in Scandinavia valued at £8 million per year. However, this estimate takes no account of the unquantifiable but highly sensitive loss of amenity involved in Scandinavia. On the other hand, the total costs to Northern Europe of achieving such reductions in emissions have been estimated by a consultant to the E C E to exceed £5,000 million per year (Barnes R A 1978). Also, it is by no means clear that such a reduction in sulphur dioxide emissions would achieve the desired reduction in acidity. The relationships between levels of emissions in one region and the acidity of rain in another, and between the acidity of rainfall and that of surface waters, are complex and far from clear.

18.45 It has been established that it is possible to mitigate the effects of acidification to some extent by the local application of lime or limestone to affected lakes and rivers, but this needs to be carefully controlled to avoid toxicity from the action of lime. The use of lime to reduce the natural acidity of lakes to improve fishing has been practised in some cases since the 19th century. This seems to us to be a potentially important method of damage limitation. It is capable of giving results far sooner than any action to limit emissions; the latter would take some decades to have any effect because of the magnitude of the control problem.

18.46 There has been considerable research both abroad and in the U K on acid rain. In the U K, the C E G B, in addition to carrying out considerable research into the transport of pollutants, is also carrying out extensive studies on the effects of acid rain on the chemistry and ecology of lake and river waters. D A F S' Freshwater Fisheries Laboratory has carried out pioneering work on the liming of lochs in Scotland. A current joint project by D A F S, the Forestry Commission and the Solway Firth Water Authority on the effects of forestry practices on the acidity of lochs will also include liming studies. Attempts have been made outside the U K to study the effects of artificially induced acidification of lakes and rivers with sulphuric acid. The results were difficult to interpret and did not properly mimic the effects of acid rain. Work in the U S A, Scandinavia and the U K has shown the importance of the watershed in the acidification of fresh water. It is now generally agreed that such studies must be carried out in the context of the watershed rather than the individual lake.

18.47 We have examined possible trends in future U K emissions in terms of their national and international implications. Based on the higher G D P growth case of 2·7% per year, discussed in Chapter 3, sulphur dioxide emissions would increase by about 10% by 2000 to 5·5 million tonnes per year; nitrogen oxides emissions by 30% to 2·5 million tonnes per year. The lower G D P growth case of 2% per year would yield a decrease in sulphur dioxide emissions of 8%, to 4·7 million tonnes per year, and an increase in nitrogen oxide emissions of 10% to 2·1 million tonnes per year.

18.48 As we have emphasised in Chapter 3, all such economic projections are surrounded by great uncertainty. Moreover, the above projections of future emissions take no account of the improvement which would result from changes in technology. F B C systems for coal retain about 20% or more of the sulphur in the ash, even without the addition of limestone to enhance sulphur absorption, whereas conventional systems retain only 10%. If much of the additional coal burn expected in industry does use such systems, the emissions of sulphur dioxide would be lower than those implied by the projections.

18.49 Secondly, burner manufacturers and electricity generating companies abroad are actively researching the development of coal burners which produce less than 50% of the nitrogen oxides produced by current systems. We understand that the C E G B and boiler manufacturers in the U K also have some interest in such burners. We would recommend a more active pursuit of research into the technology of low nitrogen oxides burner systems so that they may be available should research demonstrate the need to reduce emissions of nitrogen oxide in the U K.

18.50 All factors considered, we do not foresee, over the timescale considered by this Study, any significant increase in long range transport of air pollution from the U K. However, we attach the greatest importance to a comprehensive programme of research encompassing the relationship between the level of emissions and the acidity of rain, the technology of reducing emissions at source and remedial measures capable of implementation in areas subject to acid deposition. This research is the more important in view of the many uncertainties surrounding causes, effects and remedies in this sensitive area. We have noted the research programme being undertaken by the signatories to the Convention of the E C E on Long Range Transboundary Air Pollution. We strongly endorse the approach embodied in that Convention which calls for thorough research into mechanisms and causes as the essential basis for any appropriate cost-effective remedial action. We attach great importance to the U K contribution to this programme. We have a responsibility to find cost-effective remedies to the problems of others to which we may contribute as a result of essential U K operations. However, we share this responsibility with other major emitters. Therefore, such remedies must be found within an agreed international framework.

EFFECTS ON THE PHYSICAL WORLD

Effects on materials

18.51 The problem of dirt and damage to materials such as stonework and metalwork in inner city areas is no longer a problem of pollution from coal combustion. Present levels of particulates from fuel combustion are largely satisfactory even in inner cities, especially areas covered by Smoke Control Orders. That is not to say that there are no problems from particulates, but that the problems that do exist are localised and largely caused by industrial processes such as cement works and iron foundries. Measures taken to eliminate health effects, such as dust collection and the use of tall chimneys also minimise the effects of residual particulates by dispersing them. But careful location of potential sources of large quantities of particulates, both from the combustion and the handling of coal, will continue to be necessary.

18.52 Acids formed from sulphur oxides, nitrogen oxides, together with other pollutants and the natural carbon dioxide content of the atmosphere can attack stonework, especially limestone, marble and some sandstones, and can corrode metals, especially steel. But the relationship between exposure and damage is problematic for stone because present effects arise in many cases from the accumulation of previous pollution, often at higher levels than are now prevalent. Corrosion is also dependent on humidity and temperature. For metals there are somewhat better established dose-effect relationships for pollution from sulphur oxides, but it remains difficult to estimate how far corrosion is caused by sulphur oxides as opposed to other pollutants and stresses.

18.53 There are no reliable figures available for the cost of damage to materials by acid pollutants in the U K. Estimates made for an O E C D study (1981) of sulphur dioxide pollution showed that a 40% reduction in sulphur oxides concentrations might lead to a reduction in materials damage of £130 million per year. However, this figure related only to damage to metalwork; it did not take account of many factors, including alternative control strategies such as thicker galvanising in polluted areas, the desire to change the paint colours simply because people were tired of the old colour, or the fact that many metal structures subject to damage by acid corrosion are of a temporary nature.

18.54 Corrosion of materials is a function of local concentrations of acid. Because high concentrations are a local problem and as such not directly related to national emissions, a uniform overall reduction in emissions is not the most economic way to tackle essentially localised problems. As noted earlier, high concentrations of sulphur dioxide in cities are largely due to low and medium level emissions derived from fuel oil. Measures taken to protect human health should give some protection to materials, though there may be problems with art works such as marble sculptures in exposed sites. However, factors other than pollution can have adverse effects on materials. Soft stonework such as marble and limestone is subject to frost damage and to corrosion in unpolluted air from rain and atmospheric carbon dioxide. Therefore, it would seem preferable in such cases to provide for some measure of protection from the weather, especially rain, or for the use of more durable materials in accordance with good standard architectural practice.

Carbon dioxide and the greenhouse effect

18.55 In the atmosphere, carbon dioxide admits incoming short-wave solar radiation while tending to trap out-going long-wave (infra-red) radiation, so heating the earth's surface and the lower atmosphere. This is known as the greenhouse effect because the glass in greenhouses serves a similar function. Higher levels of carbon dioxide would in theory lead to atmospheric warming and hence climate change which could have long term consequences for economic, agricultural and settlement patterns (Council on Environmental Quality 1981). These effects are being studied by means of mathematical models which simulate the physical processes taking place in the atmosphere under the influence of solar radiation. Although these models are very sophisticated, it is not yet possible to include in them all the interactions which are desirable, particularly some important ones involving the oceans. Sir John Mason, Director-General of the Meteorological Office, in summing up the best available climate models (Mason J 1979), has said that on the basis of a possible doubling of carbon dioxide levels in the atmosphere over the next 50 years or so they predict a 2–3°C rise in global annual average surface temperatures; at the Poles in winter there would be a warming of up to 10°C; and there would be a 5–7% average increase in precipitation, reaching 20% in middle latitudes. Some parts of the globe would be colder or drier than at present. For comparison, mean annual temperatures in the U K have varied less than 1°C from the average over the past century.

18.56 Warming might be expected to reduce the need for heating in cold and temperate countries. It could also, over several centuries or even millenia, cause sea levels to rise as a result of melting the Polar ice caps. Crop yields might increase through more efficient photosynthesis and increases in the length of the growing season. On the other hand, the temperate cereal-growing belt of the Northern Hemisphere may be displaced towards less fertile soils. It is not yet possible to assess reliably the regional variations in these global trends which will determine the balance of advantages and disadvantages for different localities.

18.57 However, against a background of other changes of uncertain origin, it has not been possible to detect with certainty any increase in mean global temperature so far, which should theoretically amount to around 0·4°C. Some researchers believe that the effect is rather less than that predicted, or else the predicted effect is working against an underlying trend to lower temperatures. Nevertheless, some believe that the accumulation of carbon dioxide in the atmosphere is the most serious potential environmental problem caused by fuel combustion. For example in a 1979 report to the U S Council of Environmental Quality, four eminent scientists – George M Woodwell, Gordon J MacDonald, Roger Revelle and C David Kelling – warned that "The carbon dioxide problem is one of the most important contemporary environmental problems [which] threatens the stability of climates worldwide and therefore the stability of all nations. Man is setting in motion a series of events which seem certain to cause a significant warming of world climates over the next

decades unless mitigating steps are taken immediately. The displacement of agriculture in a world constantly threatened by hunger would alone constitute an extremely serious international disruption within the lifetime of those living".

18.58 Research into the potential influence of carbon dioxide is taking place under the auspices of the World Meteorological Organisation and other international bodies. The Meteorological Office and other U K organisations are taking part in these studies. However, the nature of the problem is such that they are unlikely to produce definite conclusions for many years. At this stage we consider that it would be premature to do more than note the potential importance of the issue and that research is being carried out to clarify the problems.

Nitrogen oxides and the ozone layer

18.59 Although coal is not responsible for any high ambient concentrations of nitrogen oxides in the U K, it is responsible for 47% of total emissions, equivalent to 0·9 million tonnes per year of nitrogen dioxide. Fears have been expressed that high levels of nitrogen oxide could lead to depletion of the ozone layer in the stratosphere which protects us from excessive amounts of ultra-violet light. This might be a problem if large quantities of the oxides were injected directly into the stratosphere as from the exhausts of high flying aircraft. But nitric oxide and nitrogen dioxide are both fairly reactive and the natural circulation and removal mechanisms of the atmosphere ensure that only small amounts of nitrogen oxides from fuel combustion are carried up into the stratosphere. The relatively unreactive nitrous oxide produced by biological processes is far more important in this respect and is by far the major source of nitrogen oxides in the stratosphere.

CONCLUSIONS

18.60 We consider that there is no longer major cause for concern on health grounds over ambient levels of smoke and sulphur dioxide from coal combustion, or indeed from combustion of any fuel. This is not to say that the situation is entirely satisfactory. There are a few areas where concentrations are too high and some improvement is needed. The extension of Smoke Control Orders should cope with the problems in most of these areas, but in a few areas already covered by such Orders it may be necessary to reduce sulphur dioxide levels by restricting the sulphur content of fuel oil used in commercial and industrial boilers. There remains uncertainty about the carcinogenic effects of the small amounts of smoke now released into the atmosphere and it would be unwise to allow them to increase unnecessarily.

18.61 The primary instrument of current improvements over the unacceptable pollution of former years has been the diligent application of control measures. These must not be relaxed. Monitoring will continue to be necessary in urban areas to ensure that satisfactory standards are maintained.

18.62 As long as pollution in urban areas is kept low enough for the protection of human health, we consider that the situation is also satisfactory as regards crops and vegetation in general in the U K. We do not consider the extent to which materials are affected by pollution at such safe levels to be cause for concern.

18.63 Nitrogen oxides from coal combustion have not been identified as the source of any pollution problems within the U K. Motor vehicles are in any case the major source of relatively high ambient concentrations of nitrogen oxides outdoors. Even higher levels occur indoors from appliances such as gas cookers, and flueless oil or propane heaters.

18.64 Sulphur dioxide together with nitrogen oxides may be implicated in the acidification of lakes and rivers in certain sensitive areas of the world, such as parts of Norway and Sweden, where the soils lack the normal capacity to neutralise any excess acidity. This acidification is linked with the complex problem of the long range transport of pollutants. Considerable uncertainty surrounds these phenomena. More research is required into the causes and mechanisms of the effects of acidification and into possible control technologies. The present state of knowledge would not warrant immediate and expensive action to limit emissions of sulphur dioxide or nitrogen oxides emissions because of their possible effects at long range. We strongly endorse the approach of the E C E Convention, involving research into mechanisms,

causes and remedies. We consider, however, that this in no way diminishes the need for continued research and development of cost-effective technologies for the reduction of emissions of sulphur dioxide and nitrogen oxides so that, should such reduction be shown to be necessary, suitable technology will be available. This research should also address the adverse environmental impacts of existing emission control technologies. More research is also desirable on techniques for ameliorating the effects of excess acidity in the problem areas.

18.65 Dust and grit from coal combustion has been in the past a major nuisance in urban areas and close to large emitters. The application of good emission control technology, which has been required by legislation over the past few years, has largely cured this problem. But continued monitoring is required to ensure that standards are maintained. Careful siting of plant, which because the scale of operation does not allow elaborate control techniques, is potentially a major emitter of grit and dust, will continue to be necessary.

18.66 The disposal of large quantities of ash from power stations and industrial plants causes no problems on health grounds. The material is largely chemically inert, and such problems as do arise can be dealt with by simple precautions. The major difficulty is that of finding suitable disposal sites given the scale involved. As with any large industrial enterprise, there are local problems arising from disturbance by noise and dust from transport and handling. Careful siting and treatment of disposal areas will continue to be necessary.

Chapter 19   The environmental effects of burning coal in urban areas

INTRODUCTION

19.1 In the previous Chapter on the effects of combustion we demonstrated that most of the air pollutants arising from coal no longer cause us major problems. Thus, over this country as a whole, current levels of total emissions do not constitute a critical consideration. As we have identified in our earlier analysis of the sources, quantities and effects of combustion residues, a combination of factors can result in urban areas being exposed to relatively higher concentrations. We therefore focus attention in this Chapter on the environmental effects of burning coal in these areas and on the measures which may be necessary to reinforce pollution control in order to accommodate prospective changes in patterns of fuel use. It must be stressed that in this context coal combustion cannot be considered in isolation. While we take full account of the implications of a major re-entry of coal into the industrial market, the problems of urban areas are a function of multi-source pollution as was emphasised in the Fifth Report of the R C E P (1976a).

MAJOR SOURCES OF POLLUTION

19.2 The three major sources of pollution from fuel use in urban areas are industry and commerce (both largely burning residual fuel oil at present), domestic premises burning coal, and motor vehicles. Coal use contributes the greatest proportion of national *emissions* of sulphur dioxide, smoke and nitrogen oxides. But in urban areas, at ground level, by far the major proportion of *concentrations* of nitrogen oxides comes from motor vehicles; and in most cases the majority of the sulphur dioxide concentration comes from fuel oil. In some conurbations subject to extensive Smoke Control Orders, the largest single source of ambient smoke levels is motor traffic. In general, however, over the U K as a whole, the major sources of smoke are domestic premises burning solid fuel, especially coal in open fires. Industrial coal use produces much less (and less harmful) smoke.

19.3 Tremendous gains in air quality have been made in the last 25 years or so. These have been reflected in the lowered incidence of some respiratory diseases which we have already discussed. In most areas, smoke and sulphur dioxide levels are now within acceptable limits. However, there are some residual problems in particular urban areas. We have examined therefore the contribution emissions from coal make to national levels, particularly of smoke and sulphur dioxide, and tried to assess how these might change if more coal is to be burned in these areas in the future. However, as we explain later, it is difficult to relate national trends of emissions to local pollution concentrations so that monitoring of the atmosphere will continue to be essential.

19.4 The E E C Air Quality Limit Values and Guide Values Directive (the 'Air Quality Directive' European Communities 1980), which came into effect in 1980, sets levels for ground level concentrations of smoke, sulphur dioxide and the combination of the two. Two sets of levels are given. The limit values are mandatory maxima which must be met throughout the E E C by 1 April 1983 and are intended to act as a protection for human health. The guide values, which are lower and non-mandatory, are intended to serve as reference points for the long-term improvements of air quality, towards which Member States are encouraged to aim. Both sets of values are set out in Tables 19.1 and 19.2. General medical and scientific opinion accepts that the limit values are adequate for the protection of human health and for the protection of animals and plants. It is difficult to comment on the suitability of the limit values for the protection of materials due to lack of information, but we have no reason to suspect that a similar situation does not exist. This Directive provides a framework for the control of air quality with respect to two of the most important pollutants, similar to that suggested by the Fifth Report of the R C E P (1976), which first suggested a two-tier system of guidelines.

19.5 Recent results from the National Survey of Smoke and Sulphur Dioxide, carried out by the Warren Spring Laboratory, indicated that there are areas within 71 district authorities (see Table 19.3) where ground level concentrations of smoke and/or sulphur dioxide either exceed the mandatory limit values now or are approaching them (that is, have recent results within 10% of the limit values). The Department of the Environment and the Welsh Office have recently issued a Circular (1981) to local authorities listing these districts; more detailed discussions to determine the precise extent of the problem will take place shortly. The map at Figure 19.1 shows the principal areas at risk. They include parts of the major industrial

Figure 19.1 Smoke control areas and principal areas at risk of breaching E E C limit values: smoke and sulphur dioxide

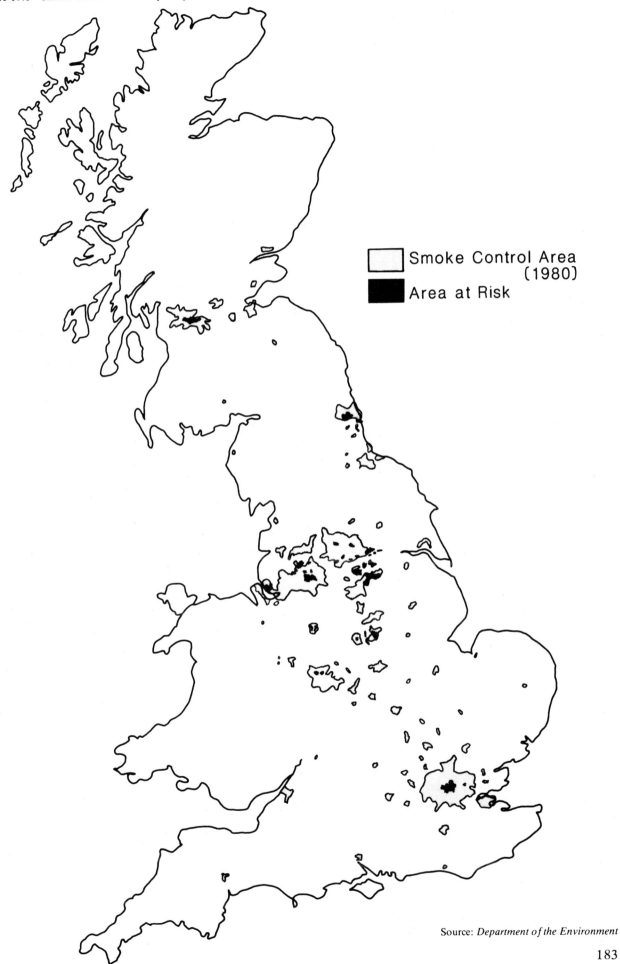

Smoke Control Area (1980)
Area at Risk

Source: *Department of the Environment*

conurbations of Glasgow, Newcastle, Greater Manchester, Liverpool, Sheffield and London.

Table 19.1 E E C limit values for smoke and sulphur dioxide

μgm⁻³

Reference period	Smoke*	Limit values for sulphur dioxide
Year (Median of daily values)	80(68)	If smoke less than 40(34):120 If smoke more than 40(34):80
Winter (Median of daily values October–March)	130(111)	If smoke less than 60(51):180 If smoke more than 60(51):130
Year (Peak) (98th percentile of daily values)**	250(213)	If smoke less than 150(128):350 If smoke more than 150(128):250

Table 19.2 E E C guide values for smoke and sulphur dioxide

μgm⁻³

Reference period	Smoke*	Sulphur Dioxide
Year (arithmetic mean of daily values)	40–60 (34–51)	40–60
24 hours (daily mean value)	100–150 (85–128)	100–150

*Limit values for smoke as stated in the Directive relate to the O E C D method of measurement: figures in brackets give equivalents for the B S I method as used in the National Survey of Smoke and Sulphur Dioxide.
** The 98th percentile is the value below which lie 98% of the daily average readings, that is only 7 days per year (2%) exceed the 98th percentile.

Table 19.3 Preliminary identification of areas within which concentrations of smoke and sulphur dioxide may exceed the limit values

List 1 below consists of those District Councils within which some evidence from the National Survey indicates that concentrations in excess of one or more of the limit values have occurred. Discussions will be held with these authorities to determine whether a continuing problem exists for which specific control action is required.

List 2 below consists of those District Councils where further investigation is needed to determine whether excess concentrations are likely to occur. Discussions will also be held with these authorities, and any additional authorities where excess concentrations are thought likely.

	List 1	List 2
EAST MIDLANDS		
Derbyshire	Amber Valley Bolsover Derby High Peak	Chesterfield Erewash
Leicestershire		Charnwood Hinckley and Bosworth
Nottinghamshire	Bassetlaw Broxtowe Mansfield	Newark Nottingham

	List 1	List 2
NORTHERN		
Tyne and Wear	Newcastle upon Tyne Sunderland	Gateshead North Tyneside South Tyneside
NORTH WEST		
Cheshire	Congleton Crewe and Nantwich Ellesmere Port Warrington	Halton
Cumbria	Copeland	Barrow-in-Furness Carlisle
Greater Manchester	Bolton Rochdale Salford Tameside Trafford Wigan	Bury Manchester Oldham
Lancashire	Chorley	Blackburn Hyndburn Rossendale
Merseyside	St Helens	Liverpool Sefton
SOUTH EAST		
London	Islington Kingston City Tower Hamlets Westminster	Barking Camden Hackney Kensington and Chelsea Southwark
Essex	Castle Point	Epping Forest
WEST MIDLANDS		
West Midlands		Sandwell Walsall Wolverhampton
WEST MIDLANDS		
Staffordshire	Cannock Chase Newcastle-u-Lyme Stoke on Trent	E Staffordshire
YORKSHIRE AND HUMBERSIDE		
North Yorkshire		York
South Yorkshire	Barnsley Doncaster Rotherham	Sheffield
West Yorkshire	Bradford Kirklees Leeds Wakefield	Calderdale

Source: *Department of the Environment and Welsh Office 1981*

19.6 A downward trend in sulphur dioxide emissions from fuel combustion since 1970 has already been shown in Table 17.6. The increase shown for 1979 was mainly attributable to the exceptionally bad winter at the beginning of that year and does not affect this trend. We understand that provisional estimates show that sulphur dioxide emissions for 1980 were less than

5 million tonnes, a considerable reduction over 1979, and that the downward trend has continued so far in 1981.

19.7 Estimates of future sulphur dioxide emissions differ to some extent but there is general consensus that no significant increase will take place. Although smoke is in many ways a more serious problem, we do not have such predictions for smoke. This is partly because detailed information is not available on the location, type and fuel consumption of domestic coal burning appliances, and partly because future emissions of smoke are not directly linked to total fuel consumption, since smoke production is much more a factor of the appliance used than the total amount of fuel burnt. Warren Spring Laboratory have produced analytical methods suitable for predicting the effect of new smoke control zones on local areas, but such methods could not be extended to large scale predictions without very considerable effort. However, since smoke pollution is derived largely from local sources, and subject to local remedies, we think that monitoring is more important than prediction.

19.8 Warren Spring Laboratory, working on the basis of the two cases in the Department of Energy's Energy Projections 1979, estimate that total sulphur dioxide emissions in 2000 will be 4·5–5·2 million tonnes. S A R U examined the geographical distribution of present and forecast sulphur dioxide emissions and estimated ground level concentrations. Their projections are based on the 1978 Green Paper on Energy Policy and on the higher G D P case in Energy Projections 1979. Their projections based on the latter do not imply sulphur dioxide levels much above those of 1976. However, the increase in industrial coal burn assumed in the Projections shows up in S A R U 's work with a rise in emissions in industrial areas such as the West Midlands and Lancashire. In view of the deterioration in economic prospects that has taken place since the Department of Energy proposed its 1979 Projections, forecasts of future emission levels based on these projections are probably over-estimates. S A R U have pointed out that their analysis is subject to a number of caveats – in particular that the assumptions of local fuel use on which it builds its projections are based on limited data, the complete information which would allow more robust prediction not being available. Moreover while this analysis, based on 10 kilometre squares, gives a picture of the country as a whole, it is too coarse to show up local problems adequately. Even with these important qualifications, the results of the S A R U work are consistent with the general view that there are urban areas where an increase in coal burn might aggravate present air pollution problems. Smoke from domestic open fires could give cause for particular concern. While the substitution of coal for fuel oil in the industrial sector may prove helpful in reducing sulphur dioxide levels, the latter will still need to be monitored carefully. We

therefore examine, taking in turn the domestic and industrial sectors, the present machinery for controlling urban air pollution and its adequacy for dealing with possible future problems.

DOMESTIC COAL BURN

19.9 As discussed in Chapters 13 and 16, the domestic coal market has declined considerably over the past three decades, partly because of legislation controlling smoke and partly because of the availability and convenience of other fuels, particularly gas. However, domestic coal is still relatively cheap and it can be an efficient source of energy. But smoke from domestic open fires is, qualitatively and quantitatively, the most objectionable form of smoke; its tar content is considerably higher than smoke from industrial premises and its likely detrimental effect on health is greater. In urban areas any discharge of products of combustion through chimneys of low or medium height will contribute to local pollution levels and could therefore be of concern, especially where sulphur dioxide and smoke levels are near to, or above, the limits in the Air Quality Directive.

Smokeless fuel and smoke reducing appliances

19.10 Domestic smoke can be reduced either by using smokeless fuel (anthracite, coke or various manufactured fuels) or by installing special appliances, described in Chapter 16, designed to reduce the amount of smoke produced. At present, smokeless fuel is the most common option, and accounts for about 5 million tonnes out of about 14 million tonnes (36%) of domestic solid fuel used each year. N C B laboratories carry out approval tests on smokeless fuels for the Department of the Environment under supervision of senior staff from Warren Spring Laboratory as part of the approval process for their use in smoke control areas. We do not see the further development of smokeless solid fuel as a viable option. Naturally smokeless coal and anthracite are generally more difficult to mine in this country, are in short supply and are likely to become more so. The manufacture of smokeless solid fuel from bituminous coal is expensive, and can cause severe pollution at the point of production. Therefore, if coal is to continue to be used for domestic heating in urban areas, more effort must be made to develop and market smoke-reducing appliances which are simple to use, reliable and foolproof under all conditions of use.

19.11 The N C B has been working for about 10 years to develop appliances which can burn bituminous coal in homes while satisfying the requirements of the Clean Air Acts. As part of the Domestic Solid Fuels Appliances Approval Scheme set up in 1970, testing of solid fuel appliances is carried out at N C B laboratories and, under observation by N C B representatives, at a few accredited laboratories of appliance manufacturers. The earlier developments were not always wholly successful. The Manchester Area Council for Clean Air and Noise Control quoted several examples of appliances which had not reduced smoke in practice, such as the N C B Housewarmers and the Everglow 1/44 – often because of faulty operation by householders. During our visit to the N C B's Coal Research Establishment, however, we saw some of the current work, which emphasises simple designs which can burn a wide range of coals. We consider such development an important means of reducing the environmental impact of domestic coal burn and welcome the N C B's work in this area. Its importance is clearly illustrated by reference to Table 17.3 which shows the dominant effect of domestic coal burn on particulate emissions. For each million tonnes switched from open fires to smoke-reducing appliances, the N C B expect a reduction in particulate emissions from 35,000 tonnes to 5,000 tonnes. We cannot overstress the importance of expert installation, regular maintenance and foolproof operation. We would like to see even greater efforts expended on the design of domestic appliances which can reduce smoke emissions using bituminous coal under a variety of conditions of use.

Combined heat and power and district heating

19.12 There are many ways of using coal to supply energy for domestic heating in urban areas. These include the generation of electricity, for immediate or off peak use, with or without heat pumps; S N G also with or without heat pumps; District Heating (D H); Combined Heat and Power together with District Heating (C H P/D H); and direct coal use. The efficiency and fuel cost to the domestic consumer of several such alternative methods have been considered in Appendix 1. None of these systems should be considered as mutually exclusive, although the economic viability of large scale D H or C H P/D H schemes may require the limitation of other heating systems within the areas served. However, it is only likely to be economic and technically feasible to provide D H networks in selected areas; and it is unlikely that many such networks could be constructed before the end of this century.

19.13 We see some advantages for coal D H schemes, in so far as they confine emissions to a single source which can be more easily controlled. The larger scale of plant involved would also permit higher standards of emission control and maintenance. Transport, storage and handling problems would be eased by the concentration of coal use and ash disposal; and householders would retain the convenience of "instant fuel". It is hoped that D H schemes would use less fuel than separate systems in each household or office block. However, D H would generally not be replacing direct coal use by individual households but would be substituting for a variety of other fuels, with gas

probably predominating. Thus coal-fired D H schemes, although potentially attractive from the energy conservation point of view, may possibly increase local pollution especially from particulate emissions.

19.14 A disadvantage of D H schemes is that, because of the high cost of installing and maintaining pipework from the boiler to the point of use, they are only really attractive for new high density housing and commerce. The installation of such pipework presents few serious difficulties in new developments. However, there could be considerable inconvenience in carrying out, in existing built-up areas, the necessary civil engineering work to supply buildings at present heated by more conventional methods. Difficulties would also arise if the owners of these buildings did not wish to take part in such a scheme.

19.15 A further problem of D H schemes is that of devising an equitable system of billing which customers both understand and trust. Heat meters have not yet been developed which are accurate and cheap enough for domestic use. The alternative of making a fixed charge is likely to encourage the wasteful use of heat and so vitiate the advantages of such schemes. We recommend therefore that the Government should coordinate research and development on heat meters, particularly for domestic use, as a matter of high priority.

19.16 The main advantage of C H P schemes is that they use energy from power generation which would otherwise be wasted. However, energy derived in this way is not necessarily cheap. Some modification of the steam plant is necessary and the output of power is reduced. Thus when applied to electricity generation more generating plant must be built and more fuel must be burnt to produce the same amount of electricity. Nevertheless, these costs can be offset by the value of the heat that would otherwise be lost. C H P is already well established in industry where about 70% of the privately generated electricity is associated with heat recovery. There are no major basic technological difficulties involved in combining C H P with D H, but it does mean coping with all the disadvantages of conventional large scale D H schemes. Much experience in C H P/D H exists abroad and there are many small D H schemes in the U K. C H P/D H has been considered in some depth in other studies including three reports published by the Department of Energy (1977c, 1979a and 1979c). Further progress on C H P schemes is dependent on detailed studies which are to be carried out in several British cities for a feasibility study to be financed by the Department of Energy. Although there are limits to the extent to which C H P/D H is feasible, we consider that, in the right circumstances, it would be a useful means of conserving energy. There would be a reduction in total emissions, but whether or not there would be an improvement locally in areas served by C H P schemes is uncertain particularly if new power stations were to be built near to towns.

Present control powers

19.17 In areas subject to smoke control orders under the Clean Air Act, smoke emissions are forbidden, subject to certain limited exceptions. As described in Chapter 4, householders continuing to use solid fuel must either install an appliance approved as being capable of burning coal smokelessly or change to an approved smokeless fuel. Where the householder has to convert his existing equipment, an Exchequer contribution to the cost can be made. About 8 million premises in over 5000 control areas are currently covered by the Clean Air Acts, and over 100 new orders are approved each year. However, most of the areas approaching the E E C limits on smoke and sulphur dioxide (see Figure 19.1 and Table 19.3) are not yet fully covered by smokeless zones and it will be necessary to extend smoke control in these areas to ensure compliance with the Air Quality Directive. The Department of the Environment and the Scottish Development Department are currently discussing with local authorities how this can be done. We note that under the Local Government Planning and Land Act 1980 the procedure for making smoke control orders has been simplified so that orders no longer need to be confirmed by the Secretary of State for the Environment. We welcome this change and hope that it may speed up the introduction of smoke control.

Adequacy of controls

19.18 In our view, an increase in domestic coal burn is highly unlikely except perhaps in rural areas. Inconvenience and problems of storing coal in built-up areas are likely to dissuade most householders, if they have options other than changing to coal, although, as we have already pointed out in Chapter 15, it is desirable to preserve consumer choice. If domestic coal burn in urban areas did increase, there might be areas in which problems of emissions would be aggravated. Air pollution levels in such areas would need to be monitored so that appropriate action might be taken if necessary. We are, of course, aware that the considerable improvements in air quality since the time of the Clean Air Acts have been a source of pride to many local authorities, who are understandably unwilling to jeopardise these achievements. Moreover, in the event of increased urban domestic coal burn, the extra staff needed to monitor smoke and enforce smoke controls on a large number of domestic dwellings could strain local authority manpower resources.

19.19 As illustrated in the map at Figure 19.1, there are a very few areas in the U K in the centres of extensive conurbations which are already covered by Smoke Control Zones but where pollution by smoke and/or sulphur dioxide exceeds or is near the limits of the E E C Air Quality Directive. In Central London

there is currently little use of solid fuel for individual domestic heating, and the major source of smoke is motor vehicles. The situation is under investigation by the Department of the Environment in other major conurbations. It might be necessary in such areas eventually to limit further, or at least prevent any increase in, the use of any fuels giving rise to emissions of either smoke or sulphur dioxide. For domestic premises the only controls available relate to smoke from solid fuel. Smoke control would not, however, avoid sulphur dioxide emissions. These would clearly need to be monitored in the event of a major increase in domestic coal burn.

INDUSTRIAL COAL BURN

19.20 The present level of industrial coal burn, and possible future developments in this market, are discussed in Chapter 13. Here we examine the implication for urban areas. Taking the middle of the range of projected industrial coal burn by 2000 which we reproduced earlier in Table 13.3, we have examined the implications of an additional coal burn of 30 million tonnes, assuming that the quality of this coal, the combustion equipment and the resulting emissions will be similar to present patterns. On these assumptions, low and medium level emissions, largely in or adjacent to urban areas, could be increased by up to 670,000 tonnes of sulphur dioxide and 90,000 tonnes of fine particulates. These estimates of emissions are likely to be maxima. Some new plant may use F B C, which can be expected to reduce sulphur dioxide emissions by 10% or more for each plant (as discussed in Chapter 16). Particulate emissions are not likely to be greatly affected by F B C. Some coal might also be used in local gasification plant to produce low or medium Btu gas and this would provide the opportunity for some sulphur removal.

19.21 Much of the 30 million tonnes of industrial coal burn will not represent increased energy consumption since a substantial proportion will be replacing fuel oil. Using figures from the lower growth case considered in the Department of Energy's 1979 Projections, it has been estimated that there will be a reduction in fuel oil consumption of 9 million tonnes. This would avoid sulphur dioxide emissions of 370,000 tonnes and particulate emissions of 14,000 tonnes. Thus the net result of the assumed changes in fuel consumption in the industrial sector by 2000, would be an increase of sulphur dioxide emissions of 300,000 tonnes (or 6% of the total from all sources in 1979) and an increase in particulate emissions of 76,000 tonnes (or 15% of the total from all sources in 1979). If other emissions were to remain unchanged, it is very likely that such increases would extend the areas which would exceed the sulphur dioxide and smoke limits of the Air Quality Directive. The Alkali and Clean Air Inspectorate pointed out to us that, if industrial coal burn were to increase, this would undoubtedly involve some increase in particulate emissions leading to more dirtiness and consequent loss of amenity. However, as we noted above, there is likely to be a continuing reduction of particulate emissions from domestic premises. It would need a switch of only 2·5 million tonnes of domestic bituminous coal (about 30%) from open fires to smoke-reducing appliances to bring about a reduction in national smoke emissions which would counter-balance an increase in particulates from such a change in industrial use.

19.22 The industrial sector can be more easily controlled than the domestic sector; it is less fragmented, and the proper installation and maintenance of plant by trained staff will be more likely. As explained in Chapter 17, most industrial installations in urban areas will be medium or small in size, typically using stoker-fired boilers with small coal and producing bottom ash. When properly operated, these installations do not produce smoke except possibly on start-up and on certain other limited occasions, as permitted by the Clean Air Regulations. Thus we conclude that the increased industrial coal burn examined should not give rise to major concern regarding smoke.

19.23 The Clean Air Acts and the Inspectorates (see paragraph 19.28) already lay down certain limiting values for grit and dust but we think that the expansion of industrial coal burn may require a review and possibly a tightening up of the limits permitted from individual installations. Meanwhile, regular surveillance of emissions and inspection of abatement equipment will be vital for proper control. This, together with the need for careful monitoring of sulphur dioxide concentrations, could strain the staff resources of local authorities. Subject to these qualifications, we conclude that the air pollution implications are not such as to prevent positive encouragement being given to increased industrial coal burn in urban areas.

19.24 Similar considerations apply to commercial premises, including office blocks, where the size of boiler involved is likely to be comparable to that in some industrial installations. But one additional consideration is that, in the City of London, there are legal controls on the sulphur content of fuel oil and there are powers under Section 76 of the Control of Pollution Act 1974 to introduce similar controls into specified areas. Low-sulphur fuel oil can at present provide a relatively low cost means of preventing high urban sulphur dioxide levels, but there may be supply constraints towards the end of the period under review. The substitution of coal for low-sulphur fuel oil would approximately double sulphur dioxide emissions.

Existing control powers

19.25 Local authorities are responsible for the control of emissions from most combustion processes which are not part of processes registrable under the Alkali Acts as discussed in Chapter 17. The principal powers available to local authorities are those of the Clean Air Acts 1956 and 1968. The nuisance provisions of the Public Health Acts 1936 and 1969 can also be used in some circumstances. The pollutants of concern are particulates, smoke, grit, dust, fume, and noxious gases, primarily sulphur dioxide.

19.26 Approval for most proposals to install new furnaces, that is, any equipment for burning fuel including boilers, ovens and dryers, as well as equipment conventionally understood as furnaces, must be sought from local authorities. The exceptions to this requirement are small and essentially domestic appliances. Such plant must be capable of operation without producing smoke and must be operated accordingly. There is a general requirement not to emit dark smoke in operation other than on certain strictly limited occasions such as start-up. Any other smoke, such as smoke from old furnaces, that does not fall within the definition of dark smoke in the Clean Air Acts, or smoke emitted by activities not covered by the Acts, such as some bonfires, can fall within the provisions of the Public Health Acts. Certain furnaces, typically of the size used for industrial premises, must have plant approved by the local authority for arresting grit and dust. Most furnaces must in any case either conform to detailed emission limits set out in regulations made under the Clean Air Acts, or use any practicable means for their control if no regulations are made. Furnaces for which there are no general emission limits are principally those in which the material being heated contributes to the emissions, such as cupolas and dryers of powdery materials. In these cases, the user must minimise emissions of grit and dust and the local authority will give advice on how best to do so. Draft proposals on general regulations for such furnaces under the Clean Air Act 1968 have been published but not as yet implemented.

19.27 The height of chimneys serving furnaces is also controlled by local authorities, primarily to provide for adequate dispersion of gaseous pollutants, in particular, sulphur dioxide. There is no direct method for local authorities to require specific controls over the emission of gaseous pollutants such as sulphur dioxide other than by regulation of the chimney height or by using the nuisance provisions of the Public Health Act.

19.28 The Alkali and Clean Air Inspectorate (in Scotland, the Industrial Pollution Inspectorate for Scotland) control those combustion activities that are registered processes (for example, power stations) or associated with registered processes (for example, boilers at a steel works) under the Alkali Act 1863, the Alkali Etc Works Regulation Act 1906 and the Health and Safety at Work Etc Act 1974. This control includes requiring the design and operation of the furnace to conform to Bpm and covers not only emissions of both particulates and gases from stacks, but also the transport and handling of fuel and ash. The requirements of Bpm can be changed from time to time as technological developments occur. Although in practice such changes are generally only applied to new plant and comprehensive refurbishment of older plant, they can be applied to existing plant if it is considered necessary.

19.29 The Clean Air Acts allow local authorities to apply to the Secretary of State for specific registered works to come under their control. These works remain on the central register but emissions of smoke, grit and dust are removed from the Alkali Inspectorate's control. The local authority controls them under the powers of the Clean Air and Public Health Act. Only a few local authorities, including Birmingham, Leeds and Manchester have done this for some registered works in their areas.

Adequacy of controls

19.30 At present, control powers seem adequate to cope with existing problems. Extension of smoke control orders and, in a limited number of areas, orders to regulate the sulphur content of fuel oil, should be sufficient to deal with individual districts which are approaching or exceeding the mandatory limits laid down in the Air Quality Directive (European Communities 1980). However, it is difficult to forecast whether these powers will cope with projected increases in coal use. As we discussed in Chapter 13, a striking feature of the evidence submitted to us is the general unawareness of local authorities and other interested bodies of the possible scale of re-entry of coal into the industrial market, in spite of the projected fourfold increase in industrial coal burn. There is a need for a continuing and systematic dialogue between Government, the N C B, industrial associations and local authorities on the preparatory action necessary to accommodate fuel substitution on the scale envisaged.

19.31 The need for such preparations is paralleled by the relative weakness of the existing pollution control system in terms of its capacity to anticipate possible pollution problems compared with its adequacy to deal with existing problems. Local authorities have no power to act in anticipation of a possible pollution problem – except in so far as they may be involved in giving or refusing planning permission for development. The powers for controlling statutory nuisance under the Public Health (Scotland) Act 1897 and the Public Health Act 1936 are now under review and proposals for change include the introduction of powers for local authorities to take action where they consider that a statutory nuisance might arise. But even if these proposals are implemented, statutory nuisance powers, which are usually confined to more direct public health hazards such as buildings in a dangerous

state of repair, are probably not appropriate to deal with the problems caused by coal burn.

19.32 The present system also provides only limited power to control the cumulative impact of emissions from new or additional sources which may operate individually within currently acceptable emission levels. The cumulative impact, added to existing emissions, could result in ambient concentrations in given areas in excess of the mandatory limits of the Air Quality Directive. This aspect of the cumulative effect of multiple sources was noted by the Fifth Report of the R C E P (1976) which recommended that local authorities should determine the total pollution capacity of a site against which to assess applications for development. However, additional powers might be needed to prevent excessive pollution from emissions arising from large numbers of existing industrial premises, if they changed from oil or gas firing to coal, even though the emissions from such premises were individually within the limits of the Clean Air Acts.

19.33 Central Government and local authorities need to monitor developments. It could well be that tighter controls may be needed. Meanwhile, in response to the scope for improvement which we have identified, we recommend that Central Government should consider the following possibilities. Local authorities might be given powers to require prior consent to an industrialist's choice of fuel or means of controlling emissions. Bpm used by the Alkali Inspectorate for controlling Scheduled Works might be extended to those processes controlled by local authorities so that they could negotiate with industries emission limits, equipment and operating practices which constituted Bpm for a particular works in a particular environment. In this context we endorse the recommendations on Registration and Consents in the Fifth Report of the R C E P. In particular, we recommend that, in exceptional circumstances, in areas proving particularly difficult to control, powers might be granted to specified local authorities to direct the use of particular fuels in particular buildings. There are precedents for this in the Manchester Corporation Act 1946 and the Bolton Corporation Act 1949, which established smokeless zones by requiring that "no smoke shall be emitted from any premises in the zone" thus enabling these authorities to ban the installation of coal fired plant in the city centres. These powers, however, lapsed at the end of 1980 and were replaced by the less stringent powers in the Clean Air Acts.

CONCLUSIONS

19.34 We established in Chapter 18 that there is no longer major cause for concern on health grounds over ambient levels of smoke and sulphur dioxide from coal combustion, nor over emissions of nitrogen oxides from coal. We consider that the mandatory limits for smoke and sulphur dioxide set out in the E E C Air Quality

Directive provide an adequate standard for the protection of human health. Appropriate measures should be taken as soon as possible to improve the air quality in all areas which do not conform to these mandatory standards. Although some areas fall only marginally within the Directive's mandatory limits, we consider the existing legislation and powers at the disposal of local authorities to be largely adequate for the control of pollution from coal combustion residues in urban areas, but it will be necessary to continue to monitor pollution. However, the scale of the possible re-entry of coal into the industrial market calls for more comprehensive preparatory action than is evident at present. Nevertheless, we consider that industrial coal burn can be positively encouraged in most circumstances.

Chapter 20   General conclusions on coal use

20.1 In the preceding Chapters on coal use we have reviewed the present and potential future use of coal in the U K. We have noted the range of views currently expressed about the likely scale of coal use in each particular market. We have not sought to propose our own independent forecasts of future patterns of coal burn. Rather, we have concentrated upon analysis of significant environmental problems which could result from a range of levels and patterns of coal use.

POTENTIAL MARKETS

20.2 There are currently four main markets for coal. By far the most important is power stations. The three other uses – for industry, for domestic purposes, and in coke ovens and manufactured fuel plants – each account for approximately 9% of the total U K coal burn. However, over the timescale covered by the Study, significant shifts could take place in the balance of coal use, although the current largest market – power stations – is likely to remain much at its present scale. Towards the turn of the century and beyond, another major market for coal could emerge in its conversion to S N G. A fundamental component of the energy strategy of the industrialised countries is the stimulation of the increased industrial use of coal over this timescale. The domestic market is generally expected to be static, and the market for coke ovens and manufactured fuel plants is expected to decline.

20.3 The evidence by the Department of Energy provided for a possible coal burn in general industry by the year 2000 in the range of 32–45 million tonnes. These projections were based on economic growth assumptions ranging from 1% to 2·7% per annum. Present fuel price relativities already signal to industrialists the desirability of switching to coal burning equipment. However, it is very difficult to predict developments in the relative prices of coal and oil over the next 20 years. Meanwhile, industrialists are deterred from making the change to coal, despite its fuel cost advantages, by the higher capital cost of coal-fired equipment. In the broader context of the energy policy imperative to reduce oil dependence, we welcome the "pump-priming" in the package of measures announced by the Government in March 1981. We see a clear role for Government to take the lead in establishing the required momentum in the market. Government support should apply to conventional and new coal burning technologies including F B C.

20.4 Major imponderables therefore surround the scale of re-entry of coal into the industrial market. However, all factors considered, we have some reservations about the attainability of the upper end of the coal burn range by 2000, provided by the Department of Energy. Nevertheless, in our analysis of both the possible environmental impact of combustion residues from coal burn, and the consequences for the environment of increased coal burn in urban areas, we have given particular attention to the possibility of a substantial increase in industrial coal burn.

20.5 The re-entry of coal into the industrial market would be facilitated by the introduction of new designs for coal handling and combustion. In particular, F B C could effect a reduction in the emission of sulphur and nitrogen oxides. Here, as with conventional boilers, their rate of introduction will depend largely on the development and marketing effort of the N C B and boiler manufacturers.

20.6 We are satisfied in general that the control measures and control technologies which relate to industrial coal burn are satisfactory. The industrial sector is inherently more easily controllable than the domestic: it is less fragmented; proper installation and maintenance under supervision by trained staff will be possible; and properly operated industrial installations are normally of a sufficient size not to produce smoke, except on starting up. Coal burn could also result in lower sulphur emission to the atmosphere than the heavy residual fuel oil it replaces. Under those circumstances, a net environmental gain could thus result from a change by industry to coal burning.

20.7 Estimation of the possible scale and timing of a coal requirement for S N G manufacture is beset by even more uncertainty. This is likely to be an entirely new market for coal, and raises questions about the application in the U K of the technologies involved on a commercial scale. An S N G programme would also provide opportunities for environmental control technologies to be incorporated as an integral part of plant design and development. We have deliberately confined our analysis of conversion to coal gasification as opposed to coal liquefaction. Firstly, coal

191

liquefaction appears to be a longer term issue in relation to prospective U K needs. Secondly, in some cases the liquefaction technology is less well developed. Thirdly, in certain major respects – such as the need to reduce the impact of chemical residues on the environment, or to secure the appropriate siting of plant – the potential problems are similar in type to those created by coal conversion to S N G.

20.8 It seems to us that S N G will be an environmentally benign form of energy at the point of its final use. In this respect, it has considerable advantages over direct coal burn. It will also have the same advantages as natural gas in terms of customer convenience, ease of use, and ease of transportation. However, there are two areas concerned in S N G production which we have examined in detail.

20.9 Firstly, there is the extent to which the technology for coal conversion will produce noxious, toxic or unpleasant chemical residues, which have to be disposed of in the environment. The great weight of evidence presented to us indicates that there should be no insurmountable problems in using currently existing technology to render any chemical residues harmless before they are released to the environment. Indeed, many aspects of S N G technology are already in application in other parts of the world, and as parts of other chemical processes, without undue environmental impact. Nevertheless, S N G plant will be very major industrial installations, and we are convinced that it will be necessary, before any particular process is brought to commercial development in the U K, that it should first have undergone a demonstration phase. By this we mean that the plant should have been built and operated either here or abroad at a larger scale than laboratory or pilot plant, although not necessarily at full commercial scale, in order to test to standards acceptable in the U K both the engineering requirements against the possibility of fugitive or accidental emissions, and the adequacy of the pollution control measures to deal with the effluents actually created by a large scale plant. In this regard, we noted that some of the proposals in the United States appear to require scaling up from pilot plants (dealing with for example 250 tonnes per day input) direct to full commercial scale (for example, 6–7,000 tonnes per day coal input) without any intermediate demonstration scale. In our view this is imprudent, and we would not recommend such a course of action in the U K. Both environmental and technical factors are too much at risk in scaling up of this magnitude.

20.10 However, two factors temper this concern. In the first place, B G C has a substantial research, development, and demonstration programme at its Westfield Development Centre; and secondly, there are a range of projects being undertaken abroad. The need for a commercial scale U K programme of S N G

manufacture is likely to occur somewhat later than in some other countries – largely because of our natural gas reserves. Thus the U K will be in a position to learn from some of the demonstration and early commercial schemes as they are applied abroad. For these reasons, therefore, while we do not urge that Government should intensify research and development of S N G plant at a demonstration scale, we recommend continued monitoring of progress in the U K, and of developments abroad, with a view to their possible application in the United Kingdom. If, in the event, experience abroad would not be directly relevant to U K circumstances, then a U K demonstration plant will be needed.

20.11 Our second area of concern relates to the provision of suitable sites for prospective S N G plants. This will be particularly important in the event of a substantial and continuing programme of S N G production, which could easily require a series of sites for the purpose. The site requirements for such large and complex plant are very constraining and from the study we have done it appears that sites with the appropriate combination of features are likely to be in short supply. There will also be a competition for suitable sites from other industrial proposals, and possible pressures from conservation interests for site preservation. Thus, although the potential commercial demand for S N G in the U K may be as much as 20 years away, we still think it prudent now to consider the types of issue likely to be raised by the siting requirements for such plant. Our analysis of this quickly led us to realise that other types of industrial and energy plant could confront similar problems. We consider this further in Chapter 21, which suggests proposals for a pilot study to assess the nature of the problems which can face both S N G and other plant.

TRANSPORT

21.12 On the transport front, we are convinced that rail is in general an environmentally preferable mode of transport to road. The particular advantages of rail transport show themselves when large quantities of coal have to be carried between fixed points. However, it is not a mode suited to the ultimate distribution of energy to very many small customers. Where local movement of coal is needed, then road transport will be the answer. In coalfield areas, road transport will be based on the landsale facilities at pits. In other areas we can envisage a reinforcement of the existing coal concentration depot network, to which the coal is moved in bulk by rail. We have not found any environmental reasons why, on a national basis, this system of distribution is not satisfactory.

20.13 However, constant care and attention will need to be paid to the design and siting of fixed road transport installations, in order to minimise unnecessary environmental impact. A considerable body of evidence to us has catalogued the adverse environmental impacts which transportation of coal by road can engender. However, we consider the solution to this is largely a matter of properly applying existing safeguards, rather than the establishment of a different system of coal transportation.

EFFECTS OF COAL COMBUSTION, AND THEIR CONTROL

20.14 There was clear apprehension in much of the evidence submitted to us that a major return to coal use would mark a return to dirt and smoke in the atmosphere which could negate many of the benefits of clean air legislation. This possibility was very much in our mind as we assessed the problems of combustion residues from coal. Before reaching conclusions on the environmental effects of coal combustion, we considered the effects of potential pollutants on human health and the natural environment.

20.15 The major adverse health effects of air pollution resulting from coal combustion have arisen from a combination of sulphur dioxide and smoke. Thus policy action since the war has been directed to dispersing sulphur dioxide from large emitters, such as power stations, through the use of tall stacks; to limiting smoke emissions from all new plants; and to limiting the amount of smoke emitted from domestic premises in controlled areas through the use of smoke control orders which have resulted in a switch to the use of smokeless fuel or other sources of energy. These orders have had a particular effect on the level of smoke from domestic installations. The combined results of these measures have been that smoke levels are now one-tenth of the values of 30 years ago in urban areas, and that sulphur dioxide levels have been halved. There is clear evidence that this reduction has resulted in the disappearance of acute effects upon health, and that the incidence of long term chronic effects has been substantially reduced. Provided that these measures continue to be enforced, we do not envisage any general or widespread difficulties with urban air pollution arising from an increased combustion of coal.

20.16 From our view of the evidence it appears that there is likely to be a "threshold level" of sulphur dioxide in the atmosphere below which significant effects on human health are not likely to occur. This is the basis on which standards have been drawn up in the E E C Air Quality Directive. The mandatory limits for smoke and sulphur dioxide set out in the Directive provide an adequate standard for the protection of human health, so far as acute effects and the development of chronic bronchitis are concerned. Particular care will need to be taken in those few areas which are currently near the accepted limit values for these pollutants. In most of these areas, the appropriate immediate policy action would be to introduce smoke control, but in a minority of cases, where smoke control

is already applied, it may be necessary to reduce sulphur emissions from burning oil. However, in the longer term, some form of legislative or regulatory restraint upon increased coal burn in a few such areas may be needed; we recommend that this possibility should be monitored. It does not at present seem likely that such "threshold levels" will occur in the case of cancer. Cancer forming agents are present in smoke from coal combustion, just as they are in smoke from tobacco. We have concluded, however, that, for the general public, current smoke pollution from coal combustion does not make any major contribution to the risk of developing cancer. Nevertheless, we regard it as desirable to keep the level of smoke pollution at the lowest practicable level, although further research could in time modify this conclusion.

20.17 We consider, therefore, that in general there is no longer major cause for concern on health grounds over ambient levels of smoke and sulphur dioxide from coal combustion, or indeed from combustion of any fuel.

20.18 We have also examined evidence on the possible impact on public health of radioactivity resulting from coal burning. We have concluded that exposure to radioactivity resulting from emissions of radioactive material from coal fired power stations is extremely small, and it is not a cause for concern. The dose equivalent received by members of the public from this source is much less than the natural variation in the annual dose received by them from all natural sources.

20.19 Provided standards of air pollution in urban areas are low enough to safeguard human health, we conclude they will also be satisfactory as regards crops or vegetation in general in the U K. Similarly, at such levels the effects on materials give no cause for concern.

20.20 Nitrogen oxide from coal combustion has not been identified as a source of any pollution problems within the U K. The major sources of relatively high ambient concentrations of nitrogen oxide out of doors are motor vehicles. Even higher levels are found indoors from appliances such as gas cookers and flueless oil or propane heaters.

20.21 Sulphur dioxide, together with nitrogen oxide, may be important factors in the acidification of lakes and rivers in certain sensitive areas of the world such as parts of Norway and Sweden, where the unusual local soils lack the capacity to neutralise any excess acidity. This problem of long range transport of pollutants is very complex. No simple solution is possible, and more research is required into the causes and mechanisms of these acidification effects, and into possible control technologies. We do not consider that enough is yet known about the problems of long range transport of air pollution to warrant any immediate action to limit sulphur dioxide or nitrogen oxide emissions because of their possible effects at long range. Considerable uncertainty also continues to surround how much of this river and lake acidity is due to the long range transport of combustion products. The approach embodied in the E C E's Convention on the matter, involving research into mechanisms, causes and remedies, is one which we strongly endorse. We do consider, however, that there is need for continued research into the technologies of removal of sulphur dioxide and nitrogen oxides from emissions, so that should the research into causes and effects show the removal of sulphur to be necessary, then suitable technology will be available without enormous economic penalties.

20.22 Dust and grit from coal combustion have been a major nuisance in urban areas in the past, and in areas close to large emitters. The application of good emission control technology, which has been required by legislation over the past few years, has largely cured this problem. But continued monitoring is required to ensure that standards are maintained; large plants which, because of the scale of operations are potentially major emitters of grit and dust, will continue to require careful siting.

20.23 The disposal of large quantities of ash from power stations and industrial plant causes no problems on health grounds. The material is largely chemically inert, and such problems as arise can be dealt with by simple precautions. The major problem is finding suitable sites for disposal of very large quantities of such ash. As with any large industrial enterprise, there are local problems arising from noise and dust created by transport and handling. Careful siting and treatment of disposal areas for such ash will continue to be necessary.

20.24 The scale and effects of the build-up of carbon dioxide in the atmosphere are surrounded by great uncertainty. Some research workers see potential problems of worldwide dimensions, with far reaching implications for the combustion of fossil fuels. The problem would not be restricted to coal combustion and use alone. It involves phenomena such as the massive scale of deforestation and ocean uptake. However, essential aspects of the mechanisms and effects are far from being fully understood and research is being undertaken worldwide into the problem. We endorse the need for a sustained programme of collaborative research, which should be directed towards assessing the scale and dimension of the problem, and the type of solutions, if any, which may need to be applied. There is no other action which can or should be taken at the present time.

CONCLUSIONS

20.25 Our overall conclusion is that there are not likely to be insuperable environmental obstacles to any

attainable increase in industrial coal burn, even if this took place in urban areas. Nevertheless, there are certain difficulties which could possibly arise; and we have indicated these in the Report. In no case do we anticipate any immediate problems arising from any increased coal burn, but in some cases continued monitoring of environmental effects is recommended against the possibility that subsequent corrective action may be necessary.

20.26 It soon became clear that there was a substantial distinction between the historic problems caused by coal burning, and what was likely to occur in future. Thus, in general, we are persuaded that the measures and standards which were introduced in the 1950s and 60s to control this historic problem, if properly applied in the future, will normally be adequate to prevent the return to an unacceptable degree of pollution of the atmosphere by the combustion of coal. However, as summarised below, we have identified certain qualifications to this general conclusion.

20.27 Domestic coal burn on simple open fires on any large scale in urban areas is likely to be unsatisfactory, leading to increased air pollution. We are not convinced that the manufacture of smokeless fuels for domestic use will be a satisfactory long term solution, in view both of the energy inefficiency in the manufacture of smokeless fuel, and of the pollution problems caused by the manufacturing plants themselves. The early designs of smoke reducing appliances which are able to burn bituminous coal have not been entirely satisfactory. We have noted the progress made to simplify designs and to ensure that they will be able to operate with greatly reduced smoke emissions under all operating conditions. In view of the need to keep the level of smoke pollution at the lowest practicable level (see paragraph 18.32), we recommend that the performance of these appliances and the effect they have on ambient smoke levels should be carefully monitored. In the meantime, we conclude that they should not be used extensively in urban areas.

20.28 We also have reservations on the adequacy of control powers over coal burn in urban areas. At present, the powers relate to controlling problems after they arise. They are also applied to individual premises on the basis that individual installations meet specified requirements. This system lacks any control over marginal users, whose emissions, although satisfactory in themselves, may add to those from existing uses in the same areas so that total emissions may result in unsatisfactory air standards in such areas. We do not see this problem arising except in "black-spot" areas as we set out in Chapter 19. Nevertheless, we recommend that this possible problem of cumulative impact of individual emissions in selective areas should be kept under review, with a view to tightening up the control powers if necessary.

PART IV

Planning and conclusions

Chapter 21 The planning process: reconciling the interests

INTRODUCTION

21.1 The preceding Chapters of the Report have analysed the possible future development of coal production and use in the United Kingdom and have attempted to assess its likely environmental impact. This development will encompass both incremental and replacement capacity and thus will involve the opening up of new capacity in greenfield sites, the extension of existing capacity and the rundown of outworn capacity. Although our general conclusion has been that overall these impacts are not likely to be so detrimental as to suggest that to make the changes in coal production and use currently envisaged would be unacceptable, we do not underestimate that there will be particular areas of concern, notably spoil disposal, which will be contentious and difficult to solve both nationally and locally.

21.2 These changes will be made up of many individual projects, each of which is likely to carry with it tensions between the national interest in energy supply, and the multitude of local interests directly affected by these projects. The achievement of the anticipated longer term increase in coal production and use, and the quality of the environment that results, will in practice be determined by the way in which these tensions are resolved.

21.3 It is the function of the town and country planning system to define policies for the best use of land, to balance competing claims, and to dissipate or resolve the tensions which arise from conflicting interests. Tensions between the general national interests are most evident at public local inquiries, which have assumed increasingly the role of reconciling these interests. In short, reconciliation of the national interest with that of the locality affected and of individual projects with the general policies to which they contribute has become the function of development control rather than development planning.

21.4 Development planning, which seeks to make the best use of scarce land resources, can only effectively respond to the requirements of the energy industries by having the earliest warning of their future land use needs and activities. In this way, the process of reconciling conflicting interests can be initiated at the earliest possible stage in the planning process and help

to reduce or remove discussion on locational aspects at any subsequent public local inquiry.

THE RANGE OF INTERESTS INVOLVED

The climate of opinion

21.5 The concept of balance and the interaction between energy policy and the environment is fundamental to our basic remit. The climate of opinion has a profound bearing upon how that interaction is perceived. There is a growing general demand by groups and individuals to participate more fully in the decision-making process. This expresses itself in the demand for more openness in Government, and, in particular, for increased public understanding, acceptance and control of decisions. It has been accompanied by a profound change in public attitudes towards the environment and by increasing attention to the consequences of major industrial projects in terms of the quality of life and of the conservation of future resources. This change also reflects a questioning of the authority of technocrats, administrators and politicians. The public no longer believes that Ministers and their advisers have a monopoly of judgement of the public good. It is frequently argued that the complexity and wide ranging number of development decisions to be taken in the next 30 years or so – among which those concerning major energy projects will be particularly sensitive – means that they cannot be left to a small number of persons whose traditional role it is to take decisions. Such views find their most articulate expression among interest groups and do not necessarily reflect the views of the public at large. Nevertheless, there is a growing concern among the general public that decisions should be taken, and be seen to be taken, reasonably, openly and fairly.

The local interest

21.6 Opposition to major projects will continue to come from local residents. The common thread running through local opposition is concern about impact on people's lives, local amenities and depreciation of property values. The general feature of major energy developments is that, while the benefits accrue to the country as a whole, the disadvantages are borne almost exclusively by local communities. This conflict is particularly acute when a new development is proposed in what has hitherto been seen as "green" areas of agriculture, recreation and open space. As our earlier analysis has demonstrated, the modernisation of the coal industry will involve the opening up of new capacity, much of which will come from new mines. However, the central feature of the coalmining scene is that the coal is where it is and must be worked there or not at all. Surface facilities have to be provided at or near the places from which access to the coal can best be achieved and the margin of choice may be small. The development of new capacity on the scale envisaged must involve the introduction of coalmining into greenfield sites in areas new to the industry. This will inevitably create opposition from local interests directly affected by such development particularly for the first time. This opposition may reflect total hostility to any change or a need for assurance that the development of new mining areas will be accompanied by adequate measures to minimise disruption and to provide compensation where such disruption cannot be avoided.

The interests of the planning authority

21.7 Against this background, the role of the planning authority is critical. Much of the evidence concentrated on the balance that must be achieved between the planning authority's control over the impact of developments on their area, and the necessary implementation of policies in the national interest. The Local Authority Associations and individual local authorities demonstrated their commitment to the system whereby they decide in the first instance whether a development should be permitted. They accept that there will be developments in the national interest which local authorities may want to oppose, but they believe as responsible and elected authorities they can be trusted to take first decisions on projects, taking into account their national importance. They state that they are already expected to do this for strategic resources other than energy. Moreover, they point out that the planning system safeguards the national interest by allowing the Secretary of State to call in major applications for his own decision, and by the right of appeal by applicants to the Secretary of State. We have much sympathy with this view.

The energy interest

21.8 As seen by the energy industries, the central problem concerning development control procedures is that of balancing energy and environmental interests in a way that does not unduly impede energy developments and efficient energy use, but which commands wide public acceptance. The aim of the planning process, therefore, should be to ensure that a balance is struck in the total evaluation of projects encompassing the energy, industrial, employment as well as health and safety, land use and environmental implications; and that the public are satisfied that in the particular circumstances a fair balance has been struck.

21.9 The U K's favourable medium term energy position does not remove the need to adjust in good order to a world in which oil and gas are likely to get increasingly scarce and expensive. Lead times in the energy sector are very long. A failure to build up smoothly the necessary technical and industrial capacity would have serious implications for the health of U K industry, for the country's wider economic performance and for the continuity of supply to domestic consumers. Some excess of energy capacity might be involved. It is worth paying this sort of insurance premium to avoid the consequences, in something as basic as energy, of a real and perhaps prolonged shortage which could not otherwise be put right for many years. However, severe opportunity costs would clearly be involved if the insurance margin were to be excessive. A gross overcapacity in energy supply would imply that there had been unnecessary sacrifice elsewhere in health, education, or defence perhaps. We must therefore try to get the balance about right, only slightly tilted in favour of energy supply.

21.10 The infrastructure of energy and other developments required to underpin industrial strategy is vital to national prosperity; thus ways must be found to enable such developments to proceed and to do so in timely manner. In planning for future energy supply, the energy industries need assurance that the town and country planning system will work effectively. They also want assurance that their investment programmes will not be dislocated by inconsistent or delayed decisions on proposed projects; and that the land use planning system can identify, safeguard, and bring forward for development sites which are necessary for the industries' purposes and which are acceptable to other public interests. It must be added that the planning authorities also need assurance that the programmes of the energy industries are related and are not in conflict.

THE COMMISSION'S APPROACH

21.11 Our primary objective has been to find effective and speedy ways of reconciling the tensions between national and local interests without introducing further complexity and delay into the existing system. At the same time any changes we advise will need to be compatible with existing provisions for local democratic control, and they will need to be responsive to the growing need to demonstrate that decisions are taken openly, fairly and reasonably.

21.12 Much of the current debate focusses too exclusively on the problems of the role played by public local inquiries. The public local inquiry provides a safeguard and an assurance to local interests that they will be taken into account in any decision that is reached. This is a positive and very important function that must be preserved.

21.13 We are fully aware that at present the public local inquiry is the only effective direct forum in which the public can make its voice heard in those Governmental decisions and procedures which culminate in actual projects. Therefore, those who wish to challenge the basic policy assumptions which underpin particular projects have to use the public local inquiry at a late stage for a purpose for which it was not intended. The major cause of the stresses now being encountered in the planning application and consent system does not derive from built-in procedural defects. The major difficulties derive from the growing burden imposed upon public local inquiries in which national and local policy issues are becoming increasingly interwoven.

21.14 We consider that the role of the local inquiry must be seen in a wider perspective. It is the culmination of a much longer process, as we have explained in Chapter 6. This earlier preparatory process can have a major effect on the nature of the proposals which come forward, and the nature of the tensions they create between the national interests and local interests and between energy and environmental issues. The increasing length and complexity of public local inquiries may well be symptomatic of shortcomings in the earlier preparatory stages, and of too great a reliance by all parties on the issues being thrashed out on the battlefield of the inquiry itself. We have therefore examined whether the possibilities for more effective use of the earlier stages of the planning system can be exploited to remove many avoidable tensions and conflicts that create difficulties at present.

21.15 Two themes have recurred throughout our Study, which have led us to this conclusion, and to the solutions we propose: firstly, the need for wider public understanding and debate of national energy policy issues; secondly, the need for mechanisms which will translate the implications of that policy into possible requirements for major developments at the local and regional level. These two themes are closely related. We attach the utmost importance to a continuing and forward-looking dialogue between developers and local authorities and interest groups. Such a dialogue must be based on a deeper understanding of the wider national issues. However, wider public debate and understanding of national energy policy can only go so far. Without the translation of energy policy into possible requirements for sites, it will not be possible to trigger the town and country planning system into making provision for them. The energy industries and planning authorities have to cooperate in agreeing on those provisions, while not losing sight of the need to carry the public, nationally and locally, along with them. It will be of vital importance to see how energy requirements fit in with other land use policies, to resolve competing land use demands and to minimise the sterilisation of land.

21.16 The aim is to move in good order from national energy policies through to proposals for individual projects in a way which minimises the tensions between the national interest and local interests. We believe that this can be achieved by exploiting the flexibility inherent in our present planning system. It does not call for radical changes in existing procedures.

THE NATIONAL DEBATE

21.17 We attach particular importance to the role of central Government in stimulating public debate and wider understanding of the broad framework of national policy. The interrelation of fuels and the repercussions of decisions in one field on others are such that sectoral strategies can only be determined separately from each other to a limited extent. The public acceptability of the requirement for major projects to implement these strategies would be considerably enhanced by a more systematic explanation by Government of the national policy framework.

21.18 We are aware of the contribution to the public debate made by the Department of Energy, notably by means of the series of Energy Policy Papers, the publication of the Department's Energy Projections and by the research into low energy futures being carried out by E T S U in conjunction with outside bodies. However, the Green Paper on Energy Policy of February 1978 still represents the fullest available exposition of national energy policy, as seen by the former Government. We recommend the early issue of a similarly wide-ranging assessment by the present Government of the evolving energy scene.

21.19 We are not concerned merely with the provision of information. There is also a need to stimulate public debate. Continuing discussion with experts outside Government and the fuel industries can make an important contribution to a better understanding of the options available and of the consequences of closing particular options. Indeed, we hope that our own Report will be seen as a useful contribution.

21.20 The role of Parliament is clearly crucial as a focus of public debate, in view of the primacy of Parliament in national decision making. We would see great benefit if Parliament held full scale debates on national energy strategy at appropriate intervals. The debate and approval by Parliament of Government White Papers setting out policy for particular issues could provide a crucial component of the policy background to public inquiries.

21.21 The Select Committee on Energy provides an important focus for Parliamentary consideration not only of the Government's overall energy strategy but also of its policy for particular fuels and industries. It also provides a channel for consideration of the views of wider interest groups. We hope that its reports will trigger major Parliamentary debates and hence stimulate wider public interest in national energy policy issues.

21.22 There are two aspects of the national energy debate to which we wish to draw particular attention. Firstly, the uncertainties that must underlie national energy policy mean that no rigid blueprint for the longer term is possible or desirable. However, long lead times for energy projects are such that national energy policy has to look forward several decades in order to determine what energy supply options should be maintained or closed and at what cost. Secondly, we would emphasise the importance of the relationship between national energy policy and the individual projects which contribute to its realisation. National energy policy cannot be a fixed blueprint for the future: rather, it is a moving canvas, which translates into projects completed and contributing to energy supply, projects over a period of up to ten years, and projects which are in different stages of planning prior to development. We recommend that a more systematic explanation by the Government of the progressive evolution of national energy policy, of its vulnerability to external events, and of the options for its implementation should be accompanied by regularly updated indications of the order of magnitude of possible future requirements for major energy projects.

21.23 One of the cardinal features of the planning of national energy supply is to maintain as many options for the future as possible without incurring unreasonable costs in doing so. We believe that this principle should be carried over to the provision of land resources for possible energy developments within the town and country planning system. We are not advocating a rigid land use blueprint. However, the presentation of national energy policy along the lines suggested above would allow potential land use requirements to be reviewed at regular intervals.

THE TOWN AND COUNTRY PLANNING SYSTEM

21.24 We have already explained in Chapter 4 the objectives of the town and country planning system and how it operates. Our primary concern in this Chapter is to identify means whereby the translation of national policy into local projects could take more account of local and regional issues without jeopardising Government policies for energy and the environment, the country's needs for energy, and the consequent investment strategies of the energy industries. We examine the scope for improved long term planning within the town and country planning system and the scope for more advance consultation and dialogue between developers, local authorities and interest groups.

21.25 The success or otherwise of the town and country planning system is critically dependent on maintaining the complementary roles of development control and development planning. This is not easy to achieve for energy projects and similar development. In development control, planning authorities have no choice but to come to a decision on every planning application made to them. In development plans, however, planning authorities can only make policies and provisions for those demands for development which they can either anticipate or are made aware of. Unless planning authorities are forewarned of possible requirements, the town and country planning system cannot make provision for energy developments. These developments have to proceed on a case by case basis through development control, often against the grain of existing policies in development plans, and leading almost inevitably to a public inquiry.

Structure plans

21.26 The main vehicle for long-term planning within the town and country planning system, is, as we have explained, *the structure plan*. The basic information on which the plan is prepared will include the existing use of land by nationalised industries. Consultations during the plan's preparation may identify proposals for developments and changes in the existing use of land by statutory undertakers. In principle, these consultations provide an opportunity for advance dialogue between the planning authority, developers and the wider public.

21.27 The energy industries have made some use of structure plans. The N C B have told us that they consider it important that the existing consultation procedure in the drafting and revising of structure plans is continued and that they hope to contribute to their revision the results of any coal exploration being undertaken in the planning area. The C E G B also ask local planning authorities to include in structure plans locations which have been identified by them as part of a reserve for long term needs. In practice, however, policies and proposals in structure and local plans tend largely to confirm the existing use of land by statutory undertakers, and the opportunity to reach agreement on the long term requirements of statutory undertakers is not realised. In short, plans tend to be retrospective, whereas they ought to be prospective.

21.28 Part of the reason seems to be that final decisions to proceed with new large projects are taken only at a late stage in the planning process of the energy industries. This arises from the need for flexibility to cope with the uncertainty surrounding future levels of demand and the long lead times associated with technological development. In view of this, planning authorities are reluctant to include large projects in structure plans because they may give rise to blight and because there may be some difficulty in justifying them when the plans are subsequently discussed in public. Moreover, because of the way in which the town and country planning system operates, the inclusion of a project will not necessarily ensure that permission to proceed will be given when final planning permission is sought. Thus, if a proposal is incorporated in a structure plan, the developer may well be called upon to defend it at the inquiry into the structure plan. In addition, he will also need to seek planning permission, and this could involve a second inquiry. Under these circumstances it is not surprising that a developer seeks to limit his involvement in the structure plan, and concentrates his effort on the planning application.

21.29 Local authorities are also able to prepare *subject plans* which set out policies towards developments in their area associated with a particular activity, such as opencast mining. Very little use has been made of this instrument as yet. Some local authorities have begun to take steps to prepare them. For example we understand that the Dorset County Planning Department have been drafting an "Oil Policy Report" which will be complementary to the County Structure Plan.

21.30 Against this background, we have concluded that long term planning of land resources suitable for energy developments needs to be improved. This would help to assure the energy industries that they will be able to proceed with their investment programmes without dislocation. It would also allow local authorities to assess the possible impacts of developments on their areas and make allowance for them in their other policies. The question we now have to address is not one of principle, but how to put that principle into practice. We have seen elsewhere the value of stated procedures for development control and public local inquiries, so that all parties are aware of the part they are expected to play. We think the same concept must apply to the preparatory stages of long term planning. In this context we have examined the steps taken in Scotland to achieve a greater degree of long term strategic planning and which we think have certain advantages.

National planning guidelines in Scotland

21.31 The S D D issued National Planning Guidelines for the coastline in 1974, and since May 1977 has issued National Planning Guidelines which are concerned with major new developments expected to involve issues of national importance. They have four major elements: statements of the nature of the siting requirements of those developments; the identification of primary search areas for planning authorities to identify a few sites for large scale development to be protected against other development; the expectation that any sites identified by planning authorities will be included in plans they prepare; and a procedure for notifying the Secretary of State for Scotland of planning applications for proposed large scale developments which raise national issues or which may prejudice the future development of any site identified by the planning authority.

21.32 A national planning guideline for petrochemical developments was issued in May 1977. The guideline gave information about the nature of petrochemical developments and the planning issues which they raised. It also stated that it would be prudent for local authorities to establish the potential for such developments in their areas and to frame their structure plans, local plans and development control policies so that sites could be readily identified and no opportunity lost through lack of preparedness. Since then a large amount of work has been carried out by local authorities in cooperation with the S D D on the feasibility of petrochemical sites in relation to the supply of feedstock and pipelines. This work has been gathered together in a way that allows a quick response to enquiries from major developers. In addition, provision has been made in a number of structure and local plans for petrochemical sites. Work has also been underaken to identify in structure and local plans a number of locations providing a supply of land with physical features suited to large scale industry. S D D have also issued a series of Planning Information Notes (1981) on the technical aspects of gasification and liquefaction of coal and their environmental implications.

Long term planning in England and Wales

21.33 We have examined the possibility of adapting to England and Wales the experience gained with the Scottish National Planning Guidelines. However, the problems of development in terms of population size and activity, length of coastline and amount of undeveloped land is much more complex in England and Wales than in Scotland. Sites for development are therefore much fewer. It is likely that a similar exercise in England and Wales on the Scottish model would end up not with guidelines but with a precise identification of only a very few feasible sites. It could be argued that it would not be appropriate for the Secretary of State to initiate such an investigation, since it might prejudice the impartiality necessary for the performance of the Secretary of State's quasi-judicial role in respect of appeals. We consider that in view of the likely shortage of the type of site involved, the quasi-judicial problem should be faced up to. Moreover, under the Scottish system it is local authorities who identify sites. We do not consider that the energy industries could be expected to commit themselves to the use of sites selected for them by local authorities. The industries must participate fully in site selection.

21.34 We nevertheless endorse the principle of the Scottish guidelines system leading towards the identification of suitable sites for major energy developments. We have commented earlier on the need for Government statements on national energy policy to be accompanied by indications of the order of magnitude of possible requirements for major energy projects. To maintain long term options for future energy supply, there is a practical need in England and Wales for more effective forward provision regarding the availability of sites. We have examined with the nationalised industries the practicalities of this problem in relation to possible needs for S N G plants. Notwithstanding uncertainties about the timing of need for such plants, we concluded that the combination of stringent site requirements and the policy or other restraints on development in large areas of the country, pointed to the considerable possibility of a scarcity of potential new sites. Moreover, there is a distinct danger that, given all the other pressures for both development and conservation in a densely populated country, those few sites which do meet these stringent requirements could in fact be used for other purposes which may not fully utilise their scarce combination of characteristics.

21.35 It became clear that this was not a problem unique to S N G plant and that it had a much wider application. Thus, we concluded that the energy industries and local authorities should undertake urgently a joint review to establish whether there is a scarcity of potential sites for major energy and other industrial developments. It would be premature to recommend that such a study should be mounted for the country as a whole, but we do believe there is a case for a pilot exercise, probably centred on Yorkshire and Humberside and the East Midlands. The exercise could then be extended to the rest of the country if the results warranted it. We so recommend.

21.36 The pilot study should seek to identify those locations which have the necessary combination of characteristics to accommodate any of a range of heavy or energy industries, in order to assess (1) whether in practice sites are likely to be in short supply (2) whether positive safeguarding is needed, and (3) whether there should be some system of formal advance notification of any other proposed developments in the vicinity of these locations, which could prejudice their ultimate use for the developments which crucially need their special site characteristics.

21.37 Such a pilot study, which should be set in hand as a matter of urgency, would constitute a specific application of the forward looking dialogue between the energy industries and local authorities, the need for which has been a recurrent theme throughout our Study. Moreover, we attach a considerable importance to the regional dimension of the study. We envisage that the Department of the Environment, as the ring-master of the town and county planning system in England and Wales, in conjunction with the Department of Energy, should take the lead in inviting the energy industries and the planning authorities in the

particular region to get together accordingly. In our view, such an exercise would not prejudice the Secretary of State's quasi-judicial role. The N C B, the C E G B, and B G C have already expressed to us a willingness to participate.

21.38 A further development of our proposal would be that once the regional review had been completed, locations could be identified in the country based structure plans as these are reviewed. We consider that our recommended approach would avoid the difficulties which presently inhibit the full use of structure plans in the siting context.

THE NEED FOR PROGRESSIVE CONSULTATION

21.39 The possibilities for consultation between developers, local authorities and interest groups extend beyond problems of the availability and location of suitable sites for future developments. The evidence of the energy industries has shown a number of examples of the way in which they approach the involvement of local authorities and local communities in their own internal planning beyond their statutory duties to consult. On the basis of this evidence, we have concluded that there is further scope for extending this type of consultation. As stated earlier, we are convinced that, for major proposals or programmes by the energy or other industries to proceed, there must be a reasonable degree of public acceptance of the proposals themselves and the reasons why they are required. The way to obtain that acceptance is by more and earlier explanation of what is involved, and, wherever possible, some public involvement when choices have to or can be made between alternatives.

21.40 It has been argued that consultation about options at any stage simply produces objections to all of them. We do not believe that any system of consultation will ensure there are no objections by those directly affected by specific proposals. However, experience suggests that consultation with responsible interests and local authorities, after an initial period of total objection, can and does begin to produce constructive dialogue. The B G C has had some notable success in pursuing a policy of early consultation. Since 1966 about sixty major projects have been completed by the B G C, almost all on greenfield sites in areas which had not been zoned for industrial use. Only two projects were taken through the planning inquiry procedure and only one application has been refused. It must be said that these projects will not have as large an impact on an area as, for example, a new deep mine. It is also the case that scope for discussion of the location of deep mines is much more limited than for the location of power stations or other similar projects.

Consultation about mining

21.41 We have discussed in Chapter 6 the concept of the "Ladder of Exploration and Project Development" as explained to us by the N C B. This clarifies the stages of planning any prospect must go through before mining can begin and the amount of information that is available at each stage of the "Ladder". We consider that this description of the sequence of mine project development is so important that we recommend the N C B to ensure that it is widely published.

21.42 Our primary concern has been to assess the extent to which the N C B have taken environmental factors into account when drawing up their forward programme of development. It must be remembered that as recently as 1974 there was virtually no exploration underway at all. Thus in the recent past there has been only the most limited choice in terms of alternative locations for major developments. It is fair to suggest that to date the Board have based their choice of where to invest in exploration and development on the concept of the next best prospect for exploitation. Once the basic geological assessment of coal in place and mining feasibility has been established, the major considerations taken into account have been related to the current economics of the engineering of the mine – including factors such as the availability of labour.

21.43 We fully recognise that, by the time a proposal is being worked up to the planning inquiry stage, the N C B are devoting a great deal of effort into minimising the impacts of that particular proposal on the surrounding area. However, we consider that the exploration programme has now reached a point at which the Board are in a position to assess alternatives giving greater weight to environmental factors than has been possible hitherto.

21.44 With the increased bank of information available to the N C B, there should be increasing opportunity for public comment on the options available before the capital sums required are committed on one particular project. We recommend that before commitment in terms of investment and effort on any one project has advanced far up the "Ladder of Exploration", consultations should take place with the local authority associations and other interested bodies, both local and national, including Government Departments and such statutory advisers as the Countryside Commissions and the N C C. There should be an opportunity for an exchange of views on the environmental desirability of different development options. We would not expect consultations on options necessarily to imply that if one goes ahead all else should be dropped, but that wherever there is flexibility for re-ordering and holding back particular projects which are less environmentally desirable, advantage should be taken of this. We are not arguing in favour of a bank of fully worked up projects. This would not be economically feasible in view of the large costs involved in developing a large scale project to the point where it can be considered at a public inquiry.

21.45 It will be important, however, that the interests to be consulted are fully aware that the information available on individual projects will be limited in the early phases of project development. In these stages, the N C B must establish the geology of the coal in place and the feasibility of mining it. The major considerations must focus primarily upon the economics of engineering the mine. Thus at the earliest stages no more than an exchange of information will be possible. Initially, this exchange will enable the N C B to inform interest groups on the results of their developing exploration programme; and those interest groups to indicate to the N C B the weight they attach to major environmental features to be found in particular areas of exploration. The N C B, as a responsible developer, would then be better able to take account of these considerations in the further development of their programme. Thereafter consultations between the parties can undergo progressive refinement. The rigour with which an acceptable balance between environmental and other considerations can be determined will increase as a project moves up the development ladder.

21.46 The Board have already expressed to us their willingness to engage in such consultations between the preliminary and significant exploration stage (see Chapter 6). Similar consultations could also take place between the significant exploration stage and the feasibility stage. For example, in some cases sufficient will be known about the location and extent of the coalfield to enable some estimate to be made of the likely limits of the area that might be affected by subsequent development.

21.47 Progressively, information will become available on the amount and characteristics of the coal in place and on recoverable reserves. As more information becomes available, general indications could emerge about the likely scale of mining and the extent to which the mine was likely to produce spoil. In parallel, more refined information will become available on the possible environmental impact of potential developments. At a very early stage it will become clear whether coalfields are located in areas of particular importance as far as landscape, nature conservation or recreation are concerned. The Nature Conservation Review (Ratcliffe D 1977) provides a valuable source of information. Later on in the process it should be possible to make some assessment of whether there would be particular local problems about spoil disposal or whether there were greater opportunities for its use for land reclamation in other areas.

Consultation on other energy developments

(a) *Opencast*
21.48 We have already discussed in some detail the role of consultation in the planning of opencast mining; our proposals are contained in Chapter 11. The

Opencast Executive and the local authority associations have agreed upon a continuous exchange of information and consultations, embodied in a Code of Practice (Association of Metropolitan Authorities et al 1980), which should enable them to work together on the preparation of 5 year rolling programmes. We welcome this development. Taken together with our recommendations for a change in the authorisation procedures, and for the addition to the Code of Practice of a set of guidelines defining criteria for opencast sites, it will help further to ensure a greater degree of mutual cooperation in the planning and preparation of opencast programmes. The opencast example illustrates how our views on advance disclosure can be translated into practice. As we discuss above, we should like to see a much closer parallel between consultations on the deep mine and the opencast side of the N C B's operations, and a greater degree of coordination between them.

(b) C E G B

21.49 The C E G B operate a process of successive refinement of projects from initial appraisal up to detailed planning. The first phase involves a study of an area by the Board to determine its development potential and constraints, and to review new prospective sites. During this phase Government Departments, statutory amenity bodies, and local authorities are notified, and discussions may take place with them. Up to two years of work may be needed to prepare a short list of sites which will be investigated in more detail in the second phase. At this stage the bodies who have already been consulted are notified again, and public announcements are made. It is normal practice to provide the public with information about the reasons for, and the extent of, the investigation. When these detailed investigations have been completed, the alternatives are evaluated by the Board. Where it is decided to nominate one or more of the sites as a reserve for long term needs, then local authorities are asked to make appropriate provision in their structure plans. Alternatively the C E G B may decide to proceed with an application for consent to develop.

21.50 The C E G B have told us of a variety of ways in which they consult with local and public interests according to local circumstances. Thus, they have been consulting with the Standing Conference on London and the South East Regional Planning about future prospects, and a report (1979) has recently been issued. In the 1977 South Hampshire structure plan, their possible interests in six sites were acknowledged, and an undertaking given to consult if alternative development pressures were likely to prejudice the development of any of those possible sites. There is currently a need for system reinforcement in the South West Peninsula, and a major consultation exercise is underway locally about the options that are open there.

21.51 These procedures, and the analogous procedures of B G C, reflect very closely our own views about the desired sequence of events in developing a project. Again, we would have made recommendations about these necessary procedures, had it not been clear during the course of our Study that they were being put into practice. Our one remaining concern relates not so much to the adequacy of these consultation procedures, but to the extent to which they are widely applied. We strongly urge developers to engage in early consultation and central Government to use their good offices to encourage this. We considered whether there should be a statutory duty to consult but concluded that it would be best to rely on increased use of current best practice in this field. It is important that the early consultations should not be too formal, so that all parties can be confident that they will not be treated as pre-hearings judging issues that can only properly be settled at a later stage.

Environmental impact assessment

21.52 It will be clear that we favour a continuing process of thorough public consultation and early assessment of the environmental implications of major projects. As the development of a project undergoes successive refinement, environmental concerns should be considered as an integral part of the process. Consultation with planning authorities is vital since part of their task is to assess and coordinate the conflicting pressures on land development. However, this environmental assessment should be seen as part of the total evaluation of the project to enable a balance to be struck between the energy, industrial, employment, and social implications of major projects and those concerning health and safety, land use and environmental considerations.

21.53 The technique of environmental impact assessment performs a particularly valuable function and is already widely deployed in the U K by the energy industries. The analysis of the siting options for the surface works at Belvoir undertaken by the N C B, and the analysis by the C E G B of alternative power station options for the South West, are but two examples of how current U K practice is applied.

21.54 We are aware that negotiations are taking place within the E E C on a draft Directive on formal environmental impact assessment. We would prefer the wider deployment of this technique within the existing flexible U K approach to the enshrinement in legislation of mandatory requirements for detailed environmental impact statements as is current practice in the United States. We consider that a mandatory formal system would risk undue additional delays and costs in planning procedures. But we would much prefer a formal system of environmental impact assessment to no system at all.

PLANNING INQUIRIES

21.55 The purpose of planning inquiries is to produce, present and assess the evidence relating to a particular development. This places the Secretary of State in a position to reach a sound decision as to whether it should be approved. The public inquiry system can be an effective method of addressing site specific issues and of providing local objectors with a means of presenting their opposition to particular developments. It can provide a vital contribution to the public acceptability of decisions to the extent that it enables it to be demonstrated that such decisions are arrived at openly, fairly and reasonably.

21.56 However, it is argued that more is being demanded of this machinery than it can provide as a result of the attempt to debate and evaluate national issues in this forum. It has been asked whether a rambling, catch-all investigation of the Belvoir kind is really the best forum in which to thrash out wider issues of national energy policy. The problems are compounded in the case of "series developments". Some would thus argue that it is highly desirable to avoid every planning application in such a series of developments becoming an occasion for examining from first principles, for example, the place of coal in U K energy policy.

21.57 The N C B and C E G B have expressed particular concern in their evidence about the possible delays arising from long, wide-ranging public inquiries and the difficulties these may cause for the implementation of their investment strategies. It is not difficult to see the cause of their concern. The Belvoir inquiry ran for 84 days and unofficial estimates put its cost at £5 million. The participants employed over a dozen Q C s and generated in excess of seven hundred documents, calling for more than one hundred expert witnesses. The cost was such that one major interest group – the Town and Country Planning Association – was forced to withdraw from the proceedings. On the breadth of the inquiry, the N U M have commented that the issues raised "range from the mating cycle of the May fly to the possibility of a revolution in Saudi Arabia". Some would argue that such inquiries can result in a disproportionate emphasis of the views of articulate minorities, at the expense of the wider public interest.

21.58 A central problem is the extent to which local inquires might be mainly limited to their intended function of addressing local issues, while enabling the wider policy issues to be considered in other fora. The crux of the problem is how to achieve this without injecting further complexity into the system. Similar problems were discussed in the Report on the Review of Highway Inquiry Procedures (Department of Transport 1978).

21.59 It is against this background that we have devoted considerable attention to the need for Government to provide for wider public understanding and debate of national energy policy issues. This would generate the essential national policy backdrop to local public inquiries. We have also emphasised the importance we attach to a continuing and forward-looking dialogue between developers, local authorities and interest groups designed to translate the implications of national policy at the local and regional level into possible requirements for major energy developments.

21.60 Some authorities have suggested that a two stage inquiry procedure, with the first considering general policy issues, would be the best solution for the expeditious handling of both national and local issues. We consider that this would add to the complexity of procedures without commensurate benefit. The timing of the two stages would be very difficult and there would be no guarantee that unresolved policy objections from the first stage would not be reintroduced, if only for tactical reasons, at the second stage.

21.61 We are not proposing that there could, or should, be a total divorce of national and local issues at the public inquiry stage. The very uncertainties of the energy scene, the impact of the contingent and the unexpected are such that national issues cannot be excluded at such inquiries. Nevertheless, implementation of our recommendations on the related themes of (i) wider public understanding and debate of the national issues, and (ii) continuing and forward-looking dialogue between developers, local authorities and interest groups, would provide what is lacking at present – namely, an informed policy backdrop on such key issues as the need for a project and its policy context. There would therefore be less need to repeat such matters in great detail at local inquiries. Thus, although we do not recommend that the terms of reference of inquiries should be restricted to exclude discussion of national issues, we do expect the need for such discussion to become much less than at present. In particular, where successive inquiries into a series of developments are required, we see no justification for later inquiries to examine the national policy issues from first principles, if such issues have been debated in Parliament. However, this assumes that the interval between inquiries will not be too long and that full account would be taken of any major developments arising between inquiries.

Public inquiries procedure

21.62 Although we have concluded that the basic form of the public inquiry should remain unchanged, we consider that there is scope for inquiry procedures to be amended to ensure that as many delays as possible are avoided. We would not want to prejudice the principles of openness, fairness and impartiality which have been at the heart of the procedural rules for inquiries since their first introduction in 1962, and are now incorporated in the Town and Country Planning (Inquiries Procedures) Rules, 1974, (S I 1974 No 419).

21.63 The energy industries have made a number of suggestions for speeding up inquiry machinery. We recommend that their suggestions are considered by the Department of the Environment, particularly the idea of a greater use of written evidence and less reliance on oral submissions and cross examination. This is not to minimise the importance of cross examination, for it performs a vital role in testing and evaluating evidence. However, it is an expensive procedure, and it tends to benefit professionals, or groups who can afford to employ professionals. We therefore recommend that it should only be used when its particular advantages in probing evidence can be fully realised.

21.64 We recommend that wherever possible proofs of evidence should be submitted as early as possible and should not subsequently be read out in evidence. We also welcome the use by Inspectors of pre-inquiry meetings to discuss with participants the issues to be considered at the public inquiry and the way in which they are to be raised. The greater the use that can be made of pre-inquiry meetings to cover the structure, scope and operation of the subsequent inquiry, then the more expeditious and ordered its proceedings are likely to be. We expect that the advanced disclosure of information by the energy industries, and the earlier consultation with interested parties, will enable participants to focus more on site specific questions.

21.65 The energy industries have also expressed concern about the timetable for reaching decisions after the conclusion of a public inquiry. We understand this concern since in some cases many months elapse before the Government reaches and publishes its decisions. There is clear scope for reducing delays at this stage of the inquiry process. As we go to print the Government's decision on the Belvoir Inquiry is still awaited 13 months after the Inquiry was completed.

CONCLUSIONS

21.66 Our basic premise has been that the planning system exists to secure the best use of scarce land resources, including development and protection of amenity, and to achieve a balance between the demands for development. The system must enable developments to take place in a properly controlled way and provide for major projects to win public acceptance. We consider that the reconciliation of conflicting interests can be better achieved by exploiting the flexibility inherent in the U K planning system and the development of better practice rather than radical changes in the planning machinery.

21.67 We consider that adequate planning procedures and controls are needed to ensure that all the implications of a development are considered in the process leading to the decision on whether it should be allowed to proceed. In particular, a developer has a duty to take the environment into account at all stages of development, not only when a particular proposal is being prepared for a planning application. Nevertheless, unduly protracted procedures are clearly undesirable and we have given careful consideration to the concern of the energy industries that important investment might be delayed, not least by the Government itself.

21.68 Our main conclusions arise from the two inter-related themes which have recurred throughout the Study. Firstly, there is a need for wider public understanding and debate of national and energy policy issues which call for more regular and systematic explanation by government of the national policy framework. Secondly, there is a need for mechanisms which will translate the implications of that policy at the local and regional level into possible requirements for major energy developments.

21.69 We attach the utmost importance to a continuing and forward-looking dialogue between developer, local authority and interest groups. Such consultation will provide for advanced disclosure and exchange of information on potential development proposals, and for environmental factors to be taken more fully into account at the earlier stages of development planning. Our recommendations for a pilot study of future siting requirements provide a specific example of what this should encompass. Unless planning authorities are forewarned of possible requirements, energy developments have to proceed on a case by case basis through development control. This leads almost inevitably to the overburdening of public local inquiries.

21.70 Many of the tensions which are encountered at these inquires, which are meant to be the culmination of a much longer process, would be eased by more effective dialogue between the relevant interests during the earlier preparatory stages of the planning process. This will serve both the public interest and that of the developer – the public in that wider discussion and consultation ensure that the developer accords adequate weight to environmental and social considerations when considering alternatives; and the developer, in that wider dialogue and consultation is necessary to obtain and sustain public acceptance of major proposals. Complete consensus is unobtainable. However, we are convinced that our proposals will provide the basis for a greater degree of consensus than that obtained hitherto in one of the most contentious areas of public policy.

Chapter 22 Conclusions and recommendations

22.1 We began by quoting from Lord Nathan's seminal report on 'Energy and the Environment', published in 1974: "a policy for the development of energy resources must be created in the context of its environmental and social consequences". In the course of our own work on 'Coal and the Environment' we have been struck by the extent to which, more than ever, energy policy cannot be made in isolation. Every decision is likely to involve a wide range of other objectives – industrial, social, political and environmental. It is not possible to find a general formula to express the right balance between different objectives. A rigid blueprint for the energy sector would be too inflexible to cater for the interactions of what can be conflicting interests.

22.2 One of the most controversial elements in the current energy debate concerns the scope for energy conservation. It raises the fundamental question whether the current balance between investment in energy supply and in energy efficiency is right. We could see considerable merit in a major study evaluating the risks inherent in a supply orientated policy compared with a policy emphasising efficiency in energy use. Such a study would need to encompass, inter alia, the implications for the fuel mix of energy pricing policy, oil and gas depletion policy and high or low electrification. It would also address the environmental impacts of energy conservation, not all of which are necessarily advantageous.

22.3 Our initial expectation was that rigorous examination of the environmental impact of coal production, use and conversion would inevitably throw up a picture highlighting their deleterious effects. In the event, we have been reassured by the more promising picture which has emerged. Most of the potential environmental problems for increased coal production and use can be overcome by the more even application of current best practice; that is, by the use of techniques and procedures which are fully understood and currently operational. This is not to diminish the potential scope for innovation. Our overall conclusion is that, subject to important qualifications concerning spoil disposal, opencast extraction and sensitive treatment of those afflicted by subsidence, there are no insuperable environmental obstacles to the role of coal as currently envisaged in the U K. The major part of our recommendations summarised below addresses the measures required to sustain this conclusion.

22.4 We have indicated areas where higher standards may be needed. We consider that these can be achieved without more complicated procedures. Flexibility of response could serve both the energy and environmental interests. We are not in favour of precise objectives, detailed schedules and specific solutions being spelt out in legislation or regulations. We are satisfied that minerals planning control legislation is generally adequate to meet current needs, and, subject to our detailed recommendations, to deal with any new problems that might arise from projected levels of coal production. We are also satisfied that the best practicable means approach to pollution control is inherently superior to control by nationally-fixed and rigid standards. The realities of pollution control require a continuing balance to be struck between the costs and benefits of pollution abatement for industry and society. The best practicable means formula provides flexibility to take account both of local circumstances and national need.

22.5 We have examined the adequacy of current air pollution control requirements to maintain air quality standards in the light of the prospects for increased coal burn. The relevant legislation, dating from the Alkali Act of 1863 to the Clean Air Acts of the past 20 years, offers an imposing range of controls. We consider it highly likely that they will afford satisfactory protection of air quality up to the turn of the century and beyond. However, continuing assessment of the adequacy of controls may be advisable given the prospect of a cumulative build-up of multi-source pollution. Even here, however, the approach of best practicable means seems to us to offer the best solution because it takes into account local circumstances, including the levels of pollution control actually achieved.

22.6 In reaching the conclusions that current legislation and controls provide satisfactory protection on air quality we have recognized that any major social activity may have some deleterious effects on health as well as social benefits. Absolute safety is as impossible to guarantee with the combustion of coal as with the generation of electricity by nuclear or other sources of power, and some few fatalities (possibly even some hundreds) may continue to be produced each year. Although every effort will be made to limit the incidence of such diseases, this is, however, one of the costs of an industrialised society and is far outweighed by the rising standard of living and reduced mortality from other causes.

22.7 Notwithstanding the possibilities which we have identified for the reconciliation of an enhanced role for coal and the protection of the environment, much needs to be done to secure greater public acceptance of increased coal production and use. A recurring theme throughout this Study has been the need for wider public debate of national policy issues and the need for more effective use of the mechanisms available to translate the implications of that policy into possible requirements for major energy developments at the local and regional level. A striking feature of the evidence submitted to us is a general unpreparedness for such developments as a major re-entry of coal into the industrial market. We have therefore underlined at many points in this Report the need for advance warning of possible developments and for early consultation between the N C B, local authorities and other interested parties. Much will also depend upon the manner in which the N C B builds upon changes, which it has introduced during the course of this Study, in its handling of environmental issues both at headquarters and in the regions. What is needed is the impetus of a sustained and coherent policy lead from the N C B 's headquarters and the implementation of that policy in the regions without stifling local initiative.

22.8 The effective practice of advance warning and early consultation does not call for radical changes in existing planning procedures. Our Study has served to highlight the flexibility inherent in the U K planning system. Thus we advocate improved practice in the operation of that system and not the introduction of radical new machinery.

22.9 For much of the period from the Second World War to the early 1970s, the coal industry experienced massive contraction and decline punctuated by changes in policy brought about by the pressure of external circumstances. In the wake of the events of 1973, the industry entered upon a new phase of its history which transformed its potential longer term prospects. In both the energy and the environmental interest, we consider that greater stability of long-term planning is the essential pre-condition of successful modernisation of the industry.

22.10 This process of modernisation affords the best prospect of striking an acceptable balance between energy and environmental interests. New low-cost high-productivity capacity can be designed to high environmental standards. The bringing on-stream of such capacity will facilitate the phasing out of obsolescent capacity in areas whose decline is associated with the worst environmental legacy of the past. The progressive introduction of such deepmined capacity would also enable the eventual limitation of opencast output so that environmental benefit could be achieved without disproportionate loss. However, the unique dependence of some regions on the coal industry will mean that the process of change, which will aid both the cleaning up of past dereliction and the generation of new employment opportunities, will need to be handled with great sensitivity.

22.11 Challenge and opportunity confronts the industry. A modernised industry designed to high environmental standards, can make an immense contribution to an energy strategy designed to diminish in the longer term the dependence of the U K economy on imported oil. The decisive determinant of achieving the required restructuring of the industry will be investment in modernisation. Only by such investment will the industry be able to grasp the opportunities afforded by the evolving energy scene. If the interests of the taxpayer, the energy consumer and the industry's workforce are to be reconciled, an integral part of modernisation must be the phasing out of heavily loss-making obsolescent capacity. This calls not only for help to individuals but for a far greater degree than hitherto of forward regional planning in anticipation of the effects of closures on mining communities. It also requires the full co-operation of both sides of the industry in the operation of its existing machinery, the local authorities, bodies such as the Welsh and Scottish Development Agencies and central Government. Failure to face this challenge would be to tether the U K to the industry's past, to postpone into the next century the clearing up of the legacy of past dereliction, and to deny to the country the immense contribution which a modernised coal industry can make in supplying the country's energy needs without unacceptable environmental costs.

SUMMARY OF DETAILED CONCLUSIONS AND RECOMMENDATIONS

22.12 We have discussed in Chapters 12 and 20 respectively the major thrust of our conclusions on coal production and coal use. We now summarise our detailed conclusions and recommendations which are listed below.

FOREWORD

Technical training and educational requirements

22.13 We RECOMMEND that an important element in the follow-up to our work should be an early examination of technical training and educational requirements. This examination should involve the nationalised industries, the unions, the universities and polytechnics, the relevant professional and technical institutions and Government Departments. (Paragraph 6)

CHAPTER 3: PROSPECTS FOR COAL

Coal reserves

22.14 We conclude that the scale of U K coal reserves satisfies any reasonable current concern about the future physical availability of coal. We attach considerable importance to the N C B's policy of maintaining, on a continuing basis, operating reserves sufficient to meet 50 years of current production. (Paragraph 3.15)

Deep mined production

22.15 We conclude that over the next 20 years the concentration of deep coal mining in the Yorkshire-Derbyshire-Nottinghamshire coalfield is likely to increase, whilst there will be some continuing decline in the older traditional mining areas of South Wales and North East England. (Paragraph 3.18)

22.16 We conclude that, to the extent that high production costs and poor environmental characteristics are associated with older, less efficient pits, the transition of coal mining to a modern industry, concentrating its activities in pits employing new technologies, would be beneficial in terms of both environmental impact and production costs. (Paragraph 3.19)

Energy policy implications

22.17 We accept that all forecasts are inevitably surrounded by great uncertainty and therefore that energy strategy cannot be translated into a blueprint. We see the role of the Department of Energy's projections as providing one among several possible technical and objective views of the future, and a quantitative backcloth against which Ministers can consider individual policy decisions. Continuing dialogue with experts outside Government and the fuel industries can make an important contribution, particularly in the difficult area of demand forecasting. (Paragraph 3.38)

22.18 We conclude that an energy policy firmly based on low growth might well prove insufficiently robust; some over-provision of energy supply is preferable to the risks attendant upon energy shortage. However, severe opportunity costs would clearly be involved if the insurance margin were to be excessive. (Paragraph 3.8)

22.19 We attach particular importance to the Government's role in stimulating public debate and wider understanding of the broader framework within which particular energy policies contribute to national objectives. (Paragraph 3.39)

22.20 We see considerable merit in a major study evaluating the risks inherent in a supply orientated policy compared with a policy emphasising efficiency in energy use. (Paragraph 3.41)

22.21 We believe that most western governments will continue to find it sensible to favour indigenous coal supplies, and to be unwilling to jeopardise sensible long-run developments by allowing customers to play the market by allowing short bursts of low-cost imports of coal during temporary recessions. Most importantly, we recognise the social unacceptability of repeating in the last decades of the century the sequence of creation and then destruction of mining communities. (Paragraph 3.42)

22.22 We conclude that the interests of the taxpayer and the energy consumer on the one hand, and the long-term prospects for employment and real wages in the mining industry on the other, can only be reconciled by investment in industrial modernisation and realistic wage settlements. These interests cannot be reconciled by policies which lead the N C B to produce the last possible tonne from obsolescent, high-cost capacity. (Paragraph 3.43)

22.23 We conclude that the range of coal consumption that it is realistic to examine is between 110–170 million tonnes per year; and the range of production is between 100–150 million tonnes per year. Even if the lower levels of production and consumption are considered, it will be essential for the Coal Board to invest in new capacity on a fairly large scale. (Paragraph 3.45)

CHAPTER 6: NEW DEEP MINE CAPACITY

The rate of build-up of new U K mining capacity

22.24 We endorse the need for a steady build-up of new capacity which will provide adequate opportunity to take environmental and other factors into proper account, but we recognise that each particular proposal must be judged on its merits in the light of competing claims for public sector investment, and likely impact on the environment and employment. (Paragraph 6.2)

The provisions of the General Development Order as they affect the N C B

22.25 We RECOMMEND that the Government should incorporate an amendment to the General Development Order 1977 to make clear that any surface development which is proposed in order to open up a new access underground should require an express grant of planning permission. (Paragraph 6.12)

N C B's environmental assessment

22.26 We conclude from our study of the N C B's development at Selby that the Board undertook a very thorough environmental impact assessment. Construction work demonstrates that the Board have achieved a high quality of architectural and landscape design. (Paragraphs 6.19 and 6.20)

Provision of local authority services consequent upon new developments

22.27 We RECOMMEND that the Department of the Environment should examine the distribution of the rate support grant as it affects the provision of additional local authority services required for new development. (Paragraph 6.32)

Acceptability of modern mines in new coalfields

22.28 We conclude that the introduction of a modern mine into a new coalfield, both in terms of the scale of increase in the population and the associated housebuilding rate, need have no greater impact than that associated with any medium-size industrial development. However, the acceptability in such areas of new mining projects will be determined by those impacts which are unique to deep mines and are not therefore associated with other forms of industrial development. (Paragraph 6.33)

CHAPTER 7: ENVIRONMENTAL EFFECTS OF DEEP MINING

Disposal of water from mines

Working mines

22.29 We conclude that it would be helpful to the water authorities to have the additional powers provided under Part II of the Control of Pollution Act 1974 available to them, and we RECOMMEND that the Government brings these into force as soon as possible. (Paragraph 7.14)

Abandoned mines

22.30 We RECOMMEND that discharges from mines abandoned by the predecessors of the N C B, or even the N C B some long time ago, should be considered as a form of dereliction comparable to abandoned pitheaps. We RECOMMEND that provision should be made through central government to meet the costs of any essential work, as is the case with derelict land clearance schemes. (Paragraph 7.19)

22.31 We RECOMMEND that when a pit is to be taken out of operation there should be close co-operation with the water authorities to assess and plan for the effects of the cessation of pumping. The N C B should carry the costs of any necessary preventive or remedial measures. Where appropriate, conditions should be imposed as part of a planning consent. We RECOMMEND that Section 46(6)(b) of the Control of Pollution Act 1974 should also be amended to ensure that the powers of the water authorities under Section 46(5) to recoup the costs of remedial works should apply in respect of mines being closed by the N C B. (Paragraph 7.20)

Coal stock piles

22.32 We RECOMMEND that an amendment should be made to the General Development Order, to make it clear that any permitted ancillary developments for a colliery relate only to the activity of that particular colliery, and cannot be used for the ancillary activities arising from the operations of any other colliery. (Paragraph 7.30)

Environmental standards at collieries

22.33 We have observed that standards of "good housekeeping" on the surface and the extent to which architectural and landscape design skills are employed varies greatly between different collieries. We conclude that the poor examples arise partly from the absence of sustained pressure from Central and Area Headquarters to achieve higher standards and partly through a lack of co-operation between the local authorities and the N C B. (Paragraphs 7.31–7.33)

CHAPTER 8: SUBSIDENCE

Prediction of subsidence

22.34 We RECOMMEND that the N C B should carry out further research into the effect of subsidence on the mass between the coal seam and the surface, especially in geological fault conditions. (Paragraph 8.6)

Sterilisation of land by subsidence

22.35 Where severe subsidence sterilises land and effectively prevents future developments, we RECOMMEND that the N C B should discuss with local interests and environmental bodies what alternative use might be made of the site. (Paragraph 8.10)

Publicity for possible subsidence

22.36 We RECOMMEND that the N C B should make every effort to ensure that individual owners and business interests are aware of likely subsidence – even when publication of notices is not obligatory. We RECOMMEND that suitable notices be posted in streets when mining is about to take place. We also RECOMMEND that the degree of publicity at Selby and Belvoir should become standard practice for all new mining projects. (Paragraph 8.19)

Preventive measures in existing structures against subsidence

22.37 We conclude that the N C B have powers, but are not obliged, to carry out preventive works to property where they consider that such works would prevent or reduce subsidence. We RECOMMEND that the possibility of such measures should at least be a matter for full public discussion in areas to be affected by mining. (Paragraph 8.30)

Precautionary measures in new buildings against subsidence

22.38 If the N C B recommend specific precautionary measures, after consultation with the local planning authority and the developer, we RECOMMEND that the local planning authority should include the appropriate planning conditions in any planning permission. (Paragraph 8.34)

Compensation for subsidence damage

22.39 We conclude that there are severe practical limits to minimising the incidence of subsidence. For the foreseeable future the Board will have to operate on the present basis – the payment of compensation for the restitution of damage after it has taken place. (Paragraph 8.35)

Speed of repair work

22.40 We RECOMMEND that all repair work should be carried out as speedily, efficiently and sensitively as possible. Where further ground movement is expected, this increases the need for rapid and unobtrusive interim repairs. (Paragraph 8.42)

22.41 WE RECOMMEND that for the future there should be additional provision for compensation for residual loss in property value modelled on Section 10 of the compulsory Purchase Act 1965, taking into account but not superseding the duties already placed on the Board by the Coal-Mining (Subsidence) Act 1957 to make reasonable repair. (Paragraph 8.45)

CHAPTER 9: SPOIL

Scale of the spoil disposal problem

22.42 We conclude that over the next 20 years the continued tipping of spoil will be one of the major environmental problems arising from deep mining and there will be major implications for land use. (Paragraphs 9.3 and 9.5)

The geographical concentration of tipping

22.43 We conclude that there are two distinct problems regarding spoil disposal in the future. The first is the environmental impact of tipping in greenfield sites; the second is that there may not be sufficient land to accommodate new spoil in an acceptable manner in the Yorkshire area, given the amount of spoil which has already been tipped there. (Paragraph 9.7)

Spoil tipping operations and controls

22.44 We RECOMMEND that landscape design proposals should be included in tip reclamation schemes from the outset, and in schemes for tipping and after-treatment of waste and that the supervisory period for restoration should be kept under review. We would like to see more imaginative and innovatory landscape design. We RECOMMEND that the N C B should be allowed to use other supervisors apart from A D A S and D A F S to oversee the restoration of land and should seek advice from the N C C on wildlife interests. (Paragraphs 9.13 and 9.17)

22.45 We conclude that where new local tipping is to take place, the N C B's methods of progressive restoration can, if properly implemented, limit its environmental impact, but there is not yet sufficient evidence about the implementation of these methods for us to reach a firm conclusion. We recognise that there may be occasions where the impact of a tip even designed to the N C B's highest standards may have such local environmental impact that it might be considered unacceptable. (Paragraph 9.14)

22.46 We conclude that the development control system provides local authorities with effective controls over new tips. (Paragraph 9.15)

22.47 We welcome the Town and Country Planning (Minerals) Bill which should provide more effective control over tips operated under older planning permissions. (Paragraph 9.17)

22.48 We conclude that the Government's proposals to amend the G D O, combined with certain provisions in the Town and Country Planning (Minerals) Bill, should meet our concern about spoil tipping aspects of the G D O. We RECOMMEND that the Government should amend the G D O so as to ensure that, where a prohibition order under the powers in the Bill has been made and confirmed in respect of mining operations at a site, it would be unlawful to continue depositing waste there and that local authorities could impose restoration conditions at such sites. (Paragraph 9.19)

Lagoons

22.49 We RECOMMEND that wherever possible the use of lagoons for the treatment of tailings should be avoided. (Paragraph 9.21)

22.50 We RECOMMEND that new deep mines should be designed wherever possible in such a way that liquid effluent would not be discharged. (Paragraph 9.21)

Alternatives to local tipping

22.51 We RECOMMEND that the N C B, in conjunction with the Research Councils, should greatly strengthen its research into backstowing even if that eventually requires substantial modification to the longwall process. (Paragraph 9.26)

22.52 We conclude that the scale of waste production in major coal mining areas is such that local land reclamation is likely to make only a very modest contribution to spoil disposal in the future, although every opportunity should be taken to make use of it. We RECOMMEND that the possibility of disposing of colliery spoil should be taken into account when considering both individual opencast programmes, and the phasing of a regional opencast programme. However, we do not consider that the need to dispose of colliery waste from deep mining can be used as a general justification for opencast mining. (Paragraph 9.30)

22.53 We conclude that there is limited scope for using remote land reclamation to dispose of colliery spoil, although it could be important in particular circumstances. Decisions concerning the disposal of colliery spoil by remote land reclamation will need to be taken in the context of defined national priorities for the use of the few sites that are available. (Paragraph 9.32 and 9.33)

22.54 We RECOMMEND that spoil should not be tipped further on beaches; that, in view of the adverse environmental impact of marine tipping, the N C B ought not to use this method of spoil disposal unless and until suitable techniques for safeguarding habitats and amenity are found; and that the ecological consequences of pipeline deposition to the seabed should be fully examined. (Paragraphs 9.34 and 9.35)

22.55 WE RECOMMEND that the Government should examine the N C B's suggestions to make greater use of colliery spoil by the introduction of an improved dual tendering system covering all appropriate building and civil engineering projects. (Paragraph 9.37)

22.56 We conclude that alternatives to local tipping can be very important in solving the problem of spoil disposal at particular collieries and within particular regions. We RECOMMEND that they should be adopted wherever feasible and appropriate, but conclude that they are unlikely to make a significant reduction in the quantity of spoil that will have to be disposed of locally. We therefore conclude that large scale local tipping is inevitable and that priority should be given to minimising its adverse environmental impact. We RECOMMEND that the N C B continue to investigate what further improvements can be made in local tipping practice. (Paragraph 9.41)

Strategic planning of spoil disposal

22.57 We conclude that the problem of spoil disposal suffers from the lack of any coherent national or regional disposal policy and that better co-ordination between the N C B, local and central government is essential. We RECOMMEND that, where the spoil disposal problem is most serious, the Department of the Environment should promote ad hoc arrangements similar to the Regional Working Parties on Aggregates.

Priority should be given to the establishment of a working group in the Yorkshire and East Midlands coalfield. (Paragraphs 9.47 and 9.48)

22.58 We RECOMMEND that the Government should establish guidelines on the use of different types of site for remote reclamation; on the availability or otherwise of Governmental financial assistance; on the priorities for different types of waste in remote disposal schemes; and on the priorities between regions in the use of remote disposal sites. (Paragraph 9.49)

CHAPTER 10: DERELICTION AND PIT CLOSURES

Nature of the dereliction problem

22.59 We endorse the view expressed in the Fourth Report of the R C E P that land dereliction may be regarded as a form of environmental degradation just as unacceptable as other forms of pollution. (Paragraph 10.2)

Methods of preventing dereliction

22.60 We RECOMMEND that the N C B continue to develop their approach of disposing of land and property not required for operational use in order to facilitate alternative industrial and commercial occupation. (Paragraph 10.7)

Existing derelict land clearance scheme

22.61 We RECOMMEND that government should exempt local authority expenditure on derelict land clearance from counting against capital expenditure allocations. (Paragraph 10.14)

The problem in Yorkshire

22.62 We conclude there is a serious need for the N C B and the Yorkshire local authorities to co-operate more effectively in drawing up a rolling programme of restoration on a regional basis, if the region is to cope with the planned increase of mining activity over the next two decades. We RECOMMEND that the government should examine the feasibility of establishing in the first instance in Yorkshire, an independent regional development agency with its own budget to deal with problems such as derelict land clearance. (Paragraphs 10.18 and 10.19)

Planning for pit closures

22.63 We conclude that there is an important linkage between pit closures, clearance of derelict land, and the generation of alternative employment opportunities. Forward planning based on early warning and the closest co-operation between all interested parties is required. (Paragraphs 10.20 and 10.21)

CHAPTER 11: OPENCAST MINING

Effects of opencast operations

22.64 We conclude that the combined effects of opencast operations can, for those badly affected, add up to a very severe diminution in the quality of life. (Paragraphs 11.36 and 11.66)

Noise in opencast mining

22.65 We RECOMMEND that the N C B should make funds available to promote further research into noise standards for opencast operation. (Paragraph 11.42)

Proximity of sites to residential property

22.66 We RECOMMEND that the Opencast Executive, the local authority associations, the Department of the Environment and the Department of Energy should discuss the feasibility of extending the current 50 yards minimum limit for opencast sites near to residential properties. (Paragraph 11.49)

Discretionary power to purchase properties in opencast cases

22.67 We RECOMMEND that the Opencast Executive should make full use of the power to make discretionary purchases of property, with the agreement of the owner, in cases where the hardship to individual property owners would be reduced. (Paragraph 11.50)

Loss of landscape value on restoration after opencast operations

22.68 We conclude that the loss of landscape quality after restoration is due to the general lack of involvement by landscape architects, the influence of farming interests restricting the use of hedges, hedgerow trees and tree planting; and design by Committee rather late in the day. (Paragraphs 11.28–11.31)

22.69 We conclude that a substantial improvement in restoration practice could be achieved if details of the restoration programme were submitted as part of the application for authorisation and were covered in the conditions attached to the grant of authorisations. We RECOMMEND that the application should include an analysis of the effect of the proposals on the food production of the area and the proportion of agricultural land to be taken out of use, a survey of the existing agricultural landscape and ecological characteristics of the site, a clear visual and verbal description of the forecast condition of the site after mining and restoration; and that the appropriate legislative changes to this end should be made as soon as possible. (Paragraphs 11.62–11.65)

22.70 We RECOMMEND that the Government's statutory advisers, the Countryside Commissions and the Nature Conservancy Council, should be consulted as a matter of course on opencast applications. (Paragraph 11.65)

Authorisation procedure for opencast applications

22.71 We RECOMMEND that planning applications for opencast coal working should be dealt with under the normal minerals planning machinery, as for all other new mineral developments, and not directly by the Secretary of State for Energy as at present. (Paragraph 11.80)

Opencast targets

22.72 We conclude that it would not be right to cut the size of the opencast target arbitrarily. However, opencast profits should not continue indefinitely to disguise and cushion deep mine losses. (Paragraph 11.82)

22.73 We RECOMMEND that as older, more unprofitable and less environmentally acceptable deep mines are closed, and more efficient and profitable operations take their place, the volume of opencast mining should be allowed to decline. In the meantime, there should be no increase in the present target of 15 million tonnes per year. (Paragraph 11.82)

22.74 We conclude that it should be possible to define more strictly the sites where opencast coal might be mined. We RECOMMEND the adoption of a series of guidelines defining such sites, that these guidelines should be brought within the recently agreed Code of Practice adopted by the Opencast Executive and the Local Authority Associations, and that they should be rigorously applied. (Paragraphs 11.83–11.85)

The financial importance of opencast operations

22.75 We RECOMMEND that the Opencast Executive is treated in the N C B 's Annual Report and Accounts as a separate accounting unit. (Paragraph 11.87)

CHAPTER 12: COAL PRODUCTION: GENERAL CONCLUSIONS

Payment of rates by N C B

22.76 We RECOMMEND that the Department of the Enviroment review the existing provisions to ensure that the rates paid by the N C B are appropriate to the environmental stress their operations can cause, as well as the benefits they can provide. (Paragraph 12.15)

Environmental concerns in the N C B 's organisation

22.77 We conclude that the Board has a responsibility to ensure that the best environmental practices available are introduced as appropriate in the day-to-day management of its operations. We RECOMMEND that this responsibility should be assigned at an appropriately senior level in the Board. (Paragraphs 12.18 and 12.21)

CHAPTER 13: COAL MARKETS

Future markets for coal

22.78 We have concluded that coal is likely to be the main source of electricity generation in the United Kingdom to the turn of the century; that there is little scope for expansion in the coke oven market; that the domestic market is unlikely to expand; but that there is substantial scope for coal substituting for oil in the industrial sector. (Paragraphs 13.7, 13.16, 13.43 and 13.45)

Future penetration of the industrial market

22.79 We RECOMMEND the Government should take the lead in stimulating continuing and systematic dialogue between the N C B, representatives of industry and local authorities on the possible scale of re-entry of coal into the industrial market. Any transition from oil to coal in the industrial sector should be gradual. (Paragraphs 13.36 and 13.37)

Future penetration of the public administration sector

22.80 We RECOMMEND that the Government give a lead in the switch from fuel oil to coal in the public administration sector. (Paragraph 13.39)

Fluidised bed combustion

22.81 We RECOMMEND that the government should stimulate the commercialisation of fluidised bed technology. There should be an integrated effort by the N C B and the Boilermakers with appropriate participation by the Departments of Industry and Energy, to formulate an agreed programme for commercialisation and to examine export potential. (Paragraph 13.40)

CHAPTER 14: TRANSPORT, HANDLING AND STORAGE

The environmentally preferable mode of coal transport

22.82 We RECOMMEND that, wherever possible, rail should be the chosen method of transport and that, in particular, power station siting (with the exception of stations with coastal or waterways access) must be compatible with the most effective use of rail transport for the provision of power station coal. Where there is a choice between direct delivery by road and a journey part by rail and part by road, there could, however, be environmental advantages and economic reasons in favour of the former. (Paragraphs 14.20 and 14.21)

Rail investment

22.83 We RECOMMEND that investment in railway track and rolling stock should proceed at a pace compatible with the plans of N C B and C E G B for the mass transport of coal. (Paragraph 14.43)

Lorry dust and dirt control

22.84. We do not favour wider or more detailed regulatory powers to control the production of dust and dirt by lorries. Nevertheless, we RECOMMEND that current best practice is put onto a more formal footing and applied in a more consistent manner. We would not advocate that sheeting and wheel washing should be obligatory in every instance but RECOMMEND that the N C B draw up a Code of Practice for all their own road vehicles leaving their sites. (Paragraph 14.41)

22.85 We RECOMMEND that the N C B associate themselves with and support research into automated load sheeting devices to make sure that full account is taken of their requirements. (Paragraph 14.41)

Coal depots

22.86 We conclude that coal depots appear to be adequately placed and of sufficient capacity to serve future industrial needs, assuming that the geographical distribution of industry does not change significantly. (Paragraph 14.12)

Coal transport by container

22.87 We RECOMMEND that further research should be carried out into the possibility of coal transport by container. (Paragraph 14.16)

Alternative modes of transport to road and rail

22.88 We conclude that the contribution of transport modes other than road and rail is likely to be small and their use uneconomic. There may be occasions when they can make a local contribution, however, and we RECOMMEND that each new coal project should be examined in case there is the possibility of supply by waterway. We conclude that coal pipelines are generally not feasible in the U K. (Paragraphs 14.44, 14.48 and 14.51)

Competition between modes of transport for coal

22.89 We support the Armitage Report's recommendation that lorries in general, and each class of lorry, should pay in taxation at least the road track costs which they impose. We support present Government policy of ensuring that there is fair competition between the various transport modes. However, we also endorse the Armitage Committee's conclusion that on environmental grounds any financial anomaly in competition between road and rail should act in favour of rail. (Paragraphs 14.55 and 14.56)

Section 8 grants

22.90 We acknowledge the importance of Section 8 grants in encouraging more environmentally acceptable transport of coal. We welcome and endorse the Armitage Committee's recommendations to increase these grants to up to 80% where the environmental benefits justify such a high rate and to increase the standard grants to 60%. We support the Government proposal to introduce legislation to extend Section 8 grants to waterway users. (Paragraph 14.58)

Storage and handling of coal and ash

22.91 We conclude that adequate storage and convenient handling of coal and ash will be extremely important to the environmental acceptability of increased coal use, and that it is likely that improved methods for storage and handling of coal and ash will be critical in stimulating conversion to coal. (Paragraph 14.59)

22.92 We conclude that there is a lack of awareness of the most up to date technology even by many of those presently using coal or supplying handling equipment. We RECOMMEND that there should be a vigorous marketing drive both to persuade present coal users to adopt this technology and to convince potential users that coal and ash handling can be done cheaply. (Paragraph 14.65)

22.93 We consider that there is room for improvement in pre-planning, particularly of emergency stocking sites, so that they are properly screened to minimise visual intrusion. N C B could give a lead by ensuring that their subsidiary, National Fuel Distributors, gives proper attention to "good housekeeping". (Paragraph 14.67)

22.94 We RECOMMEND that N C B should ensure that research and development in this crucial area of handling and storage is awarded appropriate priority within its overall research and development budget. (Paragraph 14.70)

CHAPTER 15: CONVERSION TO SUBSTITUTE FUELS

Efficiency considerations in coal conversion to S N G, and in energy use

22.95 We have concluded that S N G is sufficiently energy efficient to justify its eventual use, despite the fact that direct coal burn can be more efficient in most space heating applications. (Paragraph 15.21)

22.96 We would see major energy advantages if domestic scale heat pumps could be developed; and we RECOMMEND that Government take all necessary steps to see that such development is progressed. (Paragraph 15.21)

Expectations on the scale and timing of the U K need for S N G

22.97 We RECOMMEND that long-term contingency planning for site identification, and for project and process appraisal, should proceed on the basis of forecast need for up to 10 sites for S N G plants in the first decade of the next century. This figure should be reviewed as more accurate assessments become available. (Paragraph 15.44)

22.98 We conclude that there is some urgency in assessing siting requirements for S N G plants, despite uncertainty about how in detail the build-up of demand will take place. (Paragraph 15.61)

Demonstration plant

22.99 We RECOMMEND that B G C and N C B should undertake the necessary research into the characteristics and means of disposal of the solid residue created by S N G plant, in view of the large amount likely to be involved. (Paragraph 15.65)

22.100 We conclude that there is no certainty which of the many processes now under development would be used in any future U K programme of S N G manufacture. Serious environmental damage is unlikely from S N G plants, provided that due care is taken in their design and operation. (Paragraphs 15.70 and 15.71)

22.101 We RECOMMEND that government should make it clear to potential developers of S N G plant in the U K that commercial development of such plants will not be permitted until satisfactory experience with demonstration plants has been established. (Paragraph 15.71)

22.102 We RECOMMEND that plans should be drawn up for an early demonstration plant in the U K, if it appears likely that experience with such plant abroad will not be sufficient to assess the adequacy with which U K health and safety and environmental standards will be met. (Paragraph 15.72)

CHAPTER 17: COMBUSTION RESIDUES: SOURCES, QUANTITIES AND CONTROL

Commercial use of F B C ash

22.103 We conclude that further testing and research work is needed to assess the possibilities of commercial use of ash from fluidised bed combustion, particularly where this is mixed with spent limestone. (Paragraph 17.29)

Radiation

22.104 We conclude that, in terms of radiation dose to the public, the emission of radioactive material from coal-fired power stations is comparable with the very small amounts arising from the normal operation of present nuclear power stations including emissions from the nuclear fuel reprocessing cycle. In both cases the average dose equivalent received by members of the public is far less than the natural variations in the annual dose received from all natural sources. These conclusions apply to normal operating conditions but so far as coal is concerned the radiation cannot increase appreciably as a result of malfunctioning of the plant. (Paragraph 17.36)

Source of urban sulphur dioxide pollution

22.105 We conclude that, assuming the relative contribution of the sources examined are typical of all urban areas, low level sources are relatively important contributors of sulphur dioxide to urban pollution; high level emitters contribute less than a quarter. In addition, at least half the sulphur dioxide comes not from coal but from oil. (Paragraph 17.41)

Desulphurisation

22.106 We conclude that both the costs and the potential for application by 2000 of various desulphurisation techniques are uncertain; that the costs of sulphur control are high; and that the maximum practicable application of all the techniques would result in only modest reductions in emissions. (Paragraph 17.76)

22.107 We conclude that the use of low-sulphur oil is a valuable technique for reducing local pollution by sulphur dioxide in some urban areas, but it would have little effect on total national emissions. The displacement of residual fuel oil by coal could lead to some reduction in sulphur dioxide emissions. (Paragraph 17.77)

22.108 We conclude that further development work is needed before the costs of adding limestone to F B C systems as a means of reducing emissions can be established either for power generation or for industrial coal burn. (Paragraph 17.7)

CHAPTER 18: COMBUSTION RESIDUES: EFFECTS

Health effects

22.109 We conclude that current levels of air pollution in the United Kingdom are generally satisfactory, so far as acute and chronic effects on the respiratory system are concerned other than the production of cancer. (Paragraph 18.17)

22.110 We conclude that the present contribution of atmospheric B A P and associated materials from coal combustion to the production of lung cancer in the future is unlikely to account for more than one half per cent of cases of lung cancer among cigarette smokers, that is, for no more than 150 cases a year. Present

knowledge is incomplete and we RECOMMEND that the level of smoke pollution is kept at the lowest practicable level. (Paragraphs 18.25 and 18.32)

22.111 We consider that there is no longer major cause for concern on health grounds over ambient levels of smoke and sulphur dioxide from coal combustion. The extension of Smoke Control Orders should cope with the problems in most areas where concentrations are too high. In a few areas already covered by such Orders, it may be necessary to reduce sulphur dioxide levels by restricting the sulphur content of fuel oil used in commercial and industrial boilers. (Paragraph 18.60)

Effect on crops, vegetation and materials

22.112 We conclude that as long as pollution in urban areas is kept low enough for the protection of human health, the situation is also satisfactory as regards crops and vegetation. We do not consider the extent to which materials are affected by pollution at such levels to be cause for concern. (Paragraph 18.62)

Effects from nitrogen oxides and sulphur dioxide

22.113 We conclude that nitrogen oxides from coal combustion have not been identified as the source of any pollution problems within the U K. (Paragraph 18.63)

22.114 We conclude that more research is required into the causes and mechanisms of the effects of acidification of lakes and rivers, into possible control technologies, into cost-effective technologies for the reduction of emissions of sulphur dioxide and nitrogen oxides, into the adverse environmental impacts of existing emission control technologies and into techniques for ameliorating the effects of excess acidity in the problem areas. We endorse the approach of the Economic Commission for Europe's Convention. (Paragraphs 17.55 and 18.64)

Dust and grit

22.115 We conclude that continued monitoring is required to ensure that emission control technology maintains required standards. Careful siting of plant, which is potentially a major emitter of grit and dust, will continue to be necessary. (Paragraph 18.65)

Ash

22.116 We conclude that the disposal of large quantities of ash from power stations and industrial plants causes no problems on health grounds. The major difficulty is that of finding suitable disposal sites given the scale involved. There are local problems arising from noise and dust from transport and handling. Careful siting and treatment of disposal areas will continue to be necessary. (Paragraph 18.66)

Carbon dioxide

22.117 We consider that it would be premature at this stage to do more than note the potential importance of the possible build-up of carbon dioxide in the atmosphere and that research is being carried out to clarify the problems. No deterioration of the climate due to the build-up of carbon dioxide in the atmosphere has yet been established. If it were deemed necessary in the long run to combat the rise in atmospheric levels, any such action should be international in scope and would need to be based on the outcome of international research. (Paragraphs 17.84, 18.57 and 18.58)

CHAPTER 19: THE ENVIRONMENTAL EFFECTS OF BURNING COAL IN URBAN AREAS

Domestic coal burn

22.118 We conclude that smoke from domestic open fires is the most objectionable form of smoke. (Paragraph 19.9)

22.119 We conclude that if coal is to continue to be used for domestic heating in urban areas, more efforts must be made to develop and market smoke reducing appliances which are reliable and foolproof under all conditions of use. We RECOMMEND that, in view of the need to keep the level of smoke pollution at the lowest practicable level, the performance of smoke reducing appliances and the effect they have on ambient smoke levels should be carefully monitored. We conclude that, in the meantime, they should not be used extensively in urban areas. (Paragraphs 19.10 and 20.27)

22.120 We RECOMMEND that the Government should co-ordinate research and development on heat meters, particularly for domestic use, as a matter of high priority. (Paragraph 19.15)

Industrial coal burn

22.121 We conclude that the mandatory limits for smoke and sulphur dioxide set out in the E E C Air Quality Directive provide an adequate standard for the protection of human health. Appropriate measures should be taken as soon as possible to improve the air quality in all areas which do not conform to these mandatory standards. (Paragraph 19.34)

22.122 We consider that industrial coal burn can be positively encouraged in most circumstances. (Paragraphs 19.34)

22.123 We conclude that the prospect of a cumulative build-up of multi-source pollution may warrant continuing assessment of the adequacy of controls. We also conclude that "best practicable means" offers the best approach to multi-source emissions. (Paragraphs 19.32–19.33)

22.124 We RECOMMEND that central government should consider the possibility of empowering local authorities to require prior consent to an industrialist's choice of fuel or means of controlling emissions, and of extending the best practicable means approach to those processes controlled by local authorities. We endorse the previous recommendations on Registration and Consents in the Fifth Report of the R C E P. We RECOMMEND that in exceptional circumstances powers might be granted to specified local authorities to direct the use of particular fuels in particular buildings. (Paragraph 19.33)

CHAPTER 21: THE PLANNING PROCESS: RECONCILING THE INTERESTS

Tensions in the current decision system for major projects

22.125 We conclude that the flexibility in the present Town and Country Planning system can be exploited to minimise tensions and conflicts, and that radical changes in existing procedures are not required. (Paragraphs 21.13–21.16)

National debate on energy issues

22.126 We RECOMMEND that there should be a more systematic explanation by Government of the progressive evolution of national energy policy, and of the options for its implementation, and that this should be accompanied by regularly updated indications of the order of magnitude of possible future requirements for major energy projects. (Paragraphs 21.18, 21.21 and 21.22)

22.127 We conclude that the debate and approval by Parliament of Government White Papers setting out policy in particular areas would provide a crucial component of the policy background to public inquiries. (Paragraph 21.20)

Land availability for energy developments

22.128 We conclude that long-term planning of land resources suitable for energy developments needs to be improved. We RECOMMEND that the energy industries and the local authorities should undertake urgently a joint review, initially in Yorkshire and the East Midlands, to establish whether there is a scarcity of potential sites for major energy and other industrial developments. The Department of the Environment in conjunction with the Department of Energy should take the lead in initiating this review. (Paragraphs 21.34–21.38)

The need for progressive consultation

22.129 We conclude that if major programmes by the energy and other industries are to proceed, then there must be a reasonable degree of public acceptance which can only be obtained by more and earlier explanation of what the proposals involve and by public involvement when choices have to be made between alternatives. (Paragraph 21.39)

22.130 We RECOMMEND that N C B should publish widely their "Ladder of Exploration and Project Development" describing the sequence of mine project development. (Paragraph 21.41)

22.131 We conclude that the exploration programme of the N C B has now reached a point at which the Board is in a position to assess alternative locations for new mine developments giving greater weight and earlier consideration to environmental factors than has been possible before. We RECOMMEND that there should be consultations between the N C B and the local authority associations and other interested bodies, both local and national, before the commitment of investment in any one project has advanced too far. (Paragraphs 21.42–21.47)

Environmental impact assessment

22.132 We conclude that a mandatory system of formal environmental impact assessment would risk undue delays and costs in planning procedures. We would prefer the wider deployment of this technique within the existing flexible U K approach to project assessment. (Paragraph 21.54)

Public local inquiries

22.133 We conclude that the terms of reference of public local inquiries should not be restricted to exclude discussions of national issues, but we do expect the need for such discussion to become much less than at present once our recommendations are implemented. (Paragraph 21.61)

22.134 We RECOMMEND that the Department of the Environment should examine the possibility of the greater use of written evidence, and less reliance on oral submissions and cross-examination. Wherever possible proofs of evidence should be submitted as early as possible and should not be subsequently read out in evidence. We RECOMMEND that cross-examination should only be used when its particular advantages in probing evidence can be fully used. (Paragraphs 21.63 and 21.64)

22.135 We conclude that there is clear scope for a reduction of delays in the period after the conclusion of a public inquiry. The Government itself should reduce the delays in reaching and publishing its decisions. (Paragraphs 21.65)

Acknowledgements

1. The many organisations and individuals we have consulted during our Study are listed in Appendix 3. We are grateful to them all for the help and co-operation we have received. We must particularly express our appreciation to those bodies directly concerned with the production and use of coal in this country. The C E G B, the British Gas Corporation as well as the N C B have spared no effort to help us to understand their operations, both in written and oral evidence and in visits to their establishments. Likewise, we received particularly helpful evidence from the National Union of Mineworkers, who with the British Association of Colliery Management and the National Association of Colliery Overmen, Deputies and Shotfirers also helped us during our visits. We have received similarly invaluable help from the Association of County Councils, the Association of District Councils, the Association of Metropolitan Authorities, the Convention of Scottish Local Authorities, the Welsh Local Authority Associations, many individual local authorities in Wales and Scotland as well as England, the Welsh Development Agency and the Royal Town Planning Institute. Our work has been enriched by help and advice from numerous environmental groups. We would mention in particular our appreciation of the help we have received from the Committee for Environmental Conservation and the Council for the Preservation of Rural England. We also wish to express our thanks for the enormous volume of documentation made available to us from U S A sources.

2. During the preparatory phases of the Study, Dr Joe Gibson, the N C B Board Member for Science, and Mr Peter Moullin, the Board's Deputy Secretary, acted as technical advisers to the Commission. We were fortunate indeed to benefit from Dr Gibson's vast experience and wise counsel and for Mr Moullin's unflagging patience in dealing with our many requests for detailed information on the Board's operations. We are also indebted to Dr John Wright of the C E G B and Dr Geoffrey James and his colleagues of B G C for their lucid presentation of the very important technical contribution made by their respective organisations to our work.

3. We have also made heavy demands on the Department of the Environment, the Department of Energy and the Scottish and Welsh Offices. Indeed, given the nature of the Coal Study, it has stimulated constructive tension between the two Departments mainly concerned, namely Energy and Environment. We have seen our role as that of arbiter between them. The mutually beneficial operation of our Joint Secretariat and the Report itself is evidence of the extent to which this was successful. We have received invaluable help from Deputy Secretaries from both Departments. Representation at this level has changed during the course of the Study, but we wish to express our particular appreciation of the help we received from Sir Wilfred Burns, Mr W I McIndoe and Mr D Le B Jones. We also wish to acknowledge the help we have received throughout the Study from Mr J A Colley of the Welsh Office and Mr R G H Turnbull of the Scottish Office. However, we would emphasise that these officials did not participate in the writing of the Report or in the formulation of our conclusions and recommendations which we reached independently of all Departments concerned and of all witnesses.

4. Finally, we wish to express our sincere gratitude to all the staff of our Secretariat. They have responded to the severe demands of our wide-ranging Study with unfailing enthusiasm and efficiency. In particular, although the Secretariat contribution has been a team effort, we must thank our Joint Secretaries, Mr K C Price and Mr S T McQuillin; and Mrs M McDonald who assumed Mr McQuillin's duties when we reached the Report writing phase of our work. Inevitably, the brunt of the effort has fallen upon them and they have given unstintingly of themselves throughout in the organisation of the work, of the presentation of the voluminous evidence, and in the writing of the Report in accordance with our wishes.

Appendix 1 Thermal efficiencies and costs of domestic heating by coal-fuelled electricity, S N G and direct coal-firing

A1.1 This Appendix provides basic information used to derive comparative coal requirements and energy costs to households for various forms of domestic heating. As in Energy Papers No 20 and 35 (Department of Energy 1977c and 1979c) an average household in the year 2000 is considered to make actual use of 42.2 gigajoules (G J) (400 therms), of energy per year as space heating and hot water. The amounts and costs of energy purchased by the household to obtain 42.2 useful G J are calculated, and the equivalent primary coal burn needed to deliver that energy is shown. Calculations are carried out at December 1980 prices.

A1.2 By definition efficiency is simply energy delivered divided by energy supplied. But there are many problems in the definition of what should be counted as input and output. This Appendix takes the primary input to be the gross calorific value of the original coal burnt or converted – that is, the energy which could be obtained by burning the fuel and cooling the combustion products to the starting temperature. The energy to operate ancillary equipment and losses in transmission are also allowed for in the calculation of the efficiency of providing the secondary fuels: S N G and electricity.

A1.3 The main problem with the output is how much of the output of a device to count as useful. Here there can be no precise answer as individual circumstances vary: for example, what is the value of the heat output of an electric storage heater or a coal-fired boiler in the middle of the night?

A1.4 For many devices an appliance efficiency can be measured fairly easily. This is a figure which assumes all output is useful and ignores any losses subsequent to the initial transfer of energy. The appliance efficiency gives the maximum figure for efficiency that can be achieved by the device. For central heating boilers this is calculated from the amount of heat transferred to water continuously passed through the boiler, with a cold water input, divided by the gross calorific value of the fuel consumed. It is, however, arguable whether heat losses from the boiler casing and flue into the house should be included in the output or not. The view is taken here that in the winter they should be included

since they contribute to the end purpose of heating the house. But in the summer they should be excluded as they are of no use for the purpose of making hot water.

A1.5 The figure required to calculate the efficiency of generation of useful heat, and hence the running costs, is the system efficiency. The system efficiency takes account of losses which occur in practice after the initial heat transfer. For many reasons the importance of these losses varies considerably throughout the year and as a consequence of mode of use and lifestyle of individual households. Thus the loss of heat from the draw-off pipes and hot water storage tank may be of less consequence in the winter because some of it contributes to the general heating of the house, but is is pure loss in the summer. Storage heaters using off-peak electricity will have a poor efficiency for a household where the members are out at work all day and frequently out in the evening, because there is no one there to benefit from their heat most of the time.

A1.6 Some authors have used "intermittency factors" in calculating the cost of heating with different appliances to allow for the loss of some heat because it is produced when not needed; the device is assumed to be only intermittently producing useful heat. The system efficiency is taken to be the product of the appliance efficiency and the intermittency factor.

A1.7 For this Appendix, studies based on average cycles of use for typical domestic households are used instead. This means that in practice intermittency factors vary between 85–95% for electric fires and 65–90% for electric storage heaters, with gas and coal-fired devices lying between these two extremes. For some households, however, intermittency factors could be outside the normal ranges. They are likely to be lower for single person households where the occupant is out for much of the day; higher for households with several young children or old people needing higher standards of heating throughout the day and year.

A1.8 For the existing stock of houses, around 20% of the heating energy used over the course of a year is used to provide hot water; so around 10% of the total is used for hot water in the summer. For "low energy houses" which need less energy for space heating, the proportion used for water heating is larger and may exceed 50% over the year for exceptionally well insulated houses, so that the efficiency of hot water production in summer becomes a more important factor. For low energy houses the heat emission from the boiler casing of a heavyweight boiler can be greater than that needed to heat the kitchen in which it stands. Table A1.2 assumes 70% of the total is used for space heating and 30% for water heating – typical proportions for a new house built with good, but not outstanding, insulation.

A1.9 Table A1.1 lists the system efficiencies for average households using well controlled appliances and estimates the cost of each useful therm of heat supplied. Table A1.2 uses the data in Table A1.1 to calculate average annual fuel costs for whole-house space and water heating for the average household (defined above). Equipment is assumed to be well maintained with the hot water tanks and pipework from the boiler to the hot water tank well insulated. Draw-off pipes from the hot water tanks to the taps are assumed *not* to be insulated. The cost of S N G of 42p per therm (= £3·98 per G J) is based on the latest information available. Cost figures for two alternative prices of 36p per therm and 60p per therm are given to cover the possible range of uncertainty.

A1.10 The notes associated with the Tables explain the assumptions behind individual sets of figures and should be read with the Tables.

NOTES FOR TABLES A1.1 and A1.2

Water heating

A1.11 The main losses here are standing losses from the hot water tank and draw-off losses resulting from the cooling of water in the pipes between the tank and the tap between times of using hot water. With indirect heating systems there are also losses from the boiler itself, which are especially important in the summer when the boiler cycles on and off frequently because the hot water tank is unable to absorb heat as fast as the boiler can produce it. Such losses are particularly large for cast iron boilers with a large water capacity. Only 20% of heating energy input in average existing houses is used for hot water but in "low energy" houses which are very well insulated it could amount to over 50%. Table A1.2 assumes 30% is used for hot water, appropriate for a standard new (1980) house.

Central heating boilers and coal fires

A1.12 Heat losses from boiler casings and chimneys into the house are presumed to contribute to useful heat in the winter, but are treated as pure losses in the summer when water heating is the only requirement.

Electric heaters

A1.13 Losses in efficiency here are almost entirely due to intermittency. With direct heaters these could be practically non-existent giving virtually 100% efficiency at point of use. On the other hand, storage heaters can have very high intermittency losses in households where people are out most of the day giving efficiencies below 65%.

Heat Pumps

A1.14 Heat pumps work by using the input energy to transfer heat from a source at low temperature to a receiver at a higher temperature. The output energy is derived partly from the work done in transferring the heat and partly from the cold source. The co-efficient of performance (C O P) is the ratio of useful heat output to energy input. Electric heat pumps, which use a motor driven pump and the air as a cold source, have C O P s which vary from about 2·5 on the coldest days in the winter, to 3·5 in the hottest days in the summer. An average value of 3·0 is used in the Table; this is analogous to an appliance efficiency of 300%. Gas-fired heat pumps use an absorption cycle (though other methods such as gas engines or gas turbines are possible) and have a C O P of about 1·2, that is, an apparent appliance efficiency of 120%. The output includes heat from the cold source, the work done in the absorption cycle, and heat recovered from the exhaust gases. Although the gas fuelled C O P of 1·2 is much lower than that of 3·0 for an electric pump, this is largely compensated for by the higher delivered efficiency of S N G.

A1.15 For theoretical and practical reasons the higher the output temperature of a heat pump, the lower is the C O P. The output temperature required for domestic hot water is higher than that needed for space heating. For gas-fired heat pumps this is less of a restriction because they already use the exhaust heat for supplementary heating of the initial output so that the reduction in overall C O P resulting from producing smaller quantities of somewhat hotter water is small, and is taken account of in the average C O P of 1·2. For the production of domestic hot water by electric heat pumps it is assumed that one quarter of the energy in the hot water output is supplied by supplementary electric heating; this is equivalent to reducing the C O P from 3·0 to 2·0 for summer heating.

A1.16 Coal fuelled heat pumps using the absorption cycle are theoretically possible but are unlikely to be a practical proposition before 2000. Heat pumps are complex devices and are likely to cost from six to ten times as much as a gas boiler of equivalent output. Unless they are sized to meet the maximum requirement on the coldest days, they will need supplementing with some form of direct heating on those coldest days, which will lower the overall efficiency and hence raise the average cost of useful heat.

S N G

A1.17 The figure of £3·98 per G J (= 42 pence per therm) used here assumed a coal price at the S N G plant of 13 pence per therm, with an overall delivered efficiency of 70%. The alternative prices given cover the range of possible uncertainty as to delivered price. Existing processes are not yet able to achieve such a high delivered efficiency, but new processes being developed should achieve or exceed this figure.

Fuel prices and delivered efficiencies

A1.18 Costs are based on average domestic fuel prices for December 1980. Standing, installation and connection charges, appliance costs, maintenance and additional building costs (for example, for chimneys) are not included in the calculations, although these are clearly important in the overall assessment. Domestic gas and electricity prices vary by less than 10% from average solid fuel prices by up to 20%. The energy used for delivering coal to the domestic user is not allowed for, nor is the energy for delivering it to power stations or S N G plant. Transmission losses for S N G and electricity are allowed for and their final delivered efficiencies are those appropriate for plant expected to be in use in the year 2000.

A1.19 In practice, it is unlikely that only one source of heat will be used by any given household, except for some which use peak electicity for direct heating. In many houses a separate water heater for summer use may be justifiable.

Costs of C H P/D H for domestic heating

A1.20 C H P/D H cannot be analysed in quite the same way as other energy supplies. The original coal burn depends to some extent on essentially arbitrary assumptions about the distribution of the primary energy input between electricity generated and heat supplied.

A1.21 The figure used in Table A1.2 is derived from Energy Paper No 35 (Department of Energy 1979c) and represents the additional coal burn required to produce the same quantity of electricity whilst also supplying C H P/D H. Costs to the consumer are especially uncertain since there are no suitable U K analogues for comparison. Foreign costs are complicated by pricing distortions arising from various incentive schemes and differences in the structure of their electricity supply industries. The cost figures given are again derived from Energy Paper No 35. The range of uncertainty given here encompasses both efficiency of end use and variations in supply price which largely arise from different assumptions about discount rates.

Table A1.1 *Efficiencies of domestic heating systems from coal sources*

Heating system	Appliance efficiency %	System efficiency with allowance for intermittency %	Overall efficiency from coal %	Overall efficiency with heat pump %
Electricity – Delivered efficiency 30% from coal burn. Heat pump C O P 3·0 (equivalent to an appliance efficiency of 300%).				
Direct electric heating	100	85–95	25–28	75–84
Off-peak storage heater	100	65–90	19–27	58–81
Immersion water heater	100	65–75	19–22	38–44
Off-peak water heater	100	60–70	18–21	36–42
Instantaneous water heater	100	90–95	27–28	—
S N G – Delivered efficiency 70% from coal conversion. Heat pump C O P 1·2 (equivalent to an appliance efficiency of 120%).				
Direct gas heater	55–85	50–75	35–52	—
Central heating boilers:				
Winter – heating and water	70–85	55–75	38–52	66–73
Summer – water only:				
Lightweight boiler	75–80	35–50	24–35 }	31–52
Heavy cast iron boiler	70–80	25–35	17–24 }	
Instantaneous water heater	65–85	55–60	38–42	—
Coal – Delivered efficiency 100%.				
Open fire	25–45	20–35	20–35	—
Open fire and back boiler	30–60	25–45	25–45	—
Central heating boiler:				
Winter – heating and water	65–75	50–70	50–70	—
Summer – water only	65–70	20–30	20–30	—

Note: For manufactured smokeless fuel, the delivered and overall efficiencies from coal are variable, but can reach approximately 90% of those from coal used directly.

Source: *Department of Energy*

Table A1.2 *Annual energy requirement and cost for domestic premises for 42.2 useful G J per year*

Heating System	Delivered G J	Original coal burn in G J	Annual fuel cost at Dec 1980 prices, £
Electricity – Peak rate 4·0p/kWh = £11·11/G J. Off-peak rate 1·7p/kWh = £4·72/G J			
(1) Peak rate electricity	54·1–47·9	180·4–159·6	601–532
(2) Off-peak electricity	66·5–50·9	221·6–169·4	314–240
(3) Heat pump at peak rate C O P = 3·0	21·3–18·8	71·0– 62·7	237–209
(4) Heat pump off-peak C O P = 3·0	25·7–20·0	85·7– 66·7	121– 94
Coal – £74·10/tonne = 26p/therm = £2·46/G J. Smokeless fuels = 34p/therm = £3·20/G J			
(5) Central heating boiler with coal	103·4–72·3	103·4– 72·3	254–178
(6) Central heating boiler with naturally smokeless solid fuel	103·4–72·3	103·4– 72·3	331–231
(7) Central heating boiler with manufactured smokeless solid fuel	103·4–72·3	114·9– 80·3	331–231
S N G – at 42p/therm = £3·98/G J.			
Central heating boiler:			
(8) Lightweight boiler	82·2–60·5	118·9– 86·3	327–240
(9) Heavyweight boiler	90·5–65·8	129·3– 94·0	361–262
(10) Heat pump C O P = 1·2	51·8–43·1	74·0– 61·6	207–171
Alternative S N G prices.			
Case (8) at 36p/therm			280–205
Case (8) at 60p/therm			467–342
Case (9) at 36p/therm	Energy requirements as above		309–224
Case (9) at 60p/therm			515–374
Case (10) at 36p/therm			177–147
Case (10) at 60p/therm			295–244
C H P/D H			
(11) District heating from a C H P scheme	70–50	35–25 (in addition to the original coal burn)	450–200

Note: Of the 42·2 useful G J, 29·5 G J (70%) is for space heating, and 12·7 G J (30%) is for hot water. These figures are about average for a typical well-insulated modern house and are approximately equivalent to those used in Energy Papers Nos 20 and 35 in calculations relating to the costs of C H P and other forms of heating.

Source: *Department of Energy*

Appendix 2   Members of the commission
on energy and the environment

Chairman

Lord Flowers MA DSc FINST P FRS

Rector of Imperial College of Science and Technology
Former Chairman of the Royal Commission on
    Environmental Pollution

Members

R N Bottini Esq CBE

Part-time Member of the South Eastern Electricity Board
Retired General Secretary of the National Union of
    Agricultural and Allied Workers
Member of the Former Clean Air Council

Sir Henry Chilver DSc

Vice Chancellor, Cranfield Institute of Technology
Member of the Royal Commission on Environmental
    Pollution
Member of the Armitage Committee

Dr J G Collingwood BSc FENG FICHEME HON DSc
    FBIM

Retired Director and Head of Research Division, Unilever
    Ltd
Fellow of University College, London
Former Member of the Royal Commission on
    Environmental Pollution

A G Derbyshire Esq MA DUNIV (York) FRIBA FSIA

Partner in Robert Matthew Johnson-Marshall and Partners
    (Architects, Engineers and Planners)
Non-Executive Member of the Central Electricity
    Generating Board
External Professor of Architecture, Department of Civil
    Engineering, University of Leeds.

Professor Sir Richard Doll CBE DM DSc FRCP FRS

Warden of Green College, Oxford
Former Regius Professor of Medicine, University of Oxford
Former Member of the Royal Commission on
    Environmental Pollution

Professor Sir William Hawthorne CBE FRS

Master of Churchill College, University of Cambridge
Former Chairman of the Advisory Council on Energy
    Conservation

Professor F G T Holliday CBE FRSE

Vice Chancellor and Warden, University of Durham
Board Member of Shell U K
Trustee of the National Heritage Memorial Fund
Former Member of the Natural Environment Research
    Council
Former Chairman of the Nature Conservancy Council

G McGuire Esq OBE

Deputy National Secretary of the Youth Hostels Association
    (England and Wales)
Past President of the Ramblers Association
Vice Chairman of the Council for National Parks
Former Member of the Countryside Commission

Mrs Naomi E S McIntosh

Senior Commissioning Editor, Channel Four
Former Professor of Applied Social Research at the Open
  University
Member of the Advisory Council for Adult and Continuing
  Education
Member of the National Consumer Council
Former Chairman of the National Gas Consumers Council
Member of the former Energy Commission

Mrs Veronica Milligan BA C ENG MIEE MBIM

Partner C I V L E C Advisory Industrial Development
  Services
Member of the National Water Council
Member of the Industrial Tribunals Panel
Member of the Gwent Health Authority

Sir Austin Pearce CBE PhD

Chairman of British Aerospace
Member of the Advisory Council on Energy Conservation
Member of the former Energy Commission
Former Chairman of Esso Petroleum Co Ltd
Former Chairman of the U K Petroleum Industry
  Association Ltd

M V Posner Esq MA

Chairman of the Social Science Research Council
Fellow of Pembroke College, University of Cambridge
Member of the Advisory Council on Energy Conservation
Part-time Member of the British Railways Board

Sir Francis Tombs BSc FENG FI MECH E FIEE

Director of N M Rothschild and Sons Ltd
Chairman of the Weir Group Ltd
Former Chairman of the Electricity Council
Member of the Nature Conservancy Council
Director of Howden Group Ltd
Member of the former Energy Commission

D G T Williams Esq MA LLB

President of Wolfson College and Reader in Public Law,
  University of Cambridge
Member of the Royal Commission on Environmental
  Pollution
Member of the Council on Tribunals
Member of the former Clean Air Council

Appendix 3 Organisations and individuals who submitted evidence

Those marked * gave oral evidence at meetings of the Commission, usually following a written submission. Those who were involved in discussions during Commission visits are listed in Appendix 4.

A. Government and other organisations

Agricultural Research Council
*Association of County Councils
*Association of District Councils
*Association of Metropolitan Authorities

Barnsley Metropolitan Borough Council
*British Gas Corporation
British Petroleum
*British Railways Board
British Steel Corporation
British Waterways Board

*Central Electricity Generating Board
Chamber of Coal Traders
City of Manchester
City of Salford
Clean Air Council for Scotland
*Committee for Environmental Conservation
Confederation of British Industry
Conservation Society
*Convention of Scottish Local Authorities
*Council for the Principality
*Council for the Protection of Rural England
Countryside Commission
Countryside Commission for Scotland
Cumbria Association of Local Councils

*Department of Energy
*Department of the Environment
*Department of Transport
Dounreay Nuclear Power Development
    Establishment: Staff Associations and Trades
    Unions

Environmental Health Officers Association

Farmers Union of Wales
Federation of Civil Engineering Contractors
Fellowship of Engineering

Greater London Council Officers
Greater Manchester Council

*Health and Safety Executive (H M Alkali and Clean
    Air Inspectorate)

Imperial Chemical Industries Ltd
Institute of Energy
Institute of Petroleum
*Institute of Geological Sciences (Natural
    Environment Research Council)
Institute of Terrestial Ecology (Natural Environment
    Research Council)
International Youth Federation for Environmental
    Studies and Conservation
Institution of Geologists

Kincardine Campaign Committee

Low Temperature Coal Distillers Association of
    Great Britain Ltd

Manchester Area Council for Clean Air and Noise
    Control
Medical Research Council
Methley Environment Group
Mid-Glamorgan County Council
*Ministry of Agriculture, Fisheries and Food

*National Coal Board
National Society for Clean Air
National Farmers Union
*National Union of Mineworkers
National Water Council
Nature Conservancy Council

*Opencast Mining Intelligence Group

*Royal Society
Royal Society for the Promotion of Health
*Royal Town Planning Institute
Royal Town Planning Institute: Scottish Branch

Scottish River Purification Advisory Committee
Shell U K Limited
Strategic Conference of County Councils in Yorkshire
    and Humberside

University of Nottingham Coal Environment and
    Planning Research Programme

Wakefield Metropolitan District Council
Watt Committee on Energy
Wear Valley District Council
West Yorkshire Metropolitan County Council

B. Individuals

Dr I J Brown, formerly County Minerals Officer, West Yorkshire Metropolitan County Council
Mr I V Davies, Consultant Mining Engineer
Professor C T Shaw, Professor of Mining Engineering, Imperial College of Science and Technology, University of London
Dr D Spooner and Mr J North, Department of Geography, University of Hull

Professor Thring, Professor of Mechanical Engineering, Queen Mary College, University of London
Sir Ralph Verney, formerly Chairman of the Advisory Committee on Aggregates

A full set of evidence has been deposited with the Department of the Environment Headquarters Library, 2 Marsham Street, London SW1P 3EB and with the Department of Energy Library, Thames House South, Millbank, London SW1P 4QJ. Application to be made to the Chief Librarians for use of this material.

Appendix 4　Details of visits

The following visits were made by groups of Commissioners. Also listed are those who were involved in discussions with the Commission during the visits.

DAW MILL COLLIERY (8 March 1979)

National Coal Board

YORKSHIRE AND HUMBERSIDE COALFIELD (19–21 September 1979)

Barnsley Metropolitan Borough Council
British Association of Colliery Management (Yorkshire Region)
Central Electricity Generating Board
National Association of Colliery Overmen, Deputies and Shotfirers (Yorkshire Area)
Ministry of Agriculture, Fisheries and Food
National Coal Board
National Union of Miners (Yorkshire Area)
Nature Conservancy Council
Strategic Conference of County Councils in Yorkshire and Humberside including representatives of:
　　Humberside County Council
　　North Yorkshire County Council
　　South Yorkshire County Council
　　West Yorkshire County Council
Wakefield Metropolitan District Council.

CENTRAL ELECTRICITY RESEARCH LABORATORY (15 April 1980)

Central Electricity Generating Board
Department of the Environment
Health and Safety Executive (H M Alkali and Clean Air Inspectorate)
Royal Society
Warren Spring Laboratory

SELBY COLLIERY (1–2 May 1980)

National Coal Board

GREATER MANCHESTER (18–19 September 1981)

Antlers Ltd
Bury Borough Council
Confederation of British Industry
Department of Industry
East Lancashire Paper Mill Ltd
Greater Manchester Council
Health and Safety Executive
Manchester Area Council for Clean Air and Noise Control
National Coal Board
National Union of Mineworkers
Salford Borough Council
Trafford Borough Council
Ward and Goldstone Ltd

WEAR VALLEY, COUNTY DURHAM (22 September 1980)

Ministry of Agriculture, Fisheries and Food
N C B Opencast Executive
Wear Valley District Council

COAL RESEARCH ESTABLISHMENT (23 January 1981)

National Coal Board

SCOTTISH COALFIELD (12–13 February 1981)

British Gas Corporation
Fife Regional Council
N C B Opencast Executive
South of Scotland Electricity Board

SOUTH WALES COALFIELD (14–15 May 1981)

Council for the Principality including representatives of:
　　Islwyn Borough Council
　　Merthyr Tydfil Borough Council
　　Montgomery District Council
　　Newport Borough Council
　　Ogwr Borough Council
　　Rhondda Borough Council

Heads of the Valleys Standing Conference including representatives of:
　　Blaenau Gwent Borough Council
　　Cynon Valley Borough Council
　　Rhymney Valley District Council

National Association of Colliery Overmen, Deputies and Shotfirers (South Wales)
National Coal Board
National Union of Mineworkers

Standing Conference on Regional Policy in South
  Wales
The Wales Trades Union Congress
Welsh Counties Committee including representatives
  of:
  Clwyd County Council
  Dyfed County Council
  Gwent County Council

Mid-Glamorgan County Council
Powys County Council
West Glamorgan County Council

Welsh Confederation of British Industry
Welsh Development Agency
Welsh Water Authority

List of abbreviations

A C C	Association of County Councils		G W	Gigawatt
A C E C	Advisory Council on Energy Conservation		Ha	Hectares
			H S C	Health and Safety Commission
A D A S	Agricultural Development and Advisory Service, Ministry of Agriculture, Fisheries and Food		H S E	Health and Safety Executive
			I E A	International Energy Agency
			I C R P	International Commission for Radiological Protection
A D C	Association of District Councils		I G S	Institute of Geological Sciences
A M A	Association of Metropolitan Authorities		kj	kilojoule
			L A	Local Authority
A O N B	Area of Outstanding Natural Beauty		L A A	Local Authority Association
B A P	Benzo (a) pyrene		L N G	Liquefied Natural Gas
B G C	British Gas Corporation		L P G	Liquefied Petroleum Gas
Bpm	Best practicable means		L P A	Local Planning Authority
B R E	Building Research Establishment, Department of the Environment		m	metre
			M A F F	Ministry of Agriculture, Fisheries and Food
B S C	British Steel Corporation			
Btu	British thermal unit		mcfd	million cubic feet per day
C E G B	Central Electricity Generating Board		mg	milligram
C H E S S	Community Health Evaluation Surveillance System (U S A)		M G R	Merry-Go-Round
			M H D	Magnetohydrodynamic
C H P	Combined Heat and Power		mm	millimetre
C I S W O	Coal Industry Social Welfare Organisation		mtce	million tonnes of coal equivalent
			M W	Megawatt
CO_2	Carbon Dioxide		μg	microgram
Co En Co	Committee for Environmental Conservation		μgm^{-3}	microgram per cubic metre
			N A T O	North Atlantic Treaty Organisation
C O P	Co-efficient of performance		C C M S	Committee on the Challenges of Modern Society
C P R E	Council for the Preservation of Rural England			
			N C B	National Coal Board
C R E	Coal Research Establishment, National Coal Board		N C C	Nature Conservancy Council
			N E R C	Natural Environment Research Council
CoSLA	Convention of Scottish Local Authorities			
			N F U	National Farmers Union
D A F S	Department of Agriculture and Fisheries for Scotland		ng	nanogram
			NO	Nitric Oxide
dB(A)	Decibels measured on the A weighted scale		N_2O	Nitrous Oxide
			NO_2	Nitrogen dioxide
DH	District Heating		NO_X	Mixtures of nitrogen oxides, especially NO and NO_2
dwt	Dead weight tonnes			
E C E	United Nations Economic Commission for Europe		N R D C	National Research and Development Corporation
E E C	European Economic Community			
E T S U	Energy Technology Support Unit, Department of Energy		N R P B	National Radiological Protection Board
			N U M	National Union of Mineworkers
F B C	Fluidised bed combustion		O E C D	Organisation for Economic Co-operation and Development
F G D	Flue Gas Desulphurisation			
G D O	General Development Order		O M I G	Opencast Mining Intelligence Group
G D P	Gross Domestic Product		O P E C	Organisation of Oil Exporting Countries
G J	Gigajoules			

P A H s	Polycyclic or polynuclear aromatic hydrocarbons	SO₂	Sulphur Dioxide
P F A	Pulverised Fuel Ash	SO₃	Sulphur Trioxide
ppm	parts per million	SOₓ	Mixed sulphur oxides, SO₂ and SO₃
ppv	peak particle velocity	S C O C C	Strategic Conference of County Councils in Yorkshire and Humberside
P S A	Property Services Agency		
R C E P	Royal Commission on Environment Pollution	S D D	Scottish Development Department
		S I	Statutory Instrument
R F A C	Royal Fine Arts Commission	S S S I	Site of Special Scientific Interest
R I B A	Royal Institute of British Architects	T C P A	Town and Country Planning Act
R I C S	Royal Institute of Chartered Surveyors	T L V	Threshold Limit Value
		T O P S	Total Operations Processing System
R T P I	Royal Town Planning Institute	U K A E A	United Kingdom Atomic Energy Authority
R W P A	Regional Working Parties on Aggregates		
		W D A	Welsh Development Agency
S A R U	Systems Analysis Research Unit, Department of the Environment	W H O	World Health Organisation (United Nations)
S D D	Scottish Development Department	W O C O L	World Coal Study
S N G	Substitute Natural Gas		

Glossary of terms

AMBIENT	Used to describe air pollution concentrations in the open air as against the point of emission or indoors.
AMENITY	General term with no precise legal definition which takes account of all the factors contributing to the general quality of the environment and life in a locality.
ANGLE OF DRAW	The angle between the edge of the coal seam and the outer limit of ground movement caused by subsidence.
ANTHRACITE	A coal of high calorific value, low volatile matter, giving off little smoke.
ASH	The residue remaining after combustion whose origin is the mineral impurities contained in the fuel. Ash may also contain unburned fuel.
BACKSTOWING	The action of totally or partially filling up the waste spaces left underground by mining activity with waste material or specially quarried material to minimise subsidence.
BAGHOUSE	An installation which reduces the particulate content of gas streams by filtering them through large fabric bags.
BECQUEREL	The rate of formation of ionising radiation arising from the disintegration of one atom of a radioactive element per second. 1 becquerel = $2 \cdot 7 \times 10^{-11}$ curie. Becquerel is now the preferred term of measuring the activity of a radioactive material.
BEST PRACTICABLE MEANS (Bpm)	According to the Alkali Act 1906 a scheduled process must be provided with the best practicable means for preventing the escape of noxious or offensive gases and smoke, grit and dust to the atmosphere and for rendering such gases, where necessarily discharged, harmless and inoffensive. The "best practicable means" are often used to describe the whole approach of British anti-pollution legislation towards industrial emission, which takes account of the cost of pollution abatement and its effect on the viability of industry as well as the harmful effects pollution causes. The Alkali and Clean Air Inspectorate has the power to make bpm progressively more stringent as improved technology becomes available.
BITUMINOUS COAL	The most common form of British coal which contains or produces a relatively high proportion of volatile hydrocarbons. It gives rise to smoke when burned in the traditional domestic hearth.
BUNKER	A holder for storing solid material, such as coal and permitting it to be withdrawn in a controlled manner.
CAKING PROPERTIES	The properties manifest in some coals that, when heated up to certain temperatures the coal softens, becomes pasty and agglutinating, while the volatile matter, driven out through the sticky substance, causes it to become a cellular mass, yielding a coke or char on cooling. Coals that leave a powdery residue are non-caking; the more plastic a coal becomes on heating the stronger its caking power. The caking properties of a coal are important in the manufacture of coke and for combustion on certain types of grate.
CARBON DIOXIDE (CO_2)	Gas formed by the complete combustion of carbon. Also produced naturally by all organisms as a product of respiration. An excess of carbon dioxide can be suffocating in confined circumstances but is rarely found in hazardous concentrations in the open air.

231

CARBON MONOXIDE (C O)	Colourless, odourless highly toxic gas produced by the incomplete combustion of carbonaceous materials. It also arises from natural sources. It is absorbed in the blood in preference to oxygen with toxic effects. It is generally present only at low concentrations in the air. It was an important constituent of town gas but is not present in natural gas.
CARBONISATION	The heating of organic raw materials (usually coal) in the absence of air to obtain coke, crude gas and crude tar.
CARCINOGEN	A substance which increases the incidence of cancer in animals or man.
CAVING	Collapse of roof strata into void left by underground mining of coal.
COAL CONCENTRATION DEPOT	A trans-shipment point for coal, served in bulk by rail, for onward despatch by road to smaller industrial and domestic consumers.
COAL PREPARATION PLANT	An installation associated with collieries in which coal as mined undergoes preparation processes such as grading, cleaning, grinding, crushing, dewatering and blending to make it suitable for specific applications.
COKE	Smokeless solid fuel obtained from certain types of coal by heating in the absence of air to drive off volatile matter. See also CAKING PROPERTIES.
COMBUSTION	The chemical combination of oxygen with fuel with heat evolution so that the temperature rises. Burning.
CURIE	The rate of formation of ionising radiation of one gramme of radium in equilibrium with its decay products (which are themselves radioactive). 1 curie = 3.7×10^{10} becquerel. The preferred term is now BECQUEREL.
CYCLONE	Equipment for removing the larger particles from a gas by a combination of centrifugal and gravitational forces.
DEEP MINING	Mining in which access to the mineral deposits is obtained by means of shafts and underground workings.
DEPOSITION	The term dry deposition covers the interception and retention by the ground, vegetation, buildings etc of gases or particles reaching these surfaces by diffusion, gravitational settling or wind movement when it is not raining. In contrast wet deposition is used to describe the bringing of materials to the ground by rain.
DERELICT LAND	Generally defined as land so damaged by industrial or other development that it is incapable of beneficial use without treatment.
DUST AND GRIT	Particles or solids in the air or flue gases which fall out or deposit under their own weight. British Standard B S 3405 categorises particles between 1 to 75 μm in diameter as dust, and those above 75 μm in diameter as grit. Particles other than smoke which are less than 1 μm in diameter are commonly called fume.
EFFLUENT	Literally anything that flows out or is discharged. Usually applied to the discharge of waste material into a stream, sewer or to the air.
ELECTROSTATIC PRECIPITATOR	Equipment for removing particles from a gas by giving them an electrical charge and causing them to adhere to a plate having an opposite electrical charge.
EMISSION	The process of discharging into the atmosphere. Also applied to the material being discharged.
EMPHYSEMA	Pulmonary or lung emphysema is a swelling of the air sacs of the lungs with the sac walls later breaking down and so diminishing the areas for exchanging oxygen and carbon dioxide between the air and the blood.
ENVIRONMENTAL IMPACT ANALYSIS (E I A)	The identification of those features of the environment most likely to be affected by a project and their subsequent in-depth analysis which enables the impact of the project to be fully assessed. "Environmental Impact Statement" (E I S) refers to the statements produced by those agencies in the United States who are required to comply with the terms of the Environmental Protection Act 1969.

EXPLOITATION	All the operations involved in extracting the coal from the deposit, raising it to the surface and making it available for disposal.
EXPLORATION	Investigation to determine the extent of a coal deposit, its value and the potential for its exploitation.
DESULPHURISATION	Literally the removal of sulphur. This may be done either before the fuel is burned or afterwards (flue gas desulphurisation). Desulphurisation during combustion is possible in some conditions.
DEVELOPMENT	Activity requiring planning permission under the Town and Country Acts including the "carrying out of building, engineering, mining, or other operations in, on, or under land".
DEVELOPMENT CONTROL	Part of the Town and Country Planning System whereby all development, with certain exceptions requires planning permission.
DEVELOPMENT PLAN	In 1968 a new type of development plan system consisting of "structure" and "local" plans was introduced. Structure plans are prepared by county planning authorities (in Scotland by regional planning authorities) for approval by the appropriate Secretary of State. Local planning authorities may subsequently prepare local plans.
DISTILLATE FUEL	The term distillate fuel is used to differentiate the lighter and refined petroleum fuels – petrol, paraffin (kerosene), diesel oil and gas oil – from the less refined or residual fuel oils which retain the majority of the sulphur and other impurities. See also FUEL OIL.
DOSE	Amount taken or received. The dose rate denotes the amount taken per unit of time. In the assessment of pollution effects an important factor is the dose response relationship for any exposed individual or organism to any specified dose.
DRIFT MINING	A method of underground mining in which access to the coal seam is obtained by means of an inclined tunnel (drift) rather than by a vertical shaft. Where access is obtained by means of a level tunnel the term adit mining is preferred.
FACE	Place where the coal is being won from the seam. In longwall mining coal is cut or otherwise extracted from a generally straight wall of coal so that the face advances in parallel steps.
FIREDAMP	A gas, lighter than air and composed essentially of methane, released from coal seams and coal/rock interfaces, either continuously or sporadically.
FLUE GAS	A mixture of gas produced by combustion. Typically this mixture consists largely of carbon dioxide, water vapour, nitrogen and unused oxygen plus small amounts of sulphur dioxide, nitrogen oxides, carbon monoxide, particulates and unburnt fuel vapour.
FLUIDISED BED COMBUSTION (F B C)	A method of burning a fuel in which a bed of non-combustible particles containing the fuel is maintained in a state of suspension by the upward flow of the combustion air through the bed. The non-combustible particules are generally coal ash or a sulphur-acceptor such as limestone.
FLY ASH	Fine particles of ash carried by the flue gases arising from the combustion of pulverised coal.
FRIABILITY	Tendency to crumble.
FUEL OIL (ALSO HEAVY FUEL OIL, RESIDUAL FUEL OIL)	A term generally applied to the black oil obtained as unvaporised liquid from the distillation of crude oil, or the blends of such residues with other refinery streams. Fuel oils are normally denser and more viscous than distillate fuels such as gas oil and require pre-heating before combustion. They also usually contain more sulphur than the lighter oils. They produce a small amount of ash on combustion. See also DISTILLATE FUEL.
FUME	See DUST and GRIT.

FURNACES	Under the terms of the Clean Air Act, this means the part of any equipment designed for the combustion of fuel. It includes boilers, kilns and dryers as well as many types of device conventionally known as furnaces.
GASIFICATION	A process of manufacturing fuel gases by reacting solid fuel with a gasification medium such as air oxygen and/or steam.
GENERAL DEVELOPMENT ORDER	Statutory instrument giving deemed planning permission for certain types of development.
GREENFIELD SITE	An area usually in agricultural or open use, not yet developed for mining or industrial purposes.
GRIT	See DUST.
HALF LIFE	The half life is the time required for half the material initially present to decay. This is constant for radioactive elements so after half has decayed, the same time, one half life, is required for half of the remaining half (one quarter of the original) to decay and so on. In principle the material is never used up entirely.
INVERSION	In an inversion the normal temperature gradient in the atmosphere (decreasing temperature with height) is inverted at some height. This marks the base of an inversion layer which forms an effective barrier to the natural vertical diffusion of gases.
LAGOONS	Settlement areas used for the treatment of liquid waste from coal preparation plants.
LANDSALES	Sales of coal direct from the pit.
LEQ	A measure of the average noise level from a site in energy terms over a specified period.
LIGNITE	A lower grade of coal with lower calorific value than bituminous coal. Also known as brown coal.
LIQUEFIED NATURAL GAS	Produced by cooling natural gas to $-162°C$ to reduce its volume. L N G has to be stored and transported in heavily insulated vessels of special construction.
LIQUEFIED PETROLEUM GAS	Propane or butane gas, or a mixture of the two, liquefied by the application of pressure.
LOCAL PLAN	See DEVELOPMENT PLAN.
LONGWALL MINING	A method of mining in which coal is removed from the seam in one operation along a long working face or wall so that the face moves in a continuous line that may extend for several hundred metres.
LURGI	A process for producing gas from coal originated by the German Lurgi Company in the late 1920s.
MORBIDITY	The incidence of disease in a community or population.
MUTAGEN	A chemical which changes the genetic structure of living organisms giving rise to mutations.
NITROGEN OXIDES (N O$_x$)	Nitric oxide (N O) is a colourless toxic gas. It arises from the combination of nitrogen from the atmosphere or from coal with oxygen during combustion. It is readily oxidised in the air to produce *nitrogen dioxide* (N O$_2$) a stable brown gas which is more toxic than N O. In the atmosphere it is eventually converted to nitric acid. It may also be formed in industrial purposes such as nitric acid manufacture. *Nitrous oxide* (N $_2$ O) arises naturally from the rotting of vegetable matter and also from nitrogenous fertilisers. It is more stable than nitrogen dioxide and nitric oxide.
OPENCAST MINING	Mining of deposits of one or more seams thickness from the surface after removal of the rock above.
PARTICULATES	Small solid particles of material found in the atmosphere in addition to gases. Includes dust, grit, smoke and fume.

PNEUMOCONIOSIS	A term applied to various diseases of the lung caused by the inhalation of dust particles in such occupations as coal mining, quarrying and asbestos working; an excess of fibrous tissue forms in the lungs round the dust particles.
POLLUTER PAYS PRINCIPLE	A principle of cost allocation which means the polluter should bear the expense of carrying out measures decided by public authorities to ensure that the environment is maintained in an acceptable state.
POLLUTION	The introduction by man of substances or energy into the environment resulting in deleterious effects of such a nature as to endanger human health, harm living resources and ecosystems and material property and also to impair or interfere with amenities and other legitimate uses of the environment.
PULVERISED FUEL ASH	The ash residue resulting from combustion of pulverised coal in power stations. It includes a high proportion of fly ash.
POLYNUCLEAR AROMATIC HYDROCARBONS	Also known as polycyclic aromatic hydrocarbons. These are hydrocarbon compounds, arising from imperfect combustion, some of which are carcinogenic ie capable of causing cancer.
PROGRESSIVE RESTORATION	Method of tipping colliery spoil by which one part of a tip is restored as spoil is deposited on another part thus minimising the amount of land taken out of use for tipping at any one time.
PYROLYSIS	Breaking down a substance by the application of heat. Pyrolysis of coal produces gas, tar and coke.
RECLAMATION	Rehabilitation of land after opencast operations, spoil tipping or other activity which renders it incapable of reasonable beneficial use.
RUN-OF-MINE COAL	Coal immediately after extraction before any subsequent processing.
SCRUBBING	A cleaning operation in which gaseous or fine particulate pollutants are removed from a stream of gas by contact with a liquid.
SIEVERT	This is the unit of absorbed dose equivalent for ionising radiation. Different types of radiation have different degrees of biological effect, eg alpha particles are more damaging than beta particles or gamma rays. A common scale of equivalent effects needed to carry out quantitative studies of radiation effects is obtained by multiplying the absorbed dose in joules per kilogram (= grays) by a quality factor 1 sievert = 1 gray × quality factor. (The unit of dose equivalent formerly used was called the rem. 1 sievert = 100 rem)
SLAG	Molten mineral matter, often coal ash which resolidifies on cooling.
SMOKE	Small gas-borne solid and/or liquid particles formed by incomplete combustion of fuel below 1 μm in size.
SMOKELESS FUELS	Solid fuel whose natural properties, or whose properties resulting from treatment, are such that when burned the fuel emits only limited quantities of visible solid or liquid substances (ash, soot, tar) in the gases. Examples are Phurnacite, Coalite and Rexco.
SOOT	Aggregates of carbonaceous particles impregnated with tar formed by the incomplete combustion of carbonaceous material, particularly bituminous coal. They tend to adhere to the inside of the chimney.
SPOIL	Waste material produced in deep mining and subsequent preparation of coal for sale which includes a variety of material ranging in size from boulders to cobbles and clay particles.
STRAIN	Compression or extension of ground surface caused by subsidence.
SULPHUR OXIDES (S O$_x$)	Sulphur dioxide (S O$_2$) is a colourless gas with a choking taste and is the main product of the combustion of sulphur contained in fuels. Globally much of the sulphur dioxide in the atmosphere comes from natural sources, but in industrial regions the greater part of sulphur dioxide concentration comes from combustion of sulphur-containing fossil fuels. Sulphur dioxide oxidises in the air to form sulphur trioxide (S O$_3$).

235

STRUCTURE PLAN	See DEVELOPMENT PLAN.
SYNERGISTIC	The phenomenon in which the effect produced by two causes together is greater than the sum of the effects produced by the causes separately.
STALL AND PILLAR MINING	Traditional method of mining now largely superceded by longwall techniques in the U K.
TAILINGS	Those residues from the washing of solid fuel that are too poor in useful fuel to undergo further treatment and normally discarded. These consist of very fine particles suspended in water and are conventionally treated by settlement in lagoons.
TELECHIR	A machine closely controlled by an operator located some distance away from the work. Such devices are largely hypothetical at present. Remote handling devices used for manipulating dangerous radioactive materials are probably the best known examples of the limited telechiric devices currently in existence.
TERATOGENIC	Causing abnormal development of a foetus.
THRESHOLD LIMIT VALUES (T L V)	Threshold limit values relate to airborne concentrations of potentially harmful substances and represent conditions under which it is believed that nearly all workers may be repeatedly exposed day after day without adverse effect.
TRACE ELEMENTS	Elements, of which only trace quantities are to be found in materials or organisms.
VOID (OR GOAF)	Space behind the advancing coal face left by the mining of coal and ancillary works.

236

Conversion factors

WEIGHT	1 kilogramme (kg) = 2·2046 lb 1 pound (lb) = 0·4536 kg
	1 tonne (t) = 1,000 kg = 0·9842 long ton = 1·102 short tons (sh tn)
	1 Statute or long ton = 2,240 lb = 1·016 t = 1·120 sh tn
VOLUME	1 m³ = 35·31 ft³ 1 ft³ = 0·02832 m³
	1 litre = 1·760 U K pint 1 pint (U K pt) = 0·5682 litre
	1 Imperial gallon (U K gal) = 8 U K pt = 1·201 U S gallons (U S gal) = 4·54609 litres
	1 U S gal = 0·8327 U K gal = 6·661 U K pt = 3·785 litres
	1 barrel = 42 U S gal = 34·97 U K gal = 159·0 litres
	1 litre = 1,000 cu. centimetres = 0·22 U K gal
HEAT	1 British thermal unit (Btu) = 0·252 kilocalorie (kcal) = 1·05506 kilojoule (kJ)
	1 kcal = 4·1868 kJ = 3·9683 Btu
	1 therm = 100,000 Btu = 25,200 kcal = 105,506 kJ
	1 million therms = 25·2 Tcal = 105·506 terajoule (T J)
	1 megacalorie (Mcal) = 3,968·3 Btu = 4,186·8 kJ
	1 teracalorie (Tcal) = 39,683 therms = 4·187 T J
	1 thermie (th) = 4,185·5 kJ
	1 terajoule (T J) = 9,478·1 therms
CALORIFIC VALUE	1 kcal/kg = 1·8 Btu/lb 1 Btu/lb = 0·5556 kcal/kg
	1 kcal/m³ = 0·1124 Btu/ft³ 1 Btu/ft³ = 8·898 kcal/m³
ENERGY	1 kilowatt hour (kWh) = 1,000 watt hours 1 hp h = 1,980,000 ft lb = 0·7457 kWh
HEAT/ENERGY	1 hp h = 2,545 Btu = 641·3 kcal = 2,685 kJ
	1 kWh = 3,412 Btu = 859·845 kcal = 3,600 kJ
	1 Btu = 778 ft lb
	1 therm = 29·3 kWh
	1 GWh = 34,121 therms
POWER	1 hp = 550 ft lb/sec = 0·7457 kW
	1 kW = 1·34 hp
	1 megawatt (MW) = 1,000 kW

METRIC MULTIPLIERS	*Prefix*	*Sym-bol*	*Factor by which the unit is multiplied*
	tera	T	10^{12} = 1,000,000,000,000
	giga	G	10^{9} = 1,000,000,000
	mega	M	10^{6} = 1,000,000
	kilo	k	10^{3} = 1,000
	hecto	h	10^{2} = 100
	deca	da	10^{1} = 10
	deci	d	10^{-1} = 0·1
	centi	c	10^{-2} = 0·01
	milli	m	10^{-3} = 0·001
	micro	μ	10^{-6} = 0·000,001
	nano	n	10^{-9} = 0·000,000,001

TEMPERATURE	1 scale degree Centigrade (C) = 1·8 scale degrees Fahrenheit (F)
	For conversion of temperatures: °C = 1 (°F − 32): °F = 1°C + 32
	(C = Centigrade F = Fahrenheit)

References: Works cited and further reading

ADVISORY COMMITTEE ON AGGREGATES, 1976: Aggregates: The Way Ahead. Report . . . (Chairman) Sir Ralph Verney. HMSO, 1976, 118pp, ISBN 0117509175. Prepared for the Department of the Environment, Scottish Development Department and the Welsh Office.

ALLEN V L, 1981: Why the Miners Turned. New Society 55 (954) 26 February 1981, pp368–370.

ASHENDEN T W and MANSFIELD T A, 1978: Extreme Pollution Sensitivity of Grasses when SO_2 and NO_2 are present together in the atmosphere. Nature, *273* (5658) 11 May 1978, pp142–143.

ASSOCIATION OF METROPOLITAN AUTHORITIES, 1980: Study Group on the Interface between Road and Rail Freight. Road and Rail Freight: a study group report. 1980 [19]pp.

ASSOCIATION OF METROPOLITAN AUTHORITIES, and others 1980: Opencast Coal Mining Operations.

BARNES R A, 1978: The Long Range (Transboundary) Transport of Air Pollution: A Synthesis Report prepared for the E C E Senior Advisers on Environmental Problems. Journal of Air Pollution Control Association *29*, 1978, pp1219–1235.

BERKOVITCH I, 1977: Coal on the Switchback; the Coal Industry Since Nationalisation. Allen and Unwin. 237pp. ISBN 0046220038.

BERKOWITZ N, 1979: An introduction to coal technology. New York Academic Press, 345pp ISBN 0120919508 (Fig. 10.3.1 (b) and (c) p202).

BRITISH RAILWAYS BOARD, 1976: Safety, prepared by Environmental and Social Impact Studies (ENOSIS) for B R Strategic Planning Studies. B R, 28 leaves (Topic Report 1).

BUILDING RESEARCH ESTABLISHMENT, 1979: Noise from Opencast Mining Sites by W A UTLEY [BRE] 1979.

CABINET OFFICE, 1980: Climatic Change: Its Potential Effects on the United Kingdom and the Implications for Research. Report of the Interdepartmental Group on Climatology. Chairman Sir Kenneth Berrill, HMSO, 20pp, ISBN 0116308133.

CAMPLIN W C, 1980: Coal Fired Power Stations – the Radiological impact of effluent discharges to atmosphere. National Radiological Protection Board. HMSO 1980, 48pp (Reports R107) ISBN 0859511383.

CEDERLÖF R and others (editors) 1978: Air Pollution and Cancer: Risk Assessment Methodology and Epidemiological Evidence. Report of a Task Group. Environmental Health Perspectives 22, 1978, pp1–12.

C E G B, 1977: C E G B Research Number 5, August 1977, p6.

CENTRAL OFFICE OF INFORMATION, 1978: Environmental Planning in Britain. HMSO (R6037/78)

COAL, BRIDGE TO THE FUTURE, 1980: Report of the World Coal Study, Carroll L Wilson, Project Director, Cambridge (Mass): Ballinger Publishing Co, 247pp, ISBN 0884100995.

COAL MINING 1945: Report of the Technical Advisory Committee. Chairman C C Reid. (Cmd 6610).

COMMITTEE FOR ENVIRONMENTAL CONSERVATION, 1979: Scar on the landscape? A Report on Opencast Coal Mining and the Environment. 40pp ISBN 0903158159

CORBETT J O, 1980: The enhancement of natural radiation dosage by coal fired power generation in the United Kingdom. C E G B, 1980 (RD/B/N4760).

COUNCIL ON ENVIRONMENTAL QUALITY, 1981: Global Energy Futures and the Carbon Dioxide Problem. Washington D C. The Council, 1981.

CROOME D J, 1978: Environmental Engineering Aspects of Coalmines. Building Services Engineer 45 (10) January 1978, pp175–185.

CULLINGWORTH J B, 1976: Town and Country Planning in Britain. 6th Edition. Allen and Unwin, 287pp (The new local government series; No 8) ISBN 004352060X.

DEPARTMENT OF ENERGY, 1977a: Coal for the Future, Progress with "Plan for Coal" and Prospects to the Year 2000, 23pp.

DEPARTMENT OF ENERGY, 1977b: Coal in the U K. 19pp (Fact Sheet No 4).

DEPARTMENT OF ENERGY, 1977c: District Heating Working Party: District Heating Combined with Electricity Generation in the U K. HMSO, 120pp (Energy Paper 20) ISBN 0114106037.

DEPARTMENT OF ENERGY, ADVISORY COUNCIL ON ENERGY CONSERVATION, 1977d: Freight Transport: Short and Medium Term Considerations HMSO, 21pp (Energy Paper No 24; Advisory Council on Energy Conservation Paper 6) ISBN 011410610X.

DEPARTMENT OF ENERGY, 1978: Energy Policy: a Consultative Document HMSO, 127pp (Cmnd 7101) ISBN 0101710100.

DEPARTMENT OF ENERGY, 1979a: Heat Load Density Working Party, C H P Group: Heat Loads in British Cities HMSO, 32pp (Energy Paper No 34) ISBN 0114107602.

DEPARTMENT OF ENERGY 1979b: Energy Projections: a paper 17pp.

DEPARTMENT OF ENERGY, 1979c: C H P Group: Combined Heat and Electrical Power Generation in the U K. HMSO, 82pp (Energy Paper No 35) ISBN 0114107610.

DEPARTMENT OF ENERGY, 1980: Digest of U K Energy Statistics. HMSO, 140pp, ISBN 0114107971.

DEPARTMENT OF THE ENVIRONMENT, 1973: Towards Cleaner Air, a review of Britain's Achievements. HMSO, 36pp ISBN 0117505188.

DEPARTMENT OF THE ENVIRONMENT, 1974: Clean Air Today. HMSO, 44 pp, ISBN 0117507180.

DEPARTMENT OF THE ENVIRONMENT, PLANNING, REGIONAL AND COUNTRYSIDE DIRECTORATE, 1975a: Survey of Derelict and Despoiled Land in England 1974 (D O E PRM3/801/252).

DEPARTMENT OF THE ENVIRONMENT, 1975b: Review of the Development Control System. Final Report (by G Dobry). HMSO, 248pp, ISBN 0117508969.

DEPARTMENT OF THE ENVIRONMENT, CENTRAL UNIT ON ENVIRONMENTAL POLLUTION, 1976a: Pollution Paper No 9: Pollution Control in Great Britain – How it works: a review of legislative and administrative procedures. HMSO, 140pp, ISBN 0117511234.

DEPARTMENT OF THE ENVIRONMENT AND OTHERS, 1976b: Report of the Committee on Planning Control over Mineral Working (Chairman Sir Roger Stevens) HMSO, ISBN 0117508985*.

DEPARTMENT OF THE ENVIRONMENT, CENTRAL UNIT ON ENVIRONMENTAL POLLUTION, 1977: Pollution Paper No 11: Environmental Standards: A description of U K Practice: report of an inter-departmental working party. HMSO, 30pp, ISBN 0117511528.

DEPARTMENT OF THE ENVIRONMENT, MINERALS PLANNING DIVISION 1979a: Consultation Paper: Proposals for a Change to the General Development Order 1977 in respect of Mining Operations.

DEPARTMENT OF THE ENVIRONMENT, 1979b: Digest of Environmental Pollution Statistics, 2. HMSO, 116pp, ISBN 0117514330.

DEPARTMENT OF THE ENVIRONMENT, MINERALS PLANNING DIVISION, 1979c: The Preparation of Guidelines for Aggregates in England and Wales.

DEPARTMENT OF THE ENVIRONMENT, CENTRAL DIRECTORATE ON ENVIRONMENTAL POLLUTION, 1979d: Glossary of Air Pollution Terms. Pollution Report No 5, ISBN 0117514322.

DEPARTMENT OF THE ENVIRONMENT, 1980: Digest of Environmental Pollution Statistics, 3. HMSO, 106pp, ISBN 0117515019.

DEPARTMENT OF THE ENVIRONMENT AND WELSH OFFICE, 1981: Clean Air Procedure for Domestic Smoke Control. E C Directive on Sulphur Dioxide and Suspended Particulates. Circular 11/81 Department of the Environment. Circular 18/81 Welsh Office. ISBN 0117515167.

DEPARTMENTS OF THE ENVIRONMENT AND TRANSPORT, SYSTEMS ANALYSIS RESEARCH UNIT, 1980a: Current Geographical Distribution of Sulphur Emissions. C E N E Report No 1, April 1980.

DEPARTMENTS OF THE ENVIRONMENT AND TRANSPORT, SYSTEMS ANALYSIS RESEARCH UNIT, 1980b: Sulphur Emissions in the year 2000 Based on Department of Energy Projections. Supplement to C E N E Report No 1, April 1980.

DEPARTMENTS OF THE ENVIRONMENT AND TRANSPORT, SYSTEMS ANALYSIS RESEARCH UNIT, 1980c: Distribution of Sulphur Dioxide Concentrations in Great Britain. Current and Forecast. C E N E Report No 2, July 1980.

DEPARTMENTS OF THE ENVIRONMENT AND TRANSPORT, SYSTEMS ANALYSIS RESEARCH UNIT 1980d: Current and Forecast Impacts of Ambient Sulphur Dioxide on Population and Agricultural Targets in Great Britain. C E N E Report No 3, October 1980.

DEPARTMENTS OF THE ENVIRONMENT AND TRANSPORT, SYSTEMS ANALYSIS RESEARCH UNIT, 1980e: Further Forecast Impacts of Sulphur Dioxide on Population and Agricultural Targets in Great Britain. Supplement to C E N E Report No 3, November 1980.

DEPARTMENT OF TRANSPORT, 1978: Report of the Review of Highway Inquiry Procedures. HMSO, 16pp (Cmnd 7133) ISBN 0101713304.

DEPARTMENT OF TRANSPORT, 1980: Report of the Inquiry into Lorries, People and the Environment. Dec 1980, presented by Sir Arthur Armitage HMSO, 160pp, ISBN 0115505369.

DUNN R B, 1979: Coal in the U K. Mining Engineer. 138 (209) Feb 1979, pp527–533.

DURHAM COUNTY COUNCIL PLANNING DEPARTMENT, 1979: Reclamation in Durham.

ELLISON J McK and WALLER R E, 1978: A review of sulphur oxides and particulate matter as air pollutants with particular reference to effects on health in the United Kingdom. Environmental Research, 16, 1978, pp302–325.

ENERGY AND THE ENVIRONMENT, 1974: Report of a working party set up jointly by the Committee for Environmental Conservation, Royal Society of Arts, Institute for Fuel. [Royal Society of Arts] [1974] 45pp, ISBN 0901469025.

EUROPEAN COMMUNITIES, 1979: Council Directive of 17 December 1979 on the protection of ground water against pollution caused by certain dangerous substances (DIR 80/68). Official Journal of the European Communities L20, 26 Jan 1980, pp43–48.

EUROPEAN COMMUNITIES, 1980: Council Directive of 15 July 1980 on air quality limit values for sulphur dioxide and suspended particulates (DIR 80/779) Official Journal of the European Communities L229, 30 Aug 1980, pp30–48.

EUROPEAN COMMUNITIES COMMISSION, 1979: Summary of the study on measures taken by Member States to Utilise Mining Wastes. Luxembourg: European Commission [Eur 6634 EN].

EZRA Sir D, 1980: Coal and Energy: The need to exploit the world's most abundant fuel. 2nd edition, Benn, 1980, ISBN 0510000886.

FERRIS G F, 1978: Health effects of exposure to low levels of regulated air pollutants: a critical review. Journal of Air Pollution Control Association. 28, 1978, pp 482–497.

GORE A T and SHADDICK C W, 1958: Atmospheric pollution and mortality in the county of London. British Journal of Preventive and Social Medicine, 12, 1958, pp 104–113.

GRAINGER L, 1980: Coal Based Chemical Complexes. Philosophical Transactions of the Royal Society London, Series A, Vol 300, 1981, pp193–204.

GRIFFIN A R, 1977: The British Coalmining Industry: Retrospect and Prospect. Buxton: Moorland Publishing Co., 224pp, ISBN 0903485419.

HAMMOND E C and GARFINKEL L, 1980: General Air Pollution and Cancer in the United States. Preventative Medicine, 9, 1980, pp206–211.

HARLAND D G and ABBOTT P G, 1977: Noise and Road Traffic Outside Homes in England, Crowthorne Transport and Road Research Laboratory, 28pp (TRRL Report 770) ISSN 03051293.

HEALTH AND SAFETY COMMISSION, 1978: Report . . . . On The Hazards of Conventional Sources of Energy. HMSO, 30pp, ISBN 0118830694.

HEALTH AND SAFETY COMMISSION, 1980: Report 1979–80. HMSO, 54pp, ISBN 0118832751.

HEALTH AND SAFETY EXECUTIVE HM ALKALI AND CLEAN AIR INSPECTORATE: Health and Safety: Industrial air pollution 1975 + HMSO (annual reports).

HEALTH AND SAFETY EXECUTIVE HM INSPECTORATE OF MINES AND QUARRIES, 1979: Health and Safety: Mines. HMSO, 1980, 52pp, ISBN 0118832727.

HOUSE OF LORDS SELECT COMMITTEE ON THE EUROPEAN COMMUNITIES, 1981: Environmental Assessment of Projects . . . . . . with minutes of evidence. HMSO, 166pp, [H/L paper 69 session 1980/81] ISBN 0104069813.

INSTITUTE OF GEOLOGICAL SCIENCES, 1981: U K Mineral Statistics, 1980. HMSO, 143pp, ISBN 0118841645.

LAWTHER P J and WALLER R E 1978: Trends in urban air pollution in the U K in relation to lung cancer mortality. Environmental Health Perspective, 22, 1978, pp71–73.

LEACH G and others, 1979: A low energy strategy for the United Kingdom. The International Institute for the Environment and Development and Science Reviews Ltd., 259pp, ISBN 0905927206.

LIKENS G E and others, 1979: Acid Rain, Scientific American Vol 241 (4) Oct 1979, pp 39–47.

MASON Sir J, 1979: The meteorological effects of increasing the carbon dioxide content of the atmosphere. In Environmental Effects of Using More Coal. Proceedings of a Conference organised by the Council For Environmental Science and Engineering. London 11 and 12 December 1979. Edited by F A Robinson. Royal Society of Chemistry London (Special Publications No 37) 204pp, ISBN 0851868053.

McINNES G, 1980: Multi-element and sulphate in Particulate Surveys: Summary of Third Years Results. Warren Spring Laboratory. (LR 356 AP), ISBN 0856242098.

MARTIN A E and BRADLEY W, 1960: Mortality, Fog and atmospheric pollution. Monthly Bulletin of the Ministry of Health and the Public Health Laboratory Service. 19, May 1960, pp 56–73.

MARTIN A E, 1964: Mortality and morbidity statistics and air pollution: an investigation during the winter 1958/9. Proceedings Royal Society of Medicine. 57, 1964, pp969–975.

MINISTRY OF FUEL AND POWER, 1944: The efficient use of fuel. HMSO, 807pp.

MINISTRY OF HEALTH, 1954: Mortality and morbidity during the London Fog of December 1952 report . . . HMSO (Public Health and Medical subjects report No 95) (32–195–95).

MINISTRY OF HOUSING AND LOCAL GOVERNMENT, 1967: Chimney Heights: Second edition of the Clean Air Act memorandum, by the Ministry of Housing and Local Government, Scottish Development Department and the Welsh Office. HMSO, 10pp (75–115–0–67).

MINISTRY OF HOUSING AND LOCAL GOVERNMENT, 1970: Protection of the Environment. The Fight Against Pollution. HMSO, 32pp (Cmnd 4373) SBN 10437307.

MINISTRY OF POWER, 1965: Fuel Policy. (Cmnd 2798) HMSO.

MINISTRY OF POWER, 1967: Fuel Policy. (Cmnd 3438) HMSO.

MOSES K, 1980: Britain's Coal Resources and Reserves – the Current Position. In Assessment of Energy Resources. Watt Committee Report No 9. Watt Committee. 1981, pp39–50.

MYERS G K 1978: Cancers and Genetic Defects Resulting from the Use of Various Energy Sources. Biology and Research Physics Division, Chalk River Nuclear Laboratories U S A. June 1978, 66pp (PECL-6084).

N A T O Committee on the Challenges of Modern Society; 1980: Flue Gas Desulphurisation Pilot Study: Phase 2 Applicability Study. Study Report No 97. U S Environmental Protection Agency (EDT 114).

N C B, 1970: Spoil Heaps and Lagoons. (Technical Handbook) 232pp.

N C B 1974: Plan for Coal, 20pp.

N C B 1975: Subsidence Engineers Handbook 2nd (rev) ed 111pp, ISBN 0901429732.

N C B, 1978: Liquid Fuels from Coal, 70pp.

N C B, 1980: Fluidised Bed Combustion of Coal, 44pp.

N C B, 1981a: Report and Accounts 1980/1. 104pp, ISBN 0906394066.

N C B, 1981b: Solid Fuel Heating Plant for Industry and Commerce. ISBN 0906394058.

NOISE ADVISORY COUNCIL, 1978: Noise Implications of the Transfer of Freight from Road to Rail. Chairman D B Harrison, Working Group on noise from surface Transport. HMSO, 16pp, ISBN 0117513652.

NORTH J and SPOONER D J, 1978: The Geography of the Coal Industry in the U K in the 1970s: Changing Directions. Geojournal, 2.3.1978, pp255–272.

O E C D, 1975: The Polluter Pays Principle: Definition Analysis and Implementation. 118pp, ISBN 9264113371.

O E C D, 1977: The O E C D Programme on Long Range Transport of Air Pollutants. Measurements and Findings. O E C D Paris.

O E C D, 1981: The Cost and Benefits of Sulphur Oxide: A Methodological Study. ISBN 926421251X.

ORWELL G, 1937: The Road to Wigan Pier. Penguin 1976.

RATCLIFFE D (Editor) 1977: The Nature Conservation Review: selection of biological sites of national importance to nature conservation in Britain. Cambridge, Cambridge University Press, 1977, 2 vols, ISBN 052121159X (Vol 1); ISBN 0521214033 (Vol 2).

REID G L and Others, 1973: The nationalised fuel industries, Heinemann Educational. 264pp, ISBN 0435847759.

ROBINSON C and MARSHALL E, 1981: What Future for British Coal? Optimism or Realism on prospects to the year 2000. Institute of Economic Affairs. 104pp (Hobart Paper No 89) ISBN 0255361432.

ROGOT E and MURRAY J, 1980: Cancer Mortality among non-smokers in an insured group of U S veterans. Journal National Cancer Institute 65, 1980, pp1163–1168.

ROSE C I and HAWKSWORTH D L, 1981: Lichen re-colonization in London's cleaner air. Nature 289, pp289–292.

ROUND C, 1979: Retreat Mining with Integrated Dirt Stowing: Mining Engineer 139 (219) Dec 1979, pp495–505.

ROYAL COMMISSION ON ENVIRONMENTAL POLLUTION, 1974: Pollution Control: Progress and Problems. Chairman Sir Brian Flowers. HMSO, 99pp (Cmnd 5780) (Report 4) ISBN 0101578008.

ROYAL COMMISSION ON ENVIRONMENTAL POLLUTION, 1976a: Air Pollution Control: An Integrated Approach. Chairman Sir Brian Flowers. HMSO, 130pp (Cmnd 6371) (Report 5) ISBN 0101637101.

ROYAL COMMISSION ON ENVIRONMENTAL POLLUTION, 1976b: Nuclear Power and the Environment. Chairman Sir Brian Flowers. HMSO, 238pp (Cmnd 6618) (Report 6) ISBN 0101661800.

ROYAL COMMISSION ON ENVIRONMENTAL POLLUTION, 1979: Agriculture and Pollution. Chairman Sir Hans Kornberg. HMSO, 280pp (Cmnd 7644) (Report 7) ISBN 0101764405.

SCOTTISH DEVELOPMENT DEPARTMENT, 1981: Oil, Gas and Petrochemicals. Gas and Coal Processing. Planning information notes D series.

STANDING CONFERENCE ON LONDON AND SOUTH EAST REGIONAL PLANNING, 1979: Electricity Supply in South East England – a review (SC 1242R).

STEELE D J, 1973: Some Impressions of the Coal Industry in Poland, The Mining Engineer 132 (151) Apr 1973, pp327–339.

STRATEGIC CONFERENCE OF COUNTY COUNCILS IN YORKSHIRE AND HUMBERSIDE, 1979: Reclamation at Pyewipe using waste from the Yorkshire Coalfield.

TURNER C W, 1977: The Prevention and Underground Disposal of Extraneous Dirt Mining Engineer November 1977, pp195–203.

U S CONGRESSIONAL COMMITTEE ON SCIENCE AND TECHNOLOGY, 1976: The Environmental Protection Agency's research program with primary emphasis on the Community Health and Environmental Surveillance System (CHESS): An investigative report. U S Congressional Committee on Science and Technology, Washington D C.

U S DEPARTMENT OF ENERGY, OFFICE OF TECHNOLOGY IMPACT, 1980: Synthetic Fuels and the Environment: An Environmental and Regulatory Impacts Analysis. Technical Information Service 652pp (DOE/EV–0087).

WALKER J G and FIELD J M, 1977: Institute of Sound and Vibration Research. Technical Report No 90. University of Southampton. Preliminary Results from a Survey of the Effects of Railway Noise on Nearby Residents.

WALLER R E and others, 1969: Clean Air and Health in London. In Clean Air Conference Eastbourne 1969. Proceedings. National Society for Clean Air London. pp71–79.

WALLER R E, 1979: The effect of sulphur dioxide and related urban air pollutants on health. In International Symposium on Sulphur and the Environment, Society of Chemical Industry, London. 1979, pp171–177 ISBN 0901001589.

WARREN SPRING LABORATORY, 1972–3: National Survey on Air Pollution 1961–71. HMSO. 3 vols. ISBN 0114101493(vol 1); ISBN 0114101507(vol 2); ISBN 0114101515 (vol 3).

WELSH OFFICE, 1977: Digest of Welsh Statistics 23. HMSO, ISBN 0117900915.

WHITEHEAD Dr J C, 1978: Coal from under the North Sea. Coal and Energy Quarterly No 18, Autumn 1978, pp17–21.

WORLD HEALTH ORGANISATION, 1972: Air Quality Criteria and Guides for Urban Air Pollutants. Geneva (TRS 506) ISBN 9241205067.

WORLD HEALTH ORGANISATION, 1979: Sulphur Oxides and Suspended Particulate matter: United Nations Environment Programme. Geneva: 108pp (Environmental health criteria 8) ISBN 9241540680.

ZOLA E, 1885: Germinal. Trans from French by Ellis, M. Elek, 1972.

Relevant legislation

Alkali etc Works Regulation Act 1906 Ch 14 HMSO 17pp
Clean Air Act 1956 Ch 52 HMSO 37pp
Clean Air Act 1968 Ch 62 HMSO 18pp
Coal Industry Nationalisation Act 1946 Ch 59 HMSO
Coal Industry Act 1965 Ch 82 HMSO
Coal Industry Act 1967 Ch 91 HMSO 8pp
Coal Industry Act 1973 Ch 8 HMSO 16pp
Coal Industry Act 1975 Ch 56 HMSO 26pp
Coal Industry Act 1980 Ch 50 HMSO 6pp
Coal-Mining (Subsidence) Act 1957 Ch 59 HMSO
Control of Pollution Act 1974 Ch 40 HMSO 144pp
Countryside (Scotland) Act 1961
Countryside Act 1968 Ch 41 HMSO 64pp
Health and Safety at Work etc Act 1974 Ch 37 HMSO 122pp
Land Compensation Act 1973 Ch 26 HMSO 114pp
Local Government Act 1972 Ch 70 HMSO 464pp
Local Government (Planning and Land) Act 1980 Ch 65 HMSO 330pp
Mines and Quarries Act 1954 Ch 70 HMSO
Mines and Quarries (Tips) Act 1969 Ch 10 HMSO 48pp
Motor Vehicles (Construction and Use) Regulations 1978, SI No 1017

Noise Abatement Act 1960 Ch 68 HMSO
Opencast Act 1958 Ch 69 HMSO
Public Health (Scotland) Act 1897 Ch 38 HMSO
Public Health Act 1936 Ch 49 HMSO
Railways Act 1974
Rivers (Prevention of Pollution) Act 1951 Ch 64 HMSO
Rivers (Prevention of Pollution) Act 1961 Ch 50 HMSO
Rivers (Prevention of Pollution) (Scotland) Act 1951 Ch 66 HMSO
Town and Country Planning Act 1947 Ch 51 HMSO
Town and Country Planning Act 1971 Ch 78 HMSO 396pp
Town and Country Planning (Inquiries Procedures) Rules 1974 SI 1974 No 419 HMSO
Town and Country Planning (General Development) (Scotland) Order 1975 SI 1975 No 679 (S107) HMSO
Town and Country Planning (General Development) Order 1977 SI 1977 No 289 HMSO
Town and Country Planning (Scotland) Acts 1947–1977

Index

DEEP MINING (continued)
  drift or adit, by, 5.3–5.4
  environmental effects, 7.1 et seq, 22.25–22.28
  exploration programme, 3.20
  health and safety, 5.16–5.25
  increased costs of coal, 3.25
  longwall method
    advance mining, 5.9, figure 5.1
    coal faces, 5.8, 5.10
    differences on surface resulting from, 5.11
    early techniques, 5.6
    modern mechanised, 5.7–5.8
    origin, 5.6
    retreat system, 5.9
    ways in which operated, 5.9
  methods of working, 5.5–5.11
  new capacity, conclusions on, 22.22–22.24
  noise from, 7.22 et seq
  production
    costs, 3.19
    costs compared with opencast production costs, 3.27
    differing from opencast production, 11.87
    1980/81, in, 3.18
  prospects for, 3.18–3.25
  room and pillar, by, 5.5
  spoil, see Spoil; Spoil disposal
  stall and pillar, by, 5.5
  subsidence caused by, see Subsidence
  techniques
    effect of different types, 5.2
    minimising subsidence through, 8.22–8.27
  vertical shaft, by, 5.3–5.4
  Yorkshire–Derbyshire–Nottinghamshire coalfield, concentration
    in, 3.18, table 3.4
  see also Subsidence
DEPOTS
  coal concentration, 14.9, 14.11–14.14
  fully mechanised coal concentration, figure 14.2
  partly mechanised, figure 14.3
DERBYSHIRE
  North, see North Derbyshire
DERELICT LAND
  clearance
    block grant system, effect of, 10.19
    importance of maintaining and improving, 12.11
    level of government finance for, 10.19
    pit closures, link with, 10.20–10.21
    priority for, 10.19
  grant, 10.8–10.12, 10.13
    eligibility after land sterilisation, 8.16
  legislative machinery for reclaiming, 10.13–10.14
  local authority, transfer to, 10.9
  meaning, 10.2
  reclamation scheme, 10.9, 10.10
  reclamation schemes begun, table 10.3
DERELICTION
  clearance of, linked with alternative employment opportunities,
    10.1
  concentration, 10.3
  conclusions on, 22.45–22.49
  co-operation between N C B and local authorities, 10.15–10.18
  derelict land grant, 10.8–10.12
  environmental degradation, as, 10.2
  future, prevention of, 10.5–10.6
  land survey, 10.3
  methods of restoring land to other uses, 10.7
  new industrial development, as deterrent to, 10.2
  past, clearance of, 10.4
  pit closures, as result of, 10.1 et seq
  restoration of tips as means of avoiding, 9.15
  scale, 10.3–10.4

DESULPHURISATION
  conclusions on, 17.76
  costs of, 17.68–17.75
  flue gas, 17.63–17.67
DEVELOPING COUNTRIES
  energy problem in, 1.23
DEVELOPMENT
  meaning, 4.8
DEVELOPMENT CONTROL
  development, meaning, 4.8
  local authority powers, 4.4
  standards applied in operating system, 4.7
DEVELOPMENT PLAN
  legislation to introduce, 4.2
  local plan, 4.3
  new collieries, time for drawing up, 2.18
  structure plan, 4.3
  town and country planning, as basis of, 4.3
DISTRICT HEATING
  combined heat and power and, 19.12–19.16
DOMESTIC APPLIANCES
  closed room heaters and boilers, 16.39
  combustion, for, 16.16 et seq
  high output back boilers, 16.38, figure 16.5
  improved open fires, 16.38
  openable roomheaters, 16.40
  simple open fires, 16.37
  smoke-reducing, 16.41–16.43, figure 16.6, 19.11
DOMESTIC COAL
  burn, 19.9–19.19
  market for, 13.44
  methods of transport to premises, 14.9–14.14
  smokeless fuel, 19.10
  transport of, 14.6–14.19
  see also Transport of coal
DOMESTIC SOLID FUELS APPLIANCES APPROVAL
  SCHEME, 19.11
DONCASTER
  operating collieries in, table 3.4
  volume of spoil, 9.6
DORSET
  oil policy report, 21.29
DRAINAGE
  land, effect of subsidence, 8.11
  spoil tips, from, 7.9
  surface
    coal stock from, 7.8
    colliery premises, from, 7.8
  surplus water from underground, of, 7.7
DRAX
  trace elements, 17.31
DURHAM
  derelict land reclamation, 10.10
  dereliction, scale of, 10.3, table 10.1
DUST
  coal stocking as source of, 7.27, 7.30
  combustion residue of coal, as, 18.34
  meaning, 17.3
  nuisance from road and rail transport, 14.33
  opencast operations, from, 11.44–11.45
  tipping operations as source of, 7.27–7.28

EAST MIDLANDS
  agricultural production, effect of subsidence on, 8.14
  energy sites, review of, 21.35
  subsidence damage caused in, 8.13
EAST NOTTINGHAMSHIRE
  intensive exploration stage at, 6.15

References are to paragraph numbers

INTERNATIONAL ENERGY AGENCY, 16.49
INVESTMENT
    benefits of, since 1973, 2.18
    capital, starting point for planning, 2.22
    energy supply and energy efficiency, balance between, 3.41
    Government provision of finance for, 2.22
    improvement of productivity, in, 3.22
    new, importance to coal industry of, 3.21
    new capacity, in, 2.15, 6.1
    planned increases in production, to meet, 6.1
    programme, appropriate size of, 2.22
    rail transport, in, 14.42–14.43
    recent build up in, 2.17
    replacement of obsolete capacity, in, 3.22

KELLINGLEY
    beginning of production, 2.15
KINSLEY DRIFT
    general development order rights used to open, 6.7
    tipping operations, 7.29
    twin drifts at, 2.22
KIVERTON PARK
    appearance of, 7.31

LABOUR *see* Manpower
LAGOONING
    alternatives to, 9.21
    disadvantages of, 9.21
    safety aspects, 9.21
LANCASHIRE
    dereliction, scale of, table 10.1
LAND
    landscape value, loss of, 11.63–11.65
    loss of use during opencast working, 11.61–11.62
    reclamation after opencast operations, 11.55–11.65
    sterilisation caused by subsidence, 8.16
LAND DRAINAGE
    subsidence, effect of, 8.11
LAND RECLAMATION
    local, as alternative to tipping, 9.28–9.30
    remote, as alternative to tipping, 9.31–9.33, table 9.3
LAND USE
    spoil disposal, implications of, 9.5
LANDSCAPE
    quality of land restoration, effect of, 12.8
LEEDS
    registered works, control of, 19.29
LEICESTER
    North East, *see* Vale of Belvoir
LICENCES
    private opencast coal working, 11.21
LIQUEFACTION
    coal gasification distinguished from, 15.2–15.6
LOCAL AUTHORITY
    air pollution, investigation of, 4.24
    amenity improvements, co-operation over, 7.33
    block grant system, effect on clearance schemes, 10.19
    buildings in notified area of coal working, planning permission for, 8.32
    co-operation with N C B over dereliction, 10.15–10.18
    derelict land grant paid to, 10.8–10.12
    development control by, 4.4
    domestic sources, control of emissions from, 4.20 *et seq*
    finance, effect of new developments on, 6.32
    Government approval for planning proposals, 4.5
    industrial sources, control of emissions from, 4.20 *et seq*

    land resources, objectives for use of, in development plan, 4.3
    mineral working, responsibility for, 4.8
    planning decision,
        appeal against, 4.5
        types of, 4.4
    pollutants, powers to control release of, 4.1
    rate support grant, 6.32
    services for N C B employees, 6.30–6.31
    spoil tipping controls, 9.15 *et seq*
    standards applied in development control system, 4.7
    transfer of derelict land to, 10.9
LOCAL AUTHORITY ASSOCIATIONS, 11.20
LOCAL INTEREST
    planning process, as factor in, 21.6
LOCAL INQUIRY
    role played by, 21.12–21.14
LONGANNET POWER STATION, 14.52
LOUGHBOROUGH UNIVERSITY, 14.25
LOWER AIRE VALLEY
    landscaping after land restoration, 11.63

MALTBY
    appearance of, 7.31
    co-operation scheme of spoil disposal, 9.30
MANCHESTER
    Corporation Act 1946, 19.33
    re-entry of coal into industrial market, unawareness of, 13.36
    registered works, control of, 19.29
MANCHESTER AREA COUNCIL FOR CLEAN AIR AND NOISE CONTROL, 19.11
MANPOWER
    average age of workforce, 2.11
    decline in, 2.7 *et seq*, figure 2.3
    housing for, 6.30
    involuntary absenteeism, 2.11
    local authority services for, 6.30–6.31
    low morale, 2.11
    National Coal Board, 2.7
    recruitment
        emphasis on, 2.16
        problems of, 2.11
    redeployment of skilled miners, 2.16
    shortage, 2.8
MANUFACTURED FUEL PLANT
    future prospects for coal, 13.42–13.43
    present pattern of market demand for coal, 13.40–13.41
MARGAM, GLAMORGAN
    planning stage at, 6.15
MARINE DISPOSAL
    tipping, as alternative to, 9.34–9.35
MARKETS
    coal, *see* Coal markets
    coke ovens, 13.2
    contraction in, in response to oil and gas, 13.2
    declining role of coal, 2.5–2.6
    measures to improve coal market situation, 2.6
    solid fuel, table 16.5
    type of coal, by, 16.12–16.13
MATERIALS
    combustion residues of coal, effects of, 18.51–18.54
MEADOWGATE
    restoration of derelict land, 11.56
MERRY-GO-ROUND SYSTEM
    transport of coal, for, 14.3–14.5
METHLEY ENVIRONMENT GROUP, 11.9
MINERAL OPERATOR
    changes in legislation, 4.11

References are to paragraph numbers

References are to paragraph numbers

PLANNING AUTHORITY
  planning process, as factor in, 21.7
PLANNING INQUIRIES, 21.55 *et seq*
PNEUMOCONIOSIS
  causes, 5.23
  downward trend of, 5.22
  trend of new cases, figure 5.5
POLAND
  coal exports from, 3.31, 3.33
POLLUTANTS
  information on, 17.5
  powers to control release of, 4.1
  smoke, *see* Smoke
POLLUTER
  polluter pays principle, 1.39 *et seq*
POLLUTION
  abandoned mines, discharges from, 7.15–7.20
  air
    emissions to, 4.16
    meaning, 4.13
  alkali, production of, 4.17
  control, *see* Pollution control
  disposal of surplus water, legislative controls on, 7.11–7.14
  fuel use in urban areas, from, 19.2 *et seq*
  meaning, 4.13
  noise, *see* Noise pollution
  retrospective remedy of, 1.46
  tipping operations, avoidance in, 9.11
  water, 4.25, 17.37
  see also Dereliction
POLLUTION CONTROL
  acceptable state, notion of, 1.41
  Act of 1974, 4.24–4.26
  aim of policy, 4.15
  ash, 17.21–17.29
  clean air, *see* Clean Air Acts
  domestic coal burn, 19.17–19.19
  domestic sources, emissions from, 4.20 *et seq*
  industrial coal burn, 19.25–19.33
  industrial sources, emissions from, 4.20 *et seq*
  inspectorate responsible for, 4.18
  measures, polluter should bear expense of, 1.42
  nitrogen oxides, 17.80–17.81
  objectives, 4.14–4.15
  particulates, 17.13–17.19
  planning, relationship with, 4.12
  polluter pays principle, 1.39 *et seq*
  purpose, 17.4
  smoke, 17.8–17.9
  sulphur dioxide, 17.42–17.48
POWER LOADING MACHINE
  conditions, effect on, 5.13
POWER STATION
  coal burn
    increase in, 13.2
    2000, in, table 13.2
  coal-fired
    location of, 13.10
    schematic combustion system, figure 16.1
  combustion equipment, 16.16 *et seq*
  current pattern of coal use, 13.4–13.5
  future expectations of coal burn, 13.6–13.12
  grades of coal supplied to, table 16.2
  major, distribution of, 13.5, figure 13.2
  normal design life, 13.12
  nuclear, generating capacity from, 13.8
  pulverised fuel ash from commercial applications, table 17.5
  re-development of existing sites, 13.11
  technical developments, effect on coal's decline, 2.5
  transport of coal, 14.2–14.5

PRIVATE LICENCES
  opencast coal working, 11.21
PRIVATE OPERATORS
  opencast production in 1980/81, 3.26
PRODUCTION PLANT
  substitute natural gas, 15.49–15.71
PRODUCTIVITY
  factors responsible for increase in, 2.10
  increase in, 2.8 *et seq*
  output per manshift, 2.10
PROPERTY SERVICES AGENCY
  energy conservation, role in, 13.39
PROSPECTING
  opencast mining, 11.11–11.13
PUBLIC ADMINISTRATION SECTOR
  coal uses, 13.31–13.33
PUBLIC INQUIRY
  planning process, as part of, 21.62–21.65
  opencast coal proposals, 11.16, 11.19
PUGNEYS
  restoration of derelict land, 11.56
PULVERISED FUEL BOILERS, 16.18–16.23

RADIOACTIVITY
  trace elements, 17.32–17.36
RAIL TRANSPORT
  coal, general conclusions, 20.12–20.13, 22.67 *et seq*
  dust nuisance, 14.33
  energy efficiency, 14.24
  fair competition with other modes, 14.54–14.56
  investment in, 14.42–14.43
  noise and vibration, 14.29–14.32
  road and, of coal
    comparative impact, 14.24–14.41
    concentration depots, 14.11–14.13
    direct from colliery, 14.14
  safety and health, 14.25–14.28
  section 8 grants, 14.57–14.58
  total operations processing system, 14.10
RAILWAYS
  technical developments, effect on coal's decline, 2.5
RECOMMENDATIONS, *see* Conclusions
RECRUITMENT
  emphasis on, 2.16
REGIONAL AGGREGATES WORKING PARTIES, 9.47
REID COMMITTEE
  examination of coal production, 2.3
  surface operations, appearance of, on, 5.33
RENEWABLES
  economic contribution from, 3.5
  expenditure on, 3.5
RESEARCH AND DEVELOPMENT
  renewable sources of energy, on, 3.5
RESPIRATORY DISEASE
  non malignant, combustion residues of coal as cause, 18.2–18.17
RESTORATION
  agriculture, land for, 11.28–11.32
  forestry, land for, 11.33
ROAD TRANSPORT
  coal, general conclusions, 20.12–20.13, 22.66 *et seq*
  dust nuisance, 14.33
  energy efficiency, 14.24
  fair competition with other modes, 14.54–14.56
  noise and vibration, 14.29–14.32
  powers to reduce impact, 14.34–14.41
  rail and, of coal
    comparative impact, 14.24–14.41
    concentration depots, 14.11–14.13
    direct from colliery, 14.14

References are to paragraph numbers

TIPPING *(continued)*
control measures, 9.12
dust, as source of, 7.27, 7.28
environmental standards, improvement of, 9.11
future schemes for, 9.20
geographical concentration, 9.6
impact of, minimisation of, 9.11
impact on locality, 7.31–7.32
improvements in tip design, 9.12
inactive and disused tips, 9.6
landscape design proposals, 9.13, 9.17
leaving dirt in coal as alternative to, 9.27
limiting environmental impact of, 9.14
local, control of impact of, 9.15
local land reclamation as alternative to, 9.28–9.30
marine disposal as alternative to, 9.34–9.35
mobile earthmoving plant, 9.11
N C B rights over tips under G D O, abolition of, 9.18 *et seq*
older and new tips, distinction between in environmental terms, 9.9
older planning permissions relating to, 9.15 *et seq*
operational unreclaimed tips, environmental effects, 9.8
practices, changes in, 9.10
progressive restoration techniques, 9.12, figure 9.2
prohibition of operations, 9.19
reclaimed tips, environmental effects, 9.8
remote land reclamation as alternative to, 9.31–9.33, table 9.3
safety, importance of, 9.10
slag heap image, 9.22
suspension of operations, 9.19
tip design, 9.12
town and country planning
    controls, 9.15 *et seq*
    permission for, 4.10
waste tipping reworking, 9.16
see also Dereliction; Spoil; Spoil disposal

TOWN AND COUNTRY PLANNING
Act of 1947, 4.2, 4.10
amenity, meaning, 4.6
authorisation procedures for opencast mining, case for change of, 11.69 *et seq*
basic objective of, 4.1
buildings in notified area of coal working, 8.32
Central Electricity Generating Board, 21.49–21.51
changes in legislation, 4.11
colliery brought into operation after 1 July 1948, 6.4
Commission's approach to, 21.11–21.16
conclusions, 21.66–21.70, 22.8, 22.93–22.105
contrasting visual impact, effect of geology on, 6.24
deep boreholes, 6.14
dereliction, prevention in the future, 10.5–10.6
development control, 4.4
development plan, *see* Development plan
environmental considerations, 4.6–4.7
environmental protection, effect on, 1.12
general development order, 4.4, 4.10
generally, 21.24
greenfield sites, *see* Greenfield sites
long term, 21.33–21.38
Minerals Bill, introduction of, 6.10
mines
    existing, planning permission for, figure 4.1
    new, planning permission for, figure 4.1
mining, consultation about, 21.41–21.47
National Coal Board operations, 4.8–4.11
national planning guidelines in Scotland, 21.31–21.32
opencast
    authorisation procedures, 11.14 *et seq*
    consultation on, 21.48
    sites, planning permission for, figure 4.1
permitted development, 4.10

pit closures, planning for effects of, 7.20
planning
    decisions, types of, 4.4
    gain, 12.14
    inquiries, 21.55–21.61
progressive consultation, need for, 21.39 *et seq*
public inquiries procedure, 21.62–21.65
range of interests
    climate of opinion, 21.5
    energy interest, 21.8–21.10
    local interest, 21.6
    planning authority, 21.7
    reconciliation of, 21.1 *et seq*
reconciling the interests, 21.1 *et seq*
Scotland, 4.2
Secretary of State's powers, 4.5
spoil tipping controls, 9.15 *et seq*
structure plans, 21.26–21.30
success of system, 21.25
waste tipping reworking, 9.16

TRACE ELEMENTS
concentration in coal, 17.30–17.31
radioactivity, 17.32–17.36

TRANSPORT OF COAL
aerial ropeway, by, 14.52–14.53
bunker loading at a colliery, figure 14.1
coastal shipping, by, 14.45–14.48
conclusions, 22.65–22.74
conveyor, by, 14.52–14.53
domestic, *see* industrial and domestic *under this heading*
environmental issues associated with, 14.1 *et seq*
environmentally preferable mode, 14.20 *et seq*
general conclusions, 20.12–20.13
haul trucks, noise from, 11.39
industrial and domestic
    bulk user distinguished from, 14.6
    coal concentration depots, 14.9, 14.11–14.13
    conventional rail transport, 14.10
    developments in, 14.15–14.19
    factors increasing transport requirements, 14.7–14.8
    methods of transport to premises, 14.9–14.14
    rail and road, by, 14.11–14.13
    road direct from colliery, by, 14.14
inland waterways, by, 14.49
merry-go-round system, 14.3–14.5
opencast coal, 11.26–11.27
pipelines, by, 14.50–14.51
power stations, requirements of, 14.2–14.5
requirements of particular markets, 14.2 *et seq*
road, by, powers to reduce impact of, 14.34–14.41
road and rail
    alternative modes, 14.45–14.53
    dust and spillage, 14.33
    energy efficiency, 14.24
    fair competition between modes, 14.54–14.58
    noise and vibration, 14.29–14.32
    safety and health, 14.25–14.28
road direct from colliery, by, 14.14
See also Rail transport; Road transport

UNITED NATIONS CONFERENCE ON THE HUMAN ENVIRONMENT, 4.1
UNITED STATES OF AMERICA
coal
    conversion, 15.34–15.36
    exports, expansion of, 3.31
    prices, 3.35
    production problems, 3.32

UNITED NATIONS (continued)
    Omnibus Energy Security Act, 15.34
URBAN AREAS
    burning coal in, environmental effects of, 19.1 *et seq*

VALE OF BELVOIR
    coal preparation process, 9.21
    development of new mines, 3.23
    environmental impact of spoil tips, 9.7
    inquiry, 21.57
    N C B approach to planning and development, 6.21 *et seq*
    planning stage at, 6.15
    remote land reclamation, 9.31
    siting options for surface works, 21.53
    socio-economic impact of project, 1.20
    subsidence
        major factor, as, 8.22
        N C B liabilities for, 8.18
    tip design, 9.12
    tip size, 9.11
VALE OF BELVOIR PROTECTION GROUP, 6.25
VENICE SUMMIT
    coal production, declaration on, 1.10
    energy problem, communiqué on, 1.21
VIBRATION
    road and rail, comparative impact, 14.32

WARREN SPRING LABORATORY
    17.7, 17.39, 17.78, 18.26
WARWICKSHIRE
    South, *see* South Warwickshire
WASTE DISPOSAL
    collieries, from, 1978–79, table 9.1
    spoil disposal, *see* Spoil disposal
    strategic planning, 9.44 *et seq*
WATER
    mine drainage, volume pumped, 7.4, table 7.1
    surplus, *see* Surplus water
WATER POLLUTION
    coal combustion, resulting from, 17.37
    legislation concerning, 4.25
WEAR VALLEY
    landscaping after land restoration, 11.63
WELSH DEVELOPMENT AGENCY
    alternative employment, efforts to attract, 1.19

derelict land reclamation, 10.10
dereliction, financial assistance for, 10.8
factory space built on reclaimed land, 10.2
importance of efforts by, 10.20
new industry, reclamation of derelict land to attract, 10.18
WEST GERMANY
    coal conversion, 15.28–15.33
WEST THURROCK
    coal-oil mixture manufacture at, 16.49
WEST YORKSHIRE
    opencast mining, expansion of, 11.35
WILDLIFE
    land sterilisation, effect of, 8.16
    opencast coal proposals, effect of, 11.15
    opencast working, effect of, 11.63
    quality of land restoration, effect of, 12.8
    tip reclamation schemes, benefits of, 9.13
WITHAM
    feasibility study stage at, 6.15
WOOLLEY, coal preparation and waste tipping, 9.11
WORLD BANK
    1.23
WORLD COAL STUDY
    imported steam coal, on requirements for, 3.31
WORLD HEALTH ORGANISATION, 18.10

YORKSHIRE
    derelict land reclamation, 10.11
    N C B evasion of planning considerations over tips, 9.18
    North, *see* North Yorkshire
    preliminary exploration stage at, 6.15
    remote land reclamation, 9.31
    sites operated under G D O provisions concentrated in, 6.9
    spoil disposal in, 9.43 *et seq*
    South, *see* South Yorkshire
    tipping operations, 7.29
YORKSHIRE AND HUMBERSIDE
    energy sites, review of, 21.35
YORKSHIRE–DERBYSHIRE–NOTTINGHAMSHIRE
    COALFIELD
    active colliery spoil tipping, 3.18
    derelict spoil heaps, 3.18
    dereliction, scale of, 10.3, table 10.1
    operating collieries, 3.18, table 3.4
progressive concentration of mining industry on, 12.1
waste output from, 9.6

Printed in England for Her Majesty's Stationery Office
by Hobbs the Printers of Southampton
(1850) Dd717302 1100 8/81 G389